$33.00

Basic Plumbing with ILLUSTRATIONS

REVISED EDITION   By Howard C. Massey

Craftsman Book Company
6058 Corte del Cedro / P.O. Box 6500 / Carlsbad, CA / 92018

Acknowledgments

The author expresses his sincere appreciation to the following companies and organizations:

To the *Florida Energy Committee* and the *Environment Information Center of the Florida Conservation Foundation, Inc.* for pertinent information necessary to authenticate the thermosyphon and pumped solar water heating systems

To *Amtrol Inc.*, P.O. Box 1008, West Warwick, RI 02893
(Illustrations used for the Well-X-Trol water systems)

To *Sta-Rite Water Systems Group*, 293 Wright Street, Delavan, WI 53115
(Technical information for troubleshooting private water systems)

To *Josam Company*, Michigan City, Indiana (Fixture carrier illustrations)

To *American Standard*, P.O. Box 6820, Piscataway, NJ 08855
(Plumbing fixture photos and roughing-in measurements)

To *Mansfield Plumbing Products*, 150 First Street, Perrysville, OH 44864 (419) 938-5211.
Fax orders: (419) 938-6234 (Brass & plastic items illustrated)

To *Montgomery Ward* (Dishwasher installation illustrations)

To *Moen Incorporated*, 25300 Al Moen Drive, North Olmsted, OH 44070
(Service repair illustrations)

To *Delta Faucet Company*, P.O. Box 40980, Indianapolis, IN 46280 (Service repair illustrations)

Looking for other construction reference manuals?

Craftsman has the books to fill your needs. Call toll-free 1-800-829-8123
or write to Craftsman Book Company, P.O. Box 6500, Carlsbad, CA 92018
for a FREE CATALOG of books and videos.

Library of Congress Cataloging-in-Publication Data

Massey, Howard C.
  Basic plumbing with illustrations / by Howard C. Massey. --Rev.
ed.
      p.   cm.
  Includes index.
  ISBN 0-934041-99-7
  1. Plumbing.   I. Title.
TH6122.M38   1994   94-32305
696'.1--dc20   CIP

First edition © 1980 Craftsman Book Company
Second edition © 1994 Craftsman Book Company
Second printing 1996

Illustrations by Mike Aten and the author

Contents

1

Introduction to Plumbing

This book is a guide to good plumbing practice. Even if you've had little or no experience with plumbing systems, you should have no difficulty understanding what's explained here. The text covers the basic principles required to plan, install, and maintain common plumbing systems in residential and light commercial buildings. An understanding of the plumbing code is essential, so code requirements are discussed throughout this book. You'll find questions at the end of each chapter to help test your understanding of the ideas presented. Study the text carefully to master the fundamental principles behind each question before you turn to the answer in the back of the book.

This book will help you select the materials, pipe sizes, and methods of installation generally accepted as correct by most plumbing codes. This manual explains everything you need to know to install plumbing on nearly any residential or light commercial job. But it doesn't go into highly technical areas or specialized plumbing and piping such as hydraulic and pneumatic systems that might just be confusing.

Plumbing Codes

Nearly every city, county and state has adopted a plumbing code to protect the health, safety, and welfare of its people. Building departments enforce these codes and arrange inspections of plumbing work as it's completed. As a professional plumber, you can expect that nearly all of your work will have to meet code requirements and pass an inspection.

The plumbing code is a law intended to be enforced, not a set of directions intended to be followed. That means you shouldn't expect to learn the plumbing trade by reading the code. Still, every professional plumber (whether apprentice, journeyman, master or maintenance plumber) will have to refer to the code at least occasionally. You need a copy of the plumbing code that's enforced in the communities where you do work. If your local building department really wants to help you follow the code (rather than just enforce the code against you), they'll have copies for sale across the counter at the building department.

Please note that homeowners who install plumbing must follow the same rules as professional plumbers. Homeowners are subject to the same penalties as licensed plumbers who don't get the required permits and comply with the code. Good plumbing doesn't depend on who does the work, but on how it's done.

Plumbing codes vary. Every city, county and state can adopt any plumbing code that they want to adopt. Many adopt one of the model codes but amend certain sections. Others follow one of the model codes but not necessarily the latest version of that code. No matter what code your community has adopted, the basic principles of sanitation and safety are about the same.

You should understand clearly that this manual isn't the plumbing code. You'll have to refer to your local code from time to time. But what you learn in this book will meet code requirements nearly anywhere in the U.S. or Canada. The minor differences between model plumbing codes are emphasized throughout this book. They'll be apparent as you read and compare sections of this book with your local code.

Some codes require that only a licensed plumbing contractor do new or extensive remodeling work. However, most codes allow homeowners to do any type of plumbing work in their existing home. Some codes also permit homeowners to install plumbing in a new house if the owner can show a basic ability to do the work without professional help.

What About Plumbing Permits?

Codes require a permit for all plumbing work except repairing leaks, clearing obstructions in sewer lines or waste pipes, repairing faucets or valves, or cleaning septic tanks. When a permit is required, the permit must be posted at the site and available at all times for the plumbing inspector.

You can't cover or conceal plumbing work in any way until the plumbing inspector has checked the work for compliance with the code. As the permit holder, you have to notify the plumbing inspector when the work is ready for test and inspection. Inspections are classified into three categories for a one-story building:

1. ground roughing-in
2. tub set and interior water piping
3. final

For a two-story building, a fourth inspection is required. It's called the "topping out" inspection. All rough piping must be inspected above the first floor and up to and through the roof.

After you complete the permitted work, don't wait more than 30 days before you request the final plumbing inspection. Also the building or construction can't be used or occupied until after the final inspection is made. These inspections will be discussed in detail in later chapters so you know how to prepare for each inspection.

The Plumbing System

Cost of plumbing systems in most homes is about 10 percent of the total construction cost. There's about 300 feet of pipe under the floor and in the partitions of an average home. Each foot of this pipe will function quickly and efficiently and last many years if installed properly and maintained adequately. Proper installation is your job. Fortunately, there's a good guide to proper installation. As explained in Chapter 1, it's called the plumbing code. The code will be your best source of information on plumbing installation procedures.

The code divides every plumbing system into three basic parts:

1. the drainage and vent system

2. the water service pipe and distributing pipes

3. the plumbing fixtures

In this chapter we'll look briefly at the three systems. In later chapters, we'll discuss each one in detail.

The Drainage System

A drainage system is made up of three major classes of piping and fittings. The *first class* is all the pipes and fittings which receive used water produced within a building. The drainage system conveys that water (and waste products) to the public sewer or other approved private disposal system.

The *second class* of drainage system pipes and fixtures is the fixture traps. Traps protect the health of the building occupants. Traps provide a liquid seal that keeps sewage odors, gases, and even vermin from working their way into the building at fixture locations.

Vent pipes are the *third class* of pipes and fittings in a drainage system. These pipes admit and give off air from all parts of the drainage system so pressure inside the pipes is always the same as air pressure outside the pipes. Vent pipes are sized and arranged to relieve pressure that builds up as water is discharged into the sanitary drainage system by various types of plumbing fixtures. This free

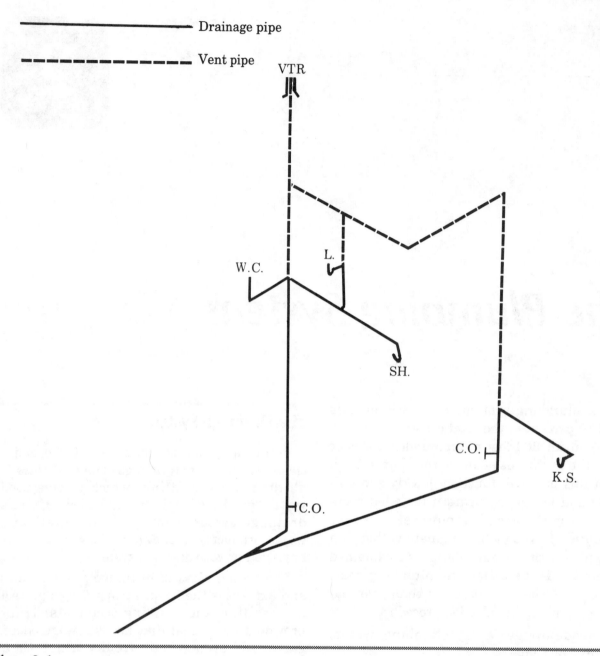

—————————————— Drainage pipe

— — — — — — — Vent pipe

Figure 2–1
Sanitary system isometric drawing

flow of air within the system keeps back-pressure and siphoning from destroying fixture trap seals.

These three classes of drainage and vent pipes and fittings are commonly called the "drain-waste-vent" system (or the DWV system). Though DWV piping is a very important part of any plumbing system, it's the part that's least understood.

Most professional plumbers think of the sanitary drainage and venting as the heart of a plumbing system. The DWV system is the first plumbing to be installed and inspected during construction. Since you install a major portion this system beneath the floor of a building, it's called "ground roughing-in" in the trade.

Figure 2–1 shows an isometric drawing of a sanitary system. The drainage system and the

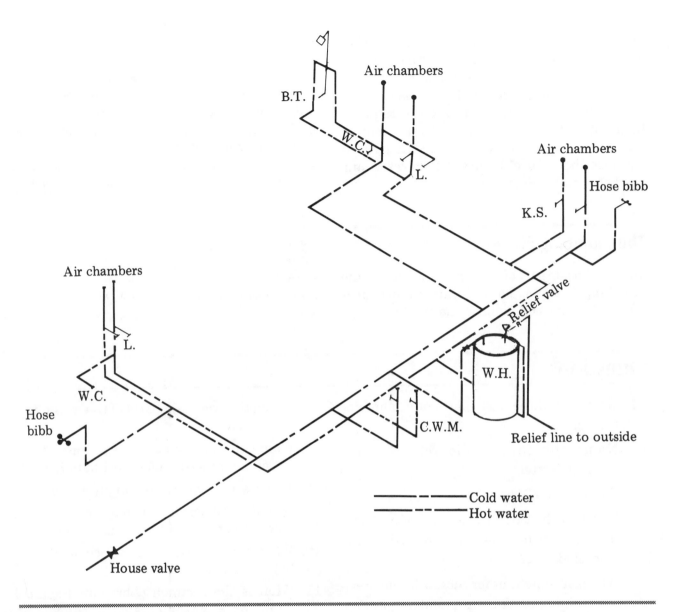

Figure 2–2
Water piping diagram

traps (the part of the system that receives liquid waste) are shown by solid lines. The dry portion (vents) is shown by dashed lines. "C.O." indicates a cleanout fitting.

If you don't have any training in plumbing, you'll need to become familiar with isometric drawings. The construction blueprints that local authorities require generally include isometric drawings. Simple jobs are easy to understand if you know how to read isometric drawings.

The Water Supply System

The second basic component of a plumbing system is the water service pipe and the distributing pipes. These pipes and fittings carry hot and cold water through separate piping to plumbing fixtures and other water outlets on private property. The water source for these pipes may be a public water distribution system carrying water to each lot or a private water system fed by a domestic well.

It's important to know the symbols for hot and cold water used in piping diagrams. Cold water is shown as a solid line broken with a single dot or dash. Hot water is shown as a solid line broken with two dots or dashes (see Figure 2–2). If a plumbing fixture requires both hot and cold water, the hot water pipe is always on the left and the cold on the right as you face the fixture. Figure 2–2 shows a simple water piping diagram.

The Plumbing Fixtures

The third basic part of a plumbing system is the plumbing fixtures. These fixtures must meet the needs of the occupants of the building.

Abbreviations are often used to identify various types of plumbing fixtures in isometric drawings and floor plans. For example, you'll see the letter "L." used to designate a lavatory. Some plans may use "LAV." The following abbreviations will be used throughout this book to identify the plumbing fixtures in isometric drawings and floor plans.

Plumbing Fixtures	Abbreviated Symbols
water closet	W.C.
bathtub	B.T.
shower	SH.
lavatory	L.
kitchen sink	K.S.
clothes washing machine	C.W.M.
water heater	W.H.

Questions

1. Name the three basic parts of a plumbing system.

2. Name the three major classes of a drainage system.

3. What is a DWV system?

4. Which of the three basic parts of a plumbing system is generally installed and inspected first?

5. Why is it important for those with no previous training or skills in plumbing practice to learn to interpret isometric drawings?

6. Name the two basic parts of the water supply system.

7. Name the two separate systems of the water supply system for most buildings.

8. What is a private water supply system?

9. In a piping diagram drawn by professionals, how are hot and cold water lines identified?

10. When facing a fixture, on which side is the hot water outlet?

11. What is the common abbreviated symbol used for a water closet? a shower? a water heater?

3

Plot Plans

The drawing of a lot that a building stands on, or will be constructed on, is called a *plot plan*. You should be familiar with plot plans and know how to interpret them. A plot plan shows:

- the shape and size of the lot

- the location and size of the building

- the setbacks from the property line

- all permanent outside construction above ground, including driveway and walkway

- electric service

- easements (if any) and streets

A plot plan also includes information on the below-ground facilities:

- location and size of the gas service

- source, location and size of the water facility

- storm drainage and any disposal facility

- type, location, and size of the sewage disposal facility

Plot plans are required by code and are an essential part of all construction plans. Plot plans are first approved by plans examiners; they are then called "as-built" plans. Plans examiners may ask for minor corrections on the plans before approval.

You must follow the plot plan exactly when installing plumbing lines shown on the plot plan. If you have to make any changes to the plan after it's approved (for example, relocating a septic tank to another area), you must first submit a revised plot plan for approval.

The three types of plot plans in Figures 3–1 to 3–3 show the three basic types of outside plumbing facilities for a building with:

1. public water and sewers

2. public water and a septic tank

3. a well and a septic tank

The house in Figure 3–1 is a typical single-family dwelling. It's connected to a public sewer lateral located at about the center of the lot. The water service is connected to a water meter on the right side of the lot. In this plot plan, the utilities serving the dwelling are in the street and not in the utility easement at the rear of the property. The sewer and water service were properly sized so the plumbing plans examiner hasn't had to make any corrections on the plan.

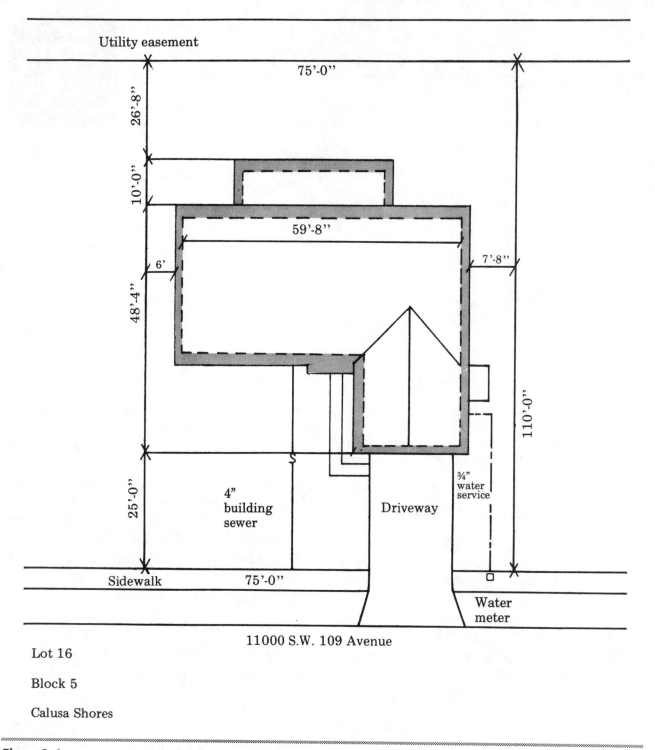

Figure 3–1
Plot plan of a building with public water and sewer

The plot plan in Figure 3–2 shows a house in a suburban area. It relies on a public water supply but has a septic tank for sewage disposal. This plot plan shows a 1200-gallon septic tank with a 480-square-foot drain field. The soil is sandy loam. The tank and drain field are located in the front yard of the dwelling and are sized for a four-bedroom dwelling. The water service is connected to a water meter located on the right side of the lot on the plan.

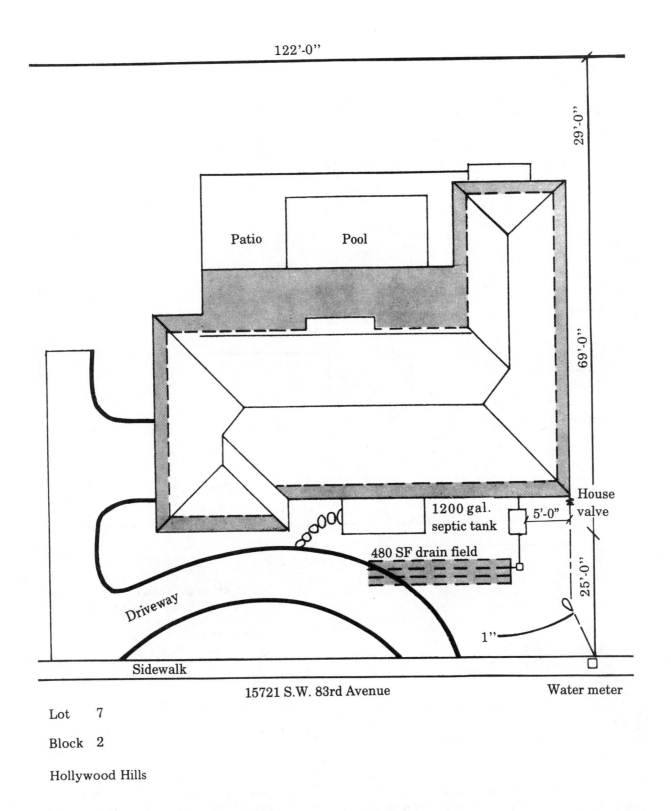

122'-0"

29'-0"

69'-0"

Patio

Pool

1200 gal. septic tank

5'-0"

House valve

480 SF drain field

25'-0"

Driveway

1"

Sidewalk

15721 S.W. 83rd Avenue

Water meter

Lot   7

Block 2

Hollywood Hills

Figure 3–2
Plot plan of a building with public water and septic tank

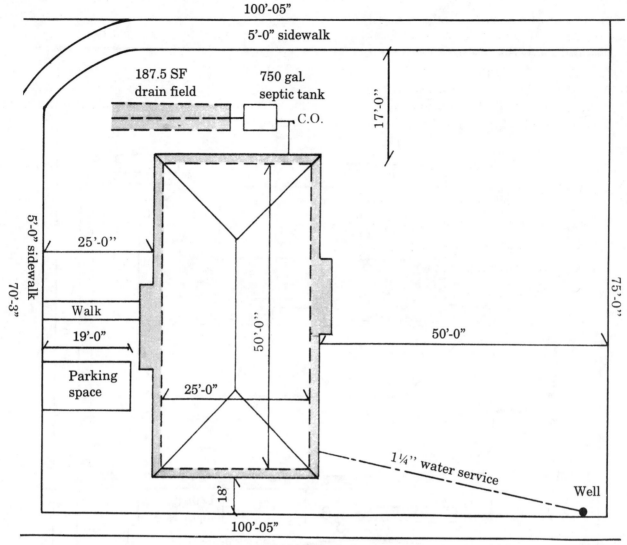

100'-05"

5'-0" sidewalk

187.5 SF drain field

750 gal. septic tank

C.O.

17'-0"

5'-0" sidewalk

70'-3"

25'-0"

Walk

19'-0"

Parking space

50'-0"

25'-0"

50'-0"

75'-0"

1¼" water service

Well

18'

100'-05"

304 N.E. 19th Drive

Lot 13
Block 7
Pinetree Estates

Septic tanks shall have a 50 ft. separation, and drain fields a 100 ft. separation, from all water supply wells.

Figure 3–3
Plot plan of a building with a well and septic tank

Also note that this service is for a three-bath house and is not sized. The plumbing plans examiner sized this service for the architect. The 5-foot separation between the septic tank and water service is also noted on the plans, as required by code.

The house in Figure 3–3 is in a rural area. It relies on a domestic well water and a septic tank for sewage disposal. The well is at the right rear of the property. The septic tank is on the left side of the residence. The water service (suction line) from the well to the house is sized at 1¼ inches because the developed length of the line is more than 40 feet. For a line less than 40 feet long, the code permits a 1-inch service (suction line).

The septic tank and drain field area are both sized properly for a two-bedroom home. The soil is fine sand. The 50-foot separation between the septic tank and the 100-foot sep-aration of the drain field from the well are noted on the plans, as required by the Department of Environmental Resource Management (DERM), or in some cases by the local health department. Since there are no notations by the plumbing plans examiner on this plot plan, you could install the facilities as shown.

Questions

1. What is the term used for approved plans?

2. If you change an approved plan, what steps do you have to take?

3. What is included in a plot plan?

4. What below-ground information is included in a plot plan?

5. When can you install facilities as shown on a given plot plan?

6. In addition to the plumbing plans examiner, who else must approve plot plans having septic systems?

7. Why does a plumber have to know how to read and interpret plot plans?

Isometric Drawings and Definitions

Every professional plumber needs to know how to read and create isometric drawings for plumbing systems. Isometrics are the language plumbing contractors, designers, foremen, plumbing estimators and installers use to communicate their ideas about plumbing systems.

An isometric drawing shows a three-dimensional view of where the pipes in a plumbing system go. The piping may be to scale, or dimensioned, or both. To give the drawing the appearance of depth, pipes in an isometric drawing meet at angles of 120 (not 90) degrees.

How to Draw the Axes of an Isometric Drawing

First, get a sheet of paper, a sharp pencil, and a 30-60 right triangle. Figure 4–1 shows how to place the triangle to draw the axes. Square the short leg of the right triangle with the right, left or bottom edge of the paper to produce angled or vertical lines. These are the axes of your isometric drawing.

Now look at Figure 4–2. Put the letter N at the top of the page to designate North and make a dot in about the center of the page. Using the 30-60 right triangle squared with

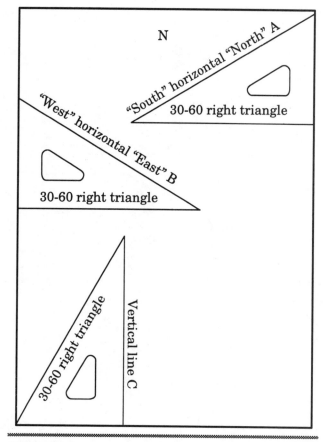

Figure 4–1

Using a 30–60 right triangle to draw angles on isometric drawings

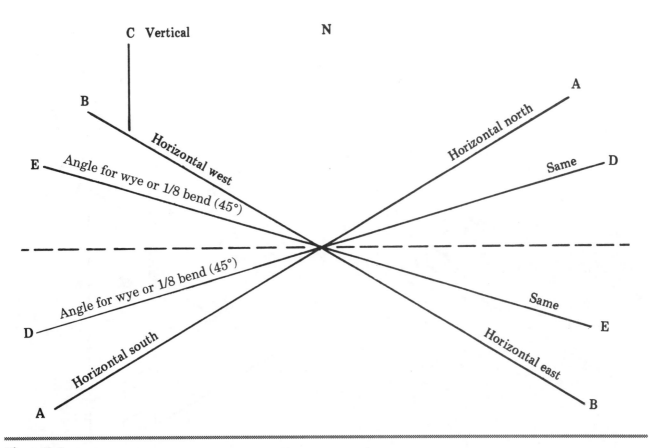

Figure 4–2
Isometric angles

the right edge of the paper, draw a line along the longest side (the hypotenuse) of the angle through the center dot. This is the north-south horizontal line (line A). Then square the triangle with the left edge of the paper and draw another angled line through the center dot. This is the east-west horizontal line (line B).

To draw the dashed line in Figure 4–2, put the short side of the triangle at the right or left edge of the paper and draw the dashed line along the long leg of the right angle of the triangle. Then draw line D halfway between the dashed line and line A (horizontal north). It's used to show the change in direction that happens with the 45-degree fitting of a wye or 1/8 bend. Repeat the procedure for line E.

To draw the vertical line (C), square the short leg of the triangle with the lower edge of the paper and draw a line using the long leg of the right angle as a straightedge. Connect line C with any of the horizontal lines, as desired.

Figure 4–3 shows a simple isometric drawing using the angles in Figure 4–2.

Fittings in an Isometric Drawing

The lines on isometric drawings stand for pipes and fittings. The symbols stand for the types of fixtures and their locations. You should learn to recognize these symbols. Figures 4–4A, 4–4B, and 4–4C show fifteen common fittings used with no-hub pipe. The symbols are the same, regardless of the type of pipe you're using.

Each fitting in the drawings is numbered to correspond to a drawing of the fitting. For example, take a look at Figure 4–4C. Fitting number 14 is a horizontal twin tap sanitary tee (also known as an "owl fitting"). This fitting connects two similar fixtures to the same waste and vent stack at the same level. In the isometric in Figure 4–4C, it connects two lavatories.

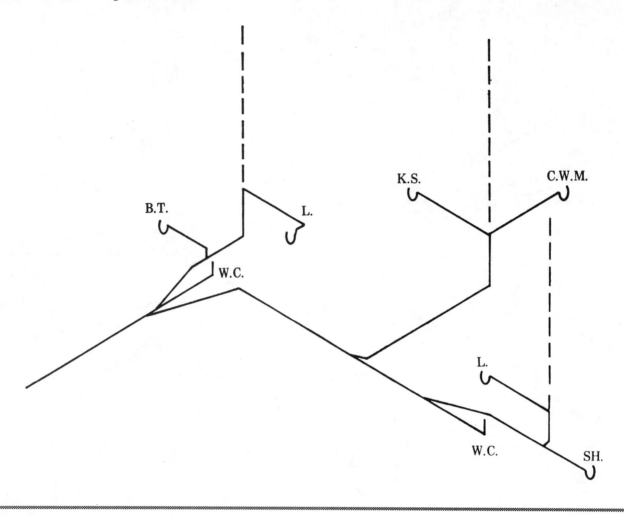

Figure 4–3
Simple isometric drawing

Terms Used on Isometric Drawings

Figures 4–5, 4–6, and 4–7 show the common terms used to describe parts of drainage and vent systems. The important parts of the system are labeled. Unfortunately, plumbing codes often use several terms to identify a single section of pipe. And a code may give a special definition of a common term. In this book, we'll try to use simple terms and group them together to cut down on possible confusion.

Figure 4–5 is an isometric drawing of a sanitary system for a typical two-bath house with a kitchen and utility room. It shows a *flat system* connected to a public sewer. This type of installation is used when vertical space available for roughing-in work is limited.

Figure 4–6 shows a small one-bath house with a kitchen and utility room. In this example the available vertical space was not limited. The type of installation shown is a *stack system* connected to a private sewage disposal unit (a septic tank). Figure 4–7 shows a typical two-story residential installation with a full bath on the second floor and kitchen and utility facilities on the first floor.

▓ Public Sewer

A *public sewer* (also called a *municipal sewer*) is usually located in a street, alley, or dedicated easement beside a parcel of privately owned property (Figure 4–5). Public sewers are installed, maintained, and controlled by the

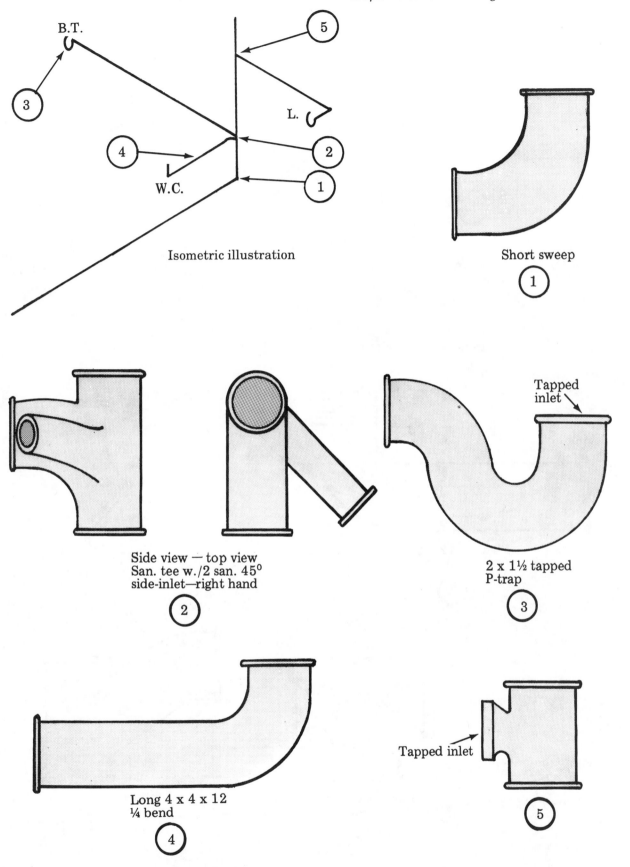

B.T.

③

⑤

L.

④

②

W.C.

①

Isometric illustration

Short sweep

①

Side view — top view
San. tee w./2 san. 45°
side-inlet—right hand

②

Tapped
inlet

2 x 1½ tapped
P-trap

③

Long 4 x 4 x 12
¼ bend

④

Tapped inlet

⑤

Figure 4–4A
Symbols and drawings of some common fittings

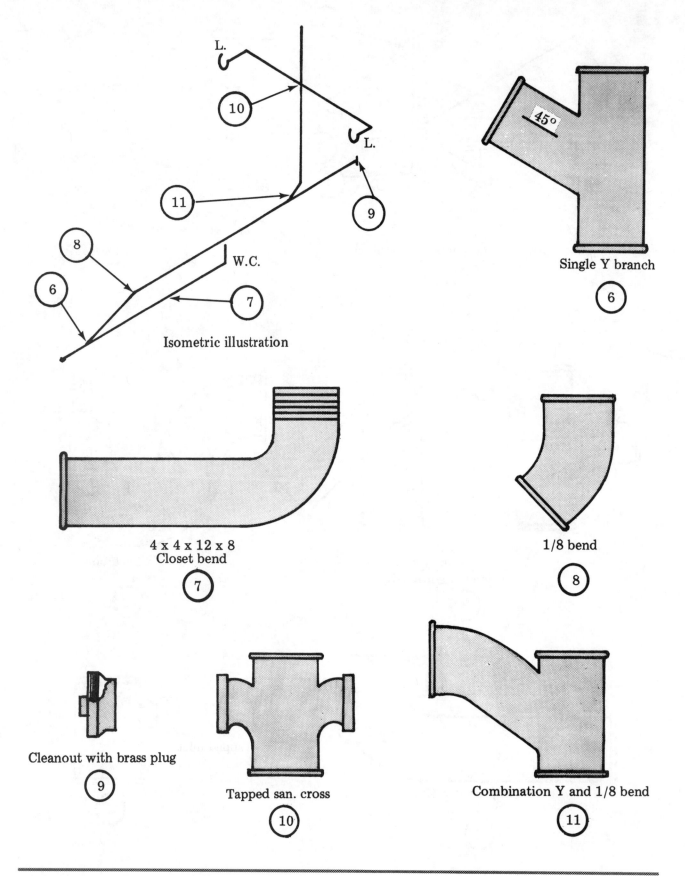

L.

10

L.

11

9

8

6

W.C.

7

Isometric illustration

Single Y branch

6

4 x 4 x 12 x 8
Closet bend

7

1/8 bend

8

Cleanout with brass plug

9

Tapped san. cross

10

Combination Y and 1/8 bend

11

Figure 4–4B
Symbols and drawings of some common fittings

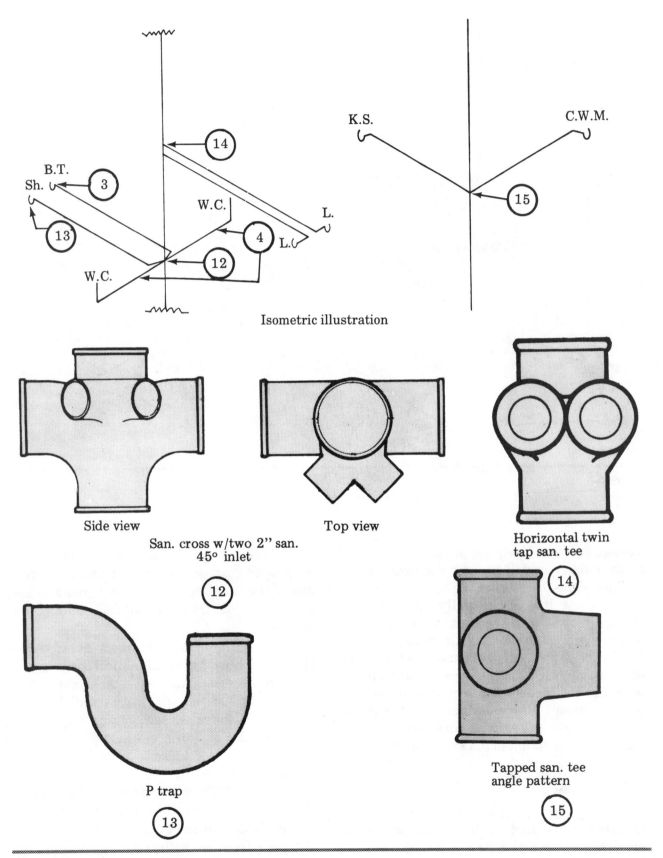

Isometric illustration

Side view

Top view

San. cross w/two 2" san.
45° inlet

Horizontal twin
tap san. tee

P trap

Tapped san. tee
angle pattern

Figure 4–4C
Symbols and drawings of some common fittings

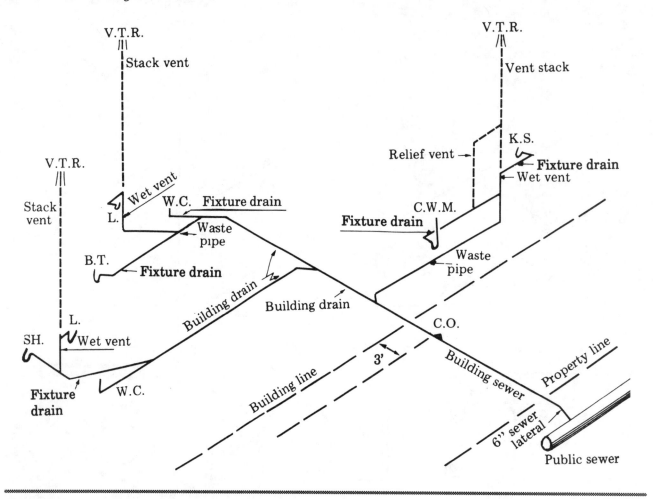

Figure 4–5
Isometric drawing of a flat system with a public sewer

local authorities. Their cost is usually supported by the public through some form of taxation.

During the installation of the sewer main, a 6-inch sewer lateral is usually extended from the main to a point several inches past the property line on each lot. This makes it easier for the property owner to connect to the sewer. After the permit for the sewer connection is issued to the homeowner or plumbing contractor, the local municipal engineering department will specify where, and how deep, to make the connection to the sewer lateral.

▓ Private Sewer

A *building sewer* (also called a *private* or *sanitary sewer*) is part of the main horizontal drainage system that takes waste from the building drain to the public sewer lateral (Figure 4–5). The building sewer begins where

it connects to the 6-inch lateral, a few inches within the property line. It ends where it meets the building drain, 3 feet (more or less in some codes) from the outside building wall or line.

The *building drain* (Figures 4–5 and 4–6) is the *main* horizontal collection system, but without the waste and vent stacks and fixture drains. It's located within the wall line of a building. It takes all sewage and other liquid waste to the building sewer which begins 3 feet (more or less in some codes) outside the building wall or line. It's also considered a *main* because the other drainage branches of the sanitary system are connected to it.

The *fixture drain* (Figures 4–5, 4–6, 4–7) is the pipe from each fixture trap to the vent which serves that fixture (referred to in some codes as *fixture branch*). It may connect directly

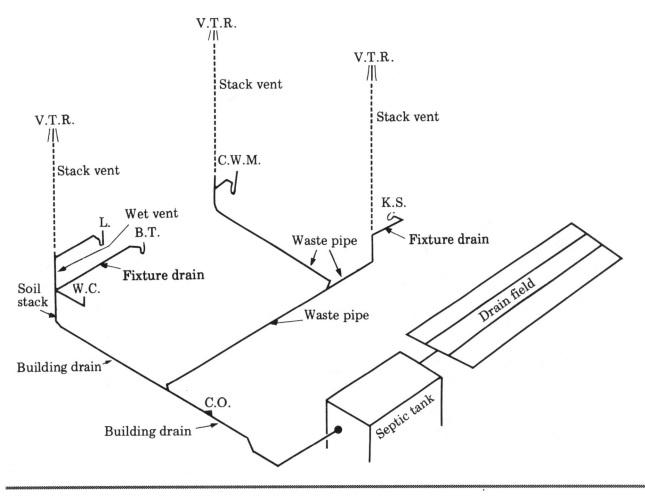

Figure 4–6
Isometric drawing of a stack system with a septic tank

to a vertical vent stack above the floor or, in the case of a shower or bathtub, to the horizontal wet vent section beneath the floor. See Figure 4–5. Plumbers also call this the "sink arm" or "lavatory arm" when it's located above the floor.

The *fixture branch* (also called *fixture drain, waste pipe,* or *wet vent*) is the drainage pipe that takes liquid waste from several fixtures to the junction of any other drain pipe (usually the building drain).

Fixture drains can't carry *body waste.* Body waste, as defined in the code, is solid fecal matter; the code makes no mention of urine. So urinals can be connected to a waste pipe.

A *soil stack* (Figures 4–6 and 4–7) is the vertical section of pipe in a plumbing system that takes waste from water closets, with or without waste from other fixtures, to the building drain. A building drain that gets waste from

water closets is generally called a *soil pipe.* The term "stack" usually refers to any vertical pipe, including the soil, waste, and vent piping of a plumbing system.

The part of a drain pipe that goes out from the side of a soil or waste stack which gets discharge from one or more fixture drains is called a *horizontal branch* (Figure 4-7).

A *common vent* (Figure 4–7) is a vertical vent that serves two fixture drains (also called *fixture branch* in some codes) that are installed at the same level. In Figure 4–7 these are the sink and clothes washing machine on the first floor and the two lavatories in the second floor bathroom.

A *continuous vent* (also called an *individual vent*) is the vertical part of a vent that's a continuation of the drain it's connected to. Look ahead to Figure 6–8 for an example.

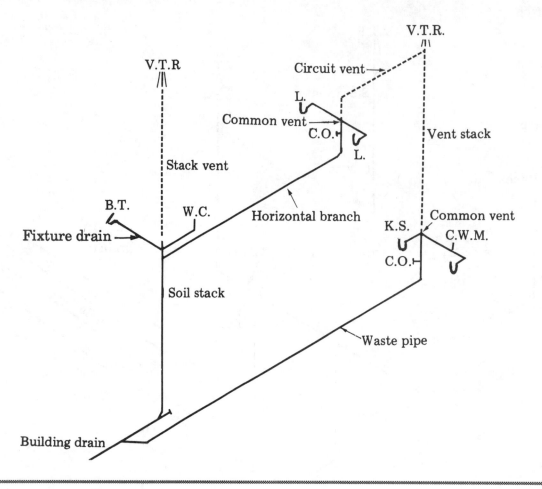

Figure 4–7
Isometric drawing of a two-story house

A *stack vent* (Figures 4–5, 4–6, 4–7) is the extension of a soil or waste stack (dry section) up through the roof of a building.

A *vent stack* (Figures 4–5, 4–7) is the vertical portion of a vent pipe. It lets air circulate to and from all parts of a drainage system.

The *relief vent* (also called a *re-vent*) lets air circulate between the drainage system and the vent of a plumbing system. See Figure 4–5.

A *circuit vent* (Figure 4–7) works like a *branch vent*. It serves two or more fixtures and rises vertically from between the last two fixture traps on a horizontal branch drain. This vent must then connect to the vent stack.

All together, the vent pipes of a building are called a *vent system*.

Questions

1. What is the purpose of isometric drawings?

2. At what angles do the pipes meet in an isometric drawing?

3. What type of triangle is used to make an isometric drawing?

4. In an isometric drawing, what do the lines represent?

5. Why is it important to be able to name the various pipes in a drainage system?

6. How is a public sewer maintained?

7. Name at least three pipes which, by definition, are not actually a part of the building drain.

8. Name three fixtures in a residence which a waste pipe may serve.

9. Name three fixtures in a residence which a soil stack or soil pipe may serve.

10. In a drainage system, what constitutes a stack?

11. A common vent serves how many fixtures installed at the same level?

12. What name is given to the continuation of the waste and vent pipes through the roof?

13. What is the primary function of a vent stack?

14. A vent system is comprised of which pipes?

5

The Drainage System

The greater the volume a pipe has to carry, the larger the pipe should be. That just makes sense. And it's also what the plumbing code requires. Under the code, minimum sizes for sewer, drain, and waste pipes vary with the volume anticipated. That volume is measured in *fixture unit values* (F.U.). Each fixture that drains into the sewer, drain, or waste pipe is assigned a value roughly in proportion to the volume of liquid the fixture can be expected to discharge in a minute. Most lavatories, for instance, are rated at 1 fixture unit (1 cubic foot or 7.5 gallons per minute).

Sizing Drain and Waste Pipes

Your local plumbing code will have a table that shows the fixture unit values for all the common plumbing fixtures. Plumbing codes vary slightly in the values they assign to each type of fixture. Don't worry about this. Just get a copy of the fixtures unit values for your area and use those values when sizing pipe in plumbing systems.

▓ Sizing a Simple Two-Bath House

Let's get some practice by sizing the pipe for the single level two-bath house shown in Figure 5–1. The first step is to find the fixture unit values of the fixtures shown in the figure. Figure 5–2 shows fixture unit values for some common residential plumbing fixtures taken from the Standard Plumbing Code.

According to the Standard Plumbing Code, the house in our example has 19 F.U. Here's how I arrived at that figure.

Fixture	F. U. Value
1 bathtub	2
1 shower	2
2 lavatories	2
2 water closets	8
1 kitchen sink	2
1 clothes washing machine	3
Total	**19**

Once we know the fixture units, we can size the drainage pipes. Once again, plumbing codes vary slightly on the trap sizes required for each fixture. But the sizing procedure is the same no matter what your code specifies.

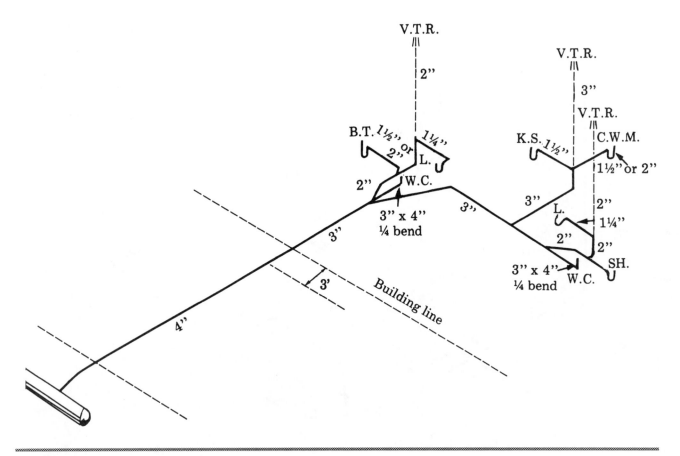

Figure 5–1
Piping layout for a single-level two-bath house

Fixture Type	Fixture Unit Value
Bathtub (with or without overhead shower)	2 F.U.
Lavatory	1 F.U.
Shower stall, domestic	2 F.U.
Kitchen sink, domestic	2 F.U.
Water closet, tank operated	4 F.U.
Clothes washing machine	3 F.U.
Laundry tray (1 or 2 compartments)	2 F.U.

Figure 5–2
Fixture unit values for common residential plumbing fixtures (Standard Plumbing Code)

Pipe Diameter (inches)	Fixture Unit Value
2	21
2½	24
3	27[1]
4	216

[1]Not over two water closets.

Figure 5–3
Maximum number of F.U. to connect to any part of the building drain or horizontal fixture drains, ¼ inch fall per foot (Standard Plumbing Code)

It's important to understand that a fixture drain can't be smaller than its fixture trap. For example, if a "shower stall, domestic" requires a 2-inch trap, the drain must also be 2 inches. It can't be 1½ inches. Figure 5–3 shows pipe diameters and fixture unit values from the Standard Plumbing Code.

These figures are based on a fall (pitch) of ¼ inch per foot of run because this is the standard fall most authorities use for drainage piping. Inspectors don't carry precise measuring devices to verify the actual fall with total accuracy. So, to avoid controversy with the installer, inspectors

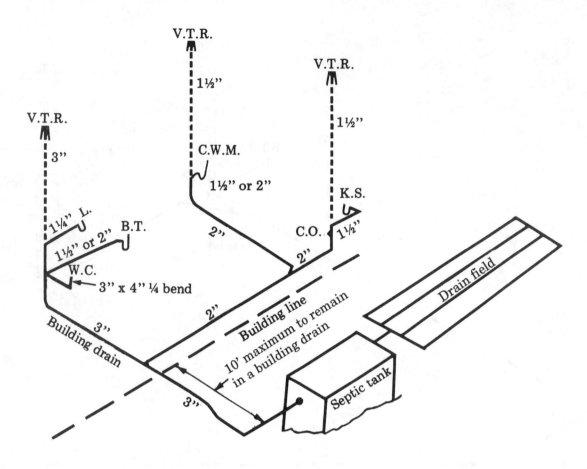

Figure 5–4
A building sewer connected to a septic tank

usually accept any fall of from ⅛ to ½ inch per foot for all horizontal waste lines.

So, using Figure 5–3, you determine that a 3-inch building drain can carry 19 fixture units (since there are only 2 water closets) to the building sewer. The minimum size of the building sewer is the same in all codes:

> "The minimum size of a building sewer shall not be smaller than 4 inches."

The drainage and waste pipe sizes for the two-bath house in Figure 5–1 should meet the minimum model code requirements wherever it's built.

Exceptions to the Rules

As with many sections of the code, there are exceptions. Let's look at a few of the more com-

mon exceptions you'll encounter when sizing the piping system.

Figure 5–4 shows a building sewer connected to a septic tank. Although this could be a building sewer, it's not classified as such because its developed length doesn't exceed 10 feet. Therefore, the portion of the sewer pipe exceeding the 3-foot limit beyond the exterior wall of this particular building (more or less in some codes) may be considered as part of the building drain. That means it can be sized at 3 inches.

Another perplexing situation which could cause some confusion is shown in Figure 5–5. A room addition and bathroom is to be installed at the side of an existing building. The plot plan shows the sewer from the new addition installed around the outside of the existing structure. It's connected to the existing sewer line in the front yard. The new

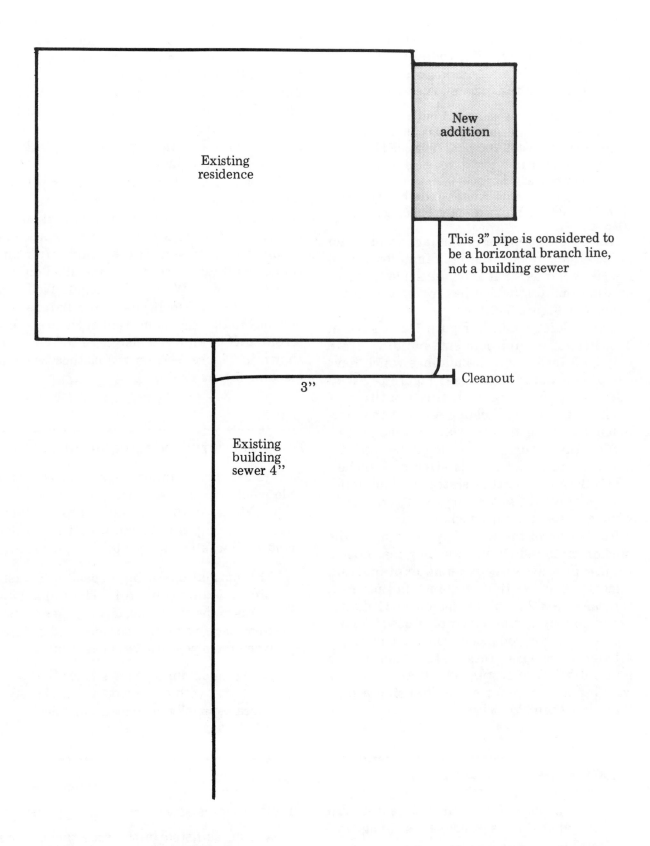

Figure 5–5
Sewer line for an addition to a building

sewer pipe is sized at only 3 inches. Normally it would be sized at the full 4 inches required for building sewers. The code has made an exception in this and similar cases, and the interpretation of the code reflects this:

> For additions to residential buildings when soil and vent lines are inaccessible and it is necessary to install the sewer line outside and around an existing structure, such line shall be considered as a horizontal branch. This horizontal branch must connect into the existing building drain, in this case the building sewer.

The code also makes a similar exception for an accessory residential building (not a commercial building) that is on the same lot as an existing building which has a single building sewer. See Figure 5–6.

The building drain in Figure 5–1 is sized at 3 inches. The minimum size building drain may be 3 inches for all buildings which have one or two water closets (see Figure 5–3 footnote). Any building with three bathrooms must have a 4-inch building drain at the junction of flow from all three water closets.

Also, the fixture drain (sometimes called a *wet vent*) in Figure 5–1 is sized at 2 inches. Why? Because the code states that the minimum size vent (wet or dry) you can use to serve a water closet is 2 inches.

The sizing of the waste pipe leading to the kitchen sink and clothes washing machine in Figure 5–1 also needs some explanation. Figure 5–3 shows that at ¼-inch fall per foot, you can use a 2-inch pipe for up to 21 fixture units. So why is the 3-inch pipe used? Again, it's because the code says you can't install a kitchen sink on a cross installation with a pump discharge fixture (such as a clothes washing machine, for example) using a pipe that's less than 2½ inches.

Most plumbing codes also require that each building have at least one minimum-size vent stack of not less than 3 or 4 inches extending through the roof. Since both bathrooms have a 2-inch wet vent and the code requires a 3-inch pipe up and through the sanitary tap cross, it's good plumbing design to satisfy the code requirements by making this vent stack the main vent for the building.

Figure 5–7 shows another plumbing design where the bathroom vent is used as the main, or minimum-size, 3- or 4-inch vent stack for the building. In this case the vent stack is 3 inches in diameter. The kitchen sink and clothes washing machine are installed on a 2-inch waste pipe. Why is that legal? Well, you can even use a 2-inch pipe if the fixture connections to the waste and vent stack are at different levels and you use a relief vent, as shown in Figure 5–7, on the clothes washing machine fixture drain.

Minimum Size Code Requirements

Always use the minimum sizes required by the code when you figure which pipe sizes to use for building drain and waste pipes. Using larger waste pipes than required isn't recommended. Here's the theory behind pipe sizing:

- The normal operating capacity of waste piping is one-third full. The other two-thirds of the pipe's capacity is to keep waste from backing up in floor-mounted fixtures when the pipe is used at peak volume.

- Using minimum pipe sizes helps the scouring action within the pipe and avoids stoppages, especially in kitchen sink lines.

Questions

1. The maximum fixture unit value is used to size what three sections of a drainage system?

2. How many gallons of water flow per minute are equal to a fixture unit?

3. Do the pipe sizes for a building drainage system, as listed in the code table, represent the *maximum* pipe sizes or the *minimum* pipe sizes?

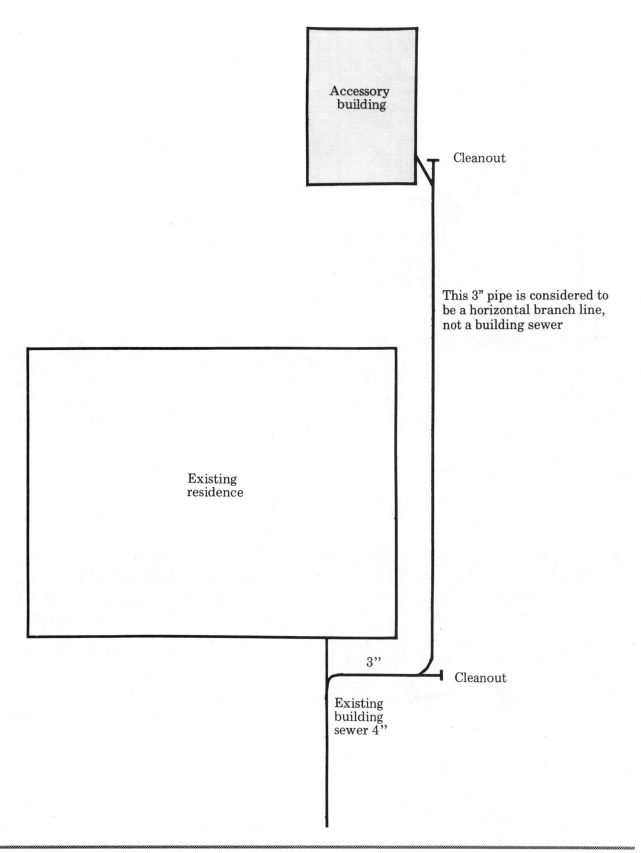

Figure 5–6
Sewer line for an accessory building

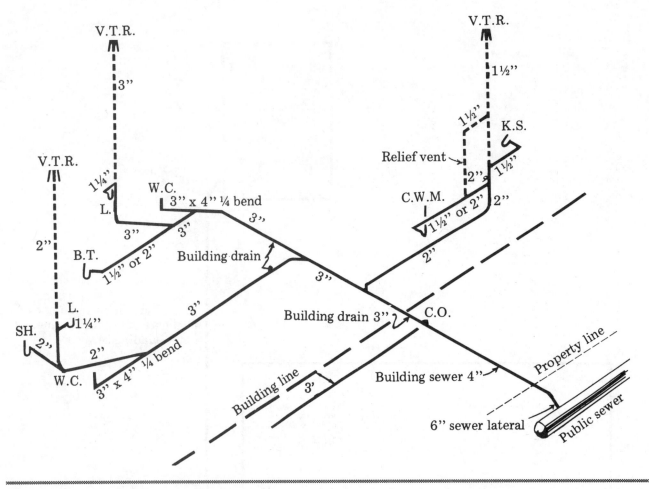

Figure 5–7
Using fixture connections at different levels

4. What part of a waste pipe is considered to be the normal capacity of waste piping?

5. Why should drainage piping be the smallest size permitted by code?

6. What are the two primary differences in model plumbing codes in determining waste pipe sizes?

7. What is the minimum size of a building sewer?

8. How far may a building drain be extended beyond the exterior wall of a building when it is connected to a septic tank and still be classified as a drain pipe?

9. What size drain pipe must be used to serve an accessory building with one bath, when both the accessory building and the main building are on the same lot and are connected to a single building sewer?

10. When is it permissible for a kitchen sink and a clothes washing machine to be installed at different levels on a 2-inch waste stack?

6

The Vent System

Vents relieve the pressure that builds up as water from plumbing fixtures is discharged into the sanitary drainage system. These pipes are sized and arranged to provide the best possible relief for each fixture and for the system. The free flow of air within the vent system keeps the back pressure or siphoning action from destroying the fixture trap seal.

In rural areas where inspections are not required, or in municipalities where inspections are lax, inadequate sizing and arrangements of vent pipes make the following problems common:

- Plumbing fixtures drain slowly, as if a partial stoppage exists.

- Water closets need several flushes to remove contents from the bowl.

- Back pressure (called *positive pressure*) within the drainage pipes may, if strong enough, force sewer gases up and through the liquid trap seals and into the building.

- Plumbing fixtures located farther from a vent pipe than is permitted by code may, when the contents are released, siphon the liquid trap seal. (This is known as *negative pressure*.)

Good venting is especially important in very cold climates where frost may form on the inside of the vent pipes. Frost can accumulate when warm, moist air flows up the vent and meets frigid outside air. Some codes require vent "increasers" or other devices to guarantee free air flow, even in the coldest weather.

Sizing the Vent System

The size of the vent pipe—like the size of its cousin, the soil and waste pipe—is determined by the maximum fixture unit load, the developed length of the vent pipe, type of plumbing fixtures to be vented, and the diameter of the soil or waste stack the vent pipe serves.

Here are the basic principles of venting as required by most plumbing codes:

- For each building with a single building sewer receiving the discharge of a water closet, there must be at least one minimum size vent stack of not less than 3 or 4 inches extending through and above the building roof.

- Use Figure 6–1 to size vent stack or stacks serving an accessory building or other building located on the same lot and

Maximum Fixture Units Permitted on	Uniform Plumbing Code	Standard Plumbing Code
a horizontal wet vent sized 2 inches	4	4
a vertical dry vent sized 1¼ inches	1	2
a vertical dry vent sized 1½ inches	8	10
a vertical dry vent sized 2 inches	24	20
a vertical dry vent sized 3 inches	84	60

Figure 6–1
Fixture unit calculations under two popular model codes

sharing one common building sewer. However, note an exception to this exception: If a water closet is located in the accessory building, the vent stack must be no smaller than 2 inches.

• No wet or dry vent serving a water closet can be less than 2 inches in diameter.

• The diameter of the vent stack must not exceed the diameter of the soil or waste stack to which it connects. For example, if the waste pipe for a kitchen sink is 2 inches, the vent stack can't be 2½ inches or larger. It can be 2 inches or smaller so long as it complies with the fixture units listed in Figure 6–1.

• The diameter of an individual vent stack can't be less than 1¼ inches or less than one-half the diameter of the drain to which it connects.

Here are two examples:

1. If a waste pipe is 2 inches and serves one lavatory, the minimum size vent is 1¼ inches.

2. If a waste pipe is 3 inches, the minimum size vent is 1½ inches.

Figure 6–1 lists fixture units and vent pipe sizes for a single-family residence with simple plumbing installations under two model codes. Note in Figure 6–1 that there are differences between the two model codes in the number of fixture units permitted. Now look back to Figure 5-1 in the previous chapter. Notice that the two bathrooms are served by a horizontal 2-inch wet vent. The kitchen sink and clothes washing machine are served by a single 3-inch

stack vent. The differences in the two model codes (Figure 6–1) will not affect the sizing of vent pipes shown in Figure 5-1.

Generally, the tables in this book will be accurate for sizing plumbing system piping in most single-family residences. But the plumbing code enforced in your community has the last word on what's required. Refer to these tables in your code for more exact information.

▓ Stack Venting

Stack venting is commonly used where plumbing fixtures are located on the same wall or where plumbing fixtures are located back to back. Each fixture drain must connect independently to the stack. This type of plumbing installation is very simple and can be used either in a single-story house or on the second floor of a two-story house. A single main pipe receives the liquid waste and also vents the group of fixtures.

The stack arrangements shown in Figures 6–2 and 6–3 are acceptable under most plumbing codes. Figure 6–2 illustrates a typical group arrangement: a water closet, lavatory, bathtub and kitchen sink. Figure 6–3 shows two bathrooms back to back, totaling two water closets, two lavatories and two bathtubs.

▓ Vertical Combination Waste and Vent Stack

The vertical combination waste and vent stack is a unique one-pipe system which both vents and receives waste from certain plumbing fixtures. Although it is used frequently in high-rise or commercial buildings, it can be used to advantage in single-family residences having more than one story. Note: Some codes require

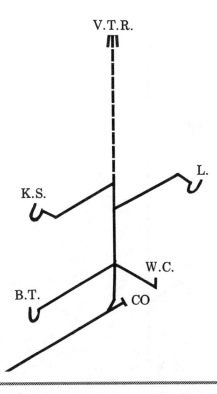

Figure 6–2
Single unit stack venting

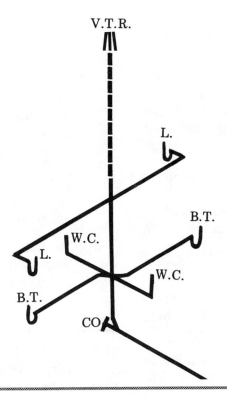

Figure 6–3
Double unit stack venting

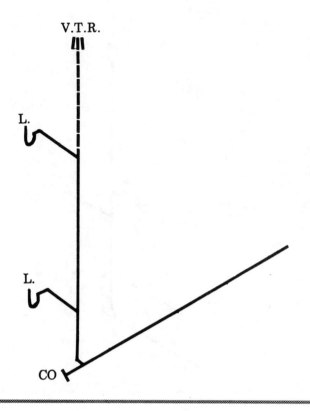

Figure 6–4
Vertical combination waste and vent stack

special approval for vertical combination waste and vent systems.

Low-unit-rated fixtures (like lavatories, bidets, bathtubs, showers and sinks) can use a combination waste and vent stack. If this arrangement is used, remember the following restrictions:

- The stack must be vertical and must extend full-size through the roof. A 3-inch combination waste and vent stack can't be reduced above the topmost fixture.

- The stack can't be connected to another existing vent pipe below the roof.

- Check the maximum number of fixture units and pipe sizes carefully. Plumbing codes vary considerably in this area.

- You can't place a kitchen sink on a 2-inch combined waste and vent stack. For example:

  1. A kitchen sink connected to a 2-inch stack on the first floor can't be extended

to a second-floor bathroom and receive the waste from, say, a lavatory.

  2. A lavatory on the first floor and a lavatory on the second floor located directly over one another could both be served by a combination waste and vent pipe. (Since they are isolated from the rest of the bathroom fixtures, it would be difficult and expensive to connect each lavatory waste pipe to the existing waste pipes across the room.)

Figure 6–4 shows a vertical combination waste and vent system installation that is acceptable under most codes.

■ Wet Venting

This special method of venting is generally used in dwellings. Wet venting provides adequate trap protection for certain plumbing fixtures. A wet vent can be vertical or horizontal, as illustrated in Figure 6–5. It can be used on a stack or a flat plumbing installation. This unique

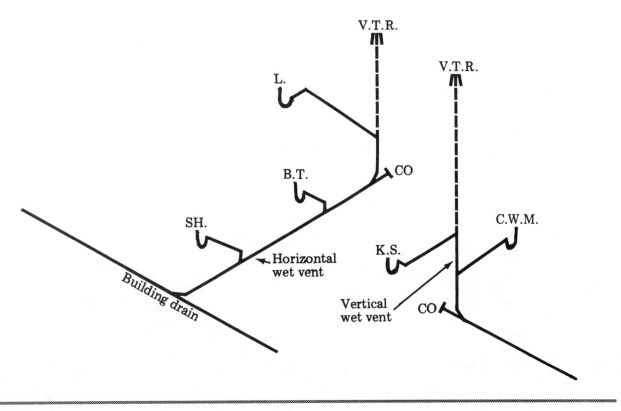

Figure 6–5
Wet venting

venting system is probably used more often than any other type and is generally accepted by all model plumbing codes. Since it is a single piping system, a wet vent is economical and can serve to vent several adjoining plumbing fixtures located on the same floor level.

As a building drain, a wet vent can convey only waste from fixtures with low unit ratings. These are lavatories, bathtubs, bidets and showers. Kitchen sinks can't use a 2-inch wet vent in some codes, but can use a 3-inch wet vent system in all model codes. *Water closets or similar fixtures may not use a wet vent system.* See Figure 6–1 for the fixture units permitted on a 2-inch horizontal wet vent.

Horizontal wet vents can't exceed 15 feet in developed length. Vertical wet vents connecting to a horizontal wet vent can't exceed 6 feet. See Figure 6–6.

Common Vent

Two methods of installing a common vent are shown in Figure 6–7. These are vertical vents serving two fixture traps installed at the

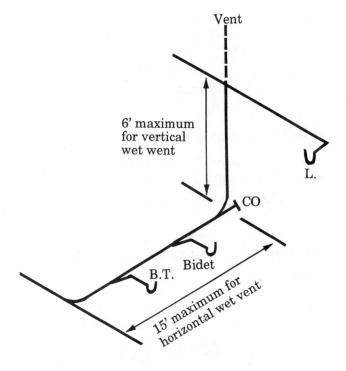

Figure 6–6
Vertical wet vent connected to horizontal wet vent

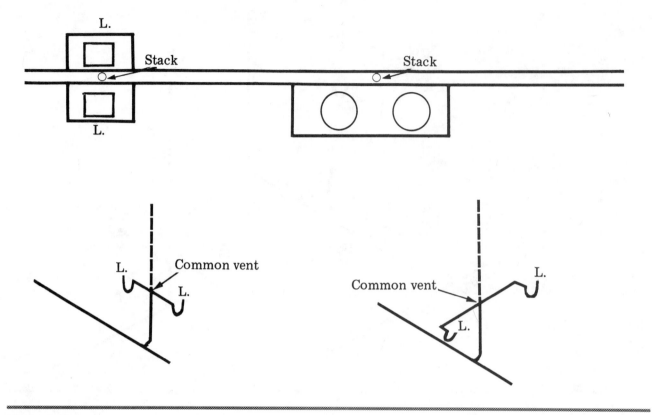

Figure 6–7
Common venting

same level. The common vent may serve two fixtures installed back to back or two fixtures installed side by side. Figure 4–7 back in Chapter 4 illustrates other examples of common vent installations.

▨ Individual Vents

This type of vent is also called a *back vent* or a *continuous vent*. Figure 6–8 illustrates these common installations.

Horizontal Distance of Fixture Trap from Vent

In the 1920s the U.S. Department of Commerce made a substantial effort to establish uniform standards for many plumbing requirements. Those standards are now reflected in most codes. Unfortunately, the various codes still do not agree on the distance and fixture drain size requirements. Refer to the tables adopted by your local authority for this section of the code. Compare your code with Figures 6–9 and 6–10.

Note carefully that the maximum distance from the trap to the vent in Figures 6–9 and 6–10 depends on three factors:

1. the size of the fixture drain
2. the size of the trap
3. the fall per foot of the fixture drain

Even though codes differ, you have to follow the tables provided in the code adopted by your community.

Some principles for horizontal runs between the fixture trap and the vent are accepted under all codes. These are summarized below and discussed in the final section of this chapter.

- The closer the trap to the vent on a minimum slope, the better.

- Every fixture trap must be protected against siphonage and back pressure and must be provided with a vent piping system that permits free admission or emission of air under normal intended use.

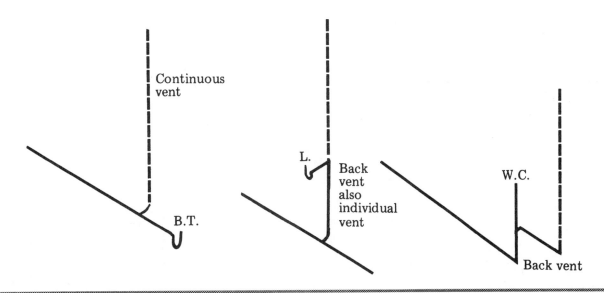

Figure 6–8
Individual vents

Size of Fixture Drain (inches)	Size of Trap (inches)	Fall per Foot (inches)	Maximum Distance Trap from Vent
1¼	1¼	¼	2'6"
1½	1½	¼	3'6"
2	2	¼	5'0"

Figure 6–9
Horizontal distance of fixture trap from vent opening (Uniform Plumbing Code)

Size of Fixture Drain (inches)	Size of Trap (inches)	Fall per Foot (inches)	Maximum Distance Trap from Vent
1¼	1¼	¼	3'6"
1½	1¼	¼	5'
2	2	¼	6'

Figure 6–10
Horizontal distance of fixture trap from vent opening (Standard Plumbing Code)

Measure the developed length of a fixture drain from the crown weir of a fixture trap to the vent pipe, as shown in Figure 6–11. This is the *developed length* and includes offsets and turns. The developed length must be within the limits prescribed in the code.

The length of the fixture drain may in some instances have to exceed the limits in Figures 6–9 and 6–10 because of fixture locations. In this case, install a relief vent to stay within the limits of your code. For example, a lavatory must be installed with a 5-foot fixture drain from the nearest vent. The drain pipe is 1¼ inches and is installed with ¼ inch per foot slope. The installation methods illustrated in Figure 6–12 comply with the requirements of

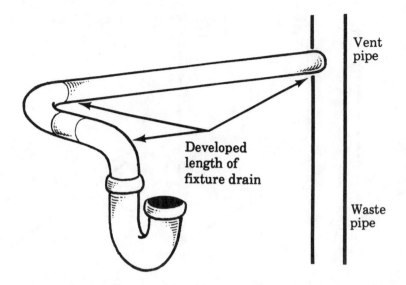

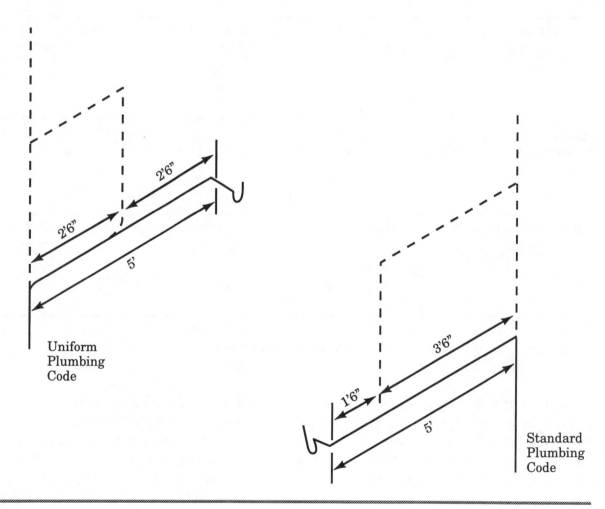

Figure 6–11
Measuring the developed length of a fixture drain

Figure 6–12
Fixture drain re-vent

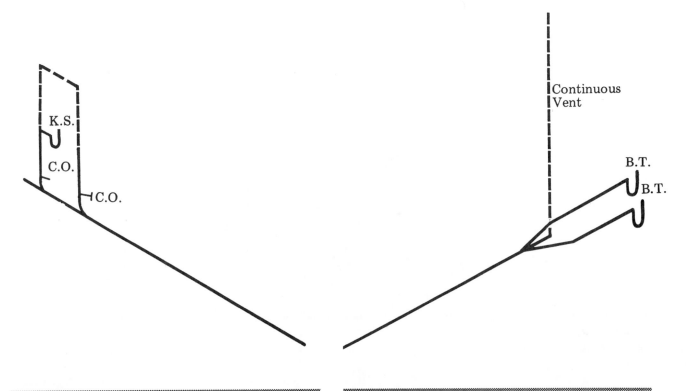

Figure 6–13
Island sink

Figure 6–14
Continuous vent

the two model codes in Figures 6–9 and 6–10. One of these methods must be used for this arrangement to be acceptable.

The isometric drawings in Figures 6–12 through 6–19 show plumbing designs approved by most codes. These piping diagrams will help you understand how waste and vent pipes are related. Note that the pipes in these drawings are not sized. But the drawings should clarify the correct location and intended use of piping shown.

Develop plumbing designs from these isometrics for practice. Then size the waste, drainage, and vent pipes according to the tables in your code. Have your isometric drawings checked for accuracy by an experienced plumber, job foreman, or plumbing inspector.

Questions

1. Name the two major reasons a fixture trap seal is destroyed when it is not properly vented.

2. Name five problems that may occur to a drainage system when vent pipes are inadequately sized and arranged.

3. Name four things that are vital in determining the sizing of vent pipes.

4. In a building having three vent stacks, what is the minimum size required by code for at least one of these three?

5. What is the minimum size vent stack for an accessory building with a full bath located on the same lot as the main building and sharing one common building sewer?

6. When is it permissible for the diameter of a vent stack to exceed the diameter of the waste stack to which it connects?

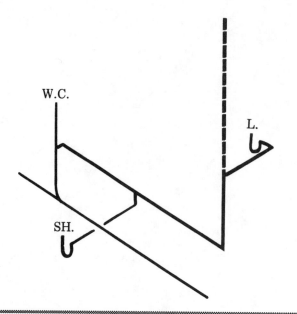

Figure 6–15
Stack wet vent installation

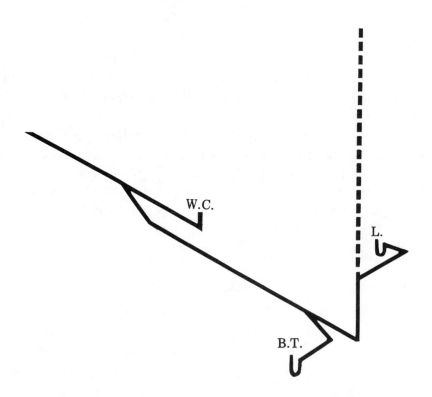

Figure 6–16
Flat wet vent installation

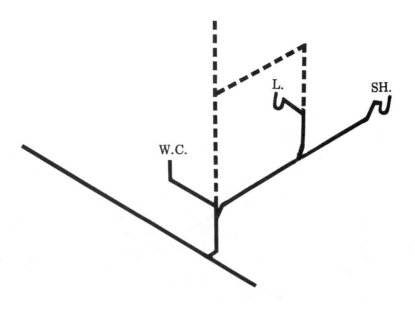

Figure 6–17
Wet vented unit

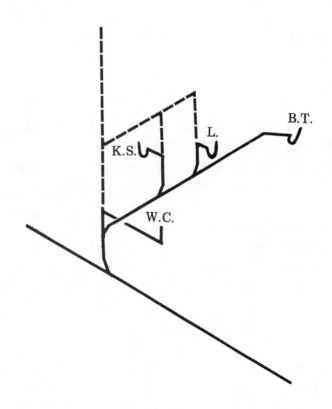

Figure 6–18
Individually vented unit

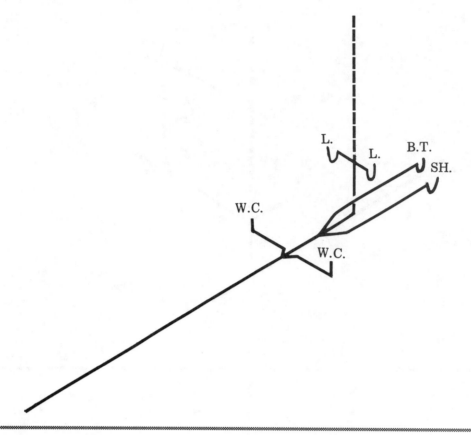

Figure 6–19
Back-to-back bathroom with horizontal wet vent installed on the flat

7. What is the smallest vent pipe in inches that can serve any plumbing fixture?

8. In stack venting, what procedure must be followed in connecting each fixture drain to the stack?

9. Stack venting a group of fixtures fulfills what two major functions?

10. What type plumbing fixtures may be served by a vertical combination waste and vent stack?

11. Can a 3-inch combination waste and vent stack connect to a 3- or 4-inch stack vent on the second floor? Why?

12. Name a common plumbing fixture that can't be installed on a 2-inch combined waste and vent stack.

13. Name three plumbing fixtures that may connect to a wet vent system.

14. Can you name the two changes in direction a wet vent may assume and still remain within the intent of the code?

15. What is the maximum distance set by code for a horizontal wet vent?

16. What is the maximum height of a vertical wet vent when it is connected to a horizontal wet vent?

17. What is the name given to a vent pipe serving two fixture traps connected at the same level?

18. What must be included when measuring the developed length of a fixture drain?

19. Most authorities agree that the closer a trap is to the vent on a minimum slope, the better it will serve the fixture. Why is this?

7

Traps

Plumbing fixtures connected directly to the sanitary drainage system must be equipped with a water seal trap. Each fixture must have its own trap. Fixtures such as water closets have integral traps fashioned within the fixture body.

The trap is designed to provide a liquid seal which prevents drainage system odors, gases and even vermin from entering the building at fixture locations. The trap must provide this liquid seal protection without restricting the flow of sewage or other waste.

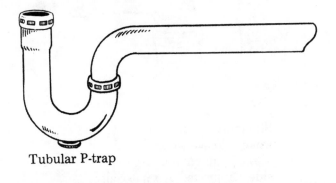

Tubular P-trap

Figure 7–1
Two-piece common tubular P-trap

Fixture Traps

Figure 7–1 shows the fixture trap most commonly used for lavatories, kitchen sinks, laundry trays, and the like. This trap is called a *P-trap* because its shape resembles an inverted capital letter P.

The code has placed many restrictions and limitations on the use of the P-trap because of its unique importance in protecting human health. These restrictions, which appear in various parts of the code, are listed below.

- Fixture traps must be self-cleaning. This means that the interior of the trap can not

have anything that will retain hair, lint, or other foreign substances.

- No trap outlet can be larger than the fixture drain to which it is connected. In other words, you can't connect a 1½-inch trap to a 1¼-inch drain with a reducer.

- Traps can't depend on movable parts to retain the water seal.

- Figure 7–2 shows traps *prohibited* by most model codes. These include bell traps, running traps, crown vented traps, pot traps, ¾ S-traps, full S-traps, and traps with

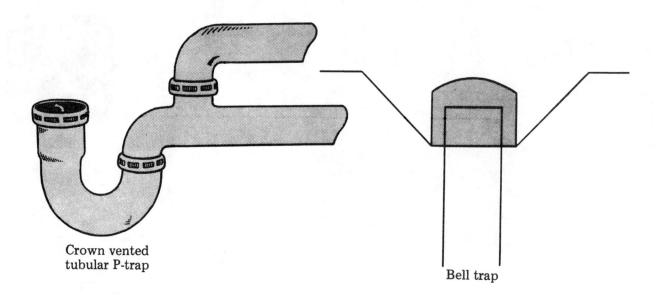

Crown vented
tubular P-trap

Bell trap

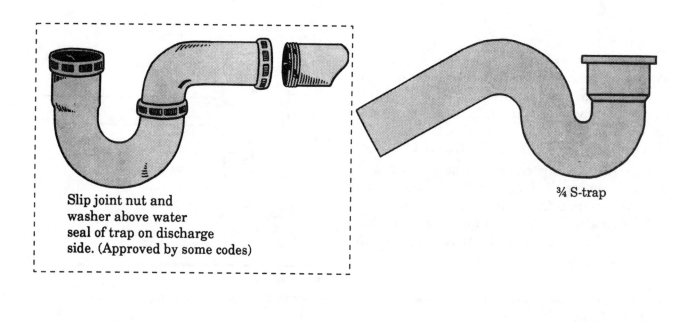

Slip joint nut and
washer above water
seal of trap on discharge
side. (Approved by some codes)

¾ S-trap

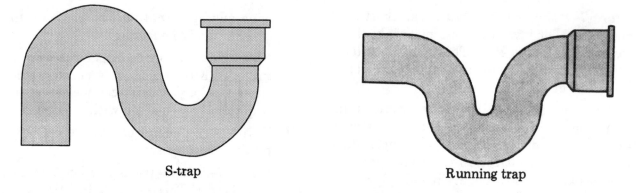

S-trap

Running trap

Figure 7–2
Traps prohibited by most codes

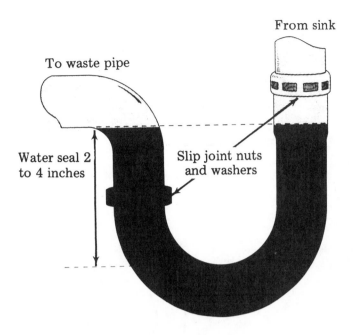

Figure 7–3
Trap illustrating proper water seal depth

slip-joint nuts and washers on the discharge side of the trap above the water seal. You may wonder why some of these prohibited traps (such as the full S-trap) are still sold at plumbing supply houses. Many older homes have full S-traps that were installed when these traps were still acceptable under the local code. Today, they may be used for replacement purposes only, never for new construction.

- Each fixture trap must have a water seal of not less than 2 inches or more than 4 inches. See Figure 7–3.

- All traps must be installed level in relation to their water seals. See Figure 7–4. This is necessary to prevent negative action or self-siphonage of liquid in the trap.

- With one exception, each plumbing fixture must be separately trapped by a water seal trap. The exception is water closets or similar fixtures with an integral trap. These can't be separately trapped. See Figure 7–5.

- Sinks, laundry trays or similar fixtures having two or three compartments may be connected to a single trap with a continu-ous waste (providing the compartments are adjacent to one another and one compartment is not more than 6 inches deeper than the other). See Figure 7–6.

- No fixture can be double trapped. This means that the liquid waste discharged from a fixture may not go through one trap and then a second trap before discharging into the building stack or drain. The water closet in Figure 7–5 is double trapped as shown and, therefore, would not be acceptable.

- Some codes permit two or three single compartment sinks or lavatories to use a single trap, provided that (a) where three fixtures are installed the trap must be located on the center fixture, (b) the fixtures must be set at the same level, and (c) the fixtures must not be spaced more than 30 inches center to center. See Figure 7–7.

- There is a maximum allowable vertical drop from a fixture waste outlet to the trap water seal. This maximum drop further prevents self-siphonage of the fixture trap water seal. The shorter the distance between these two points, the more efficient

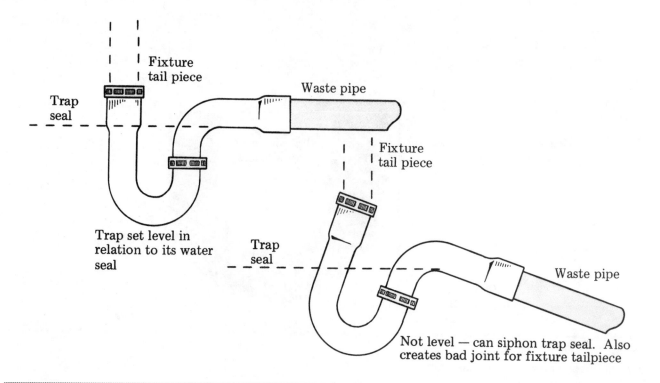

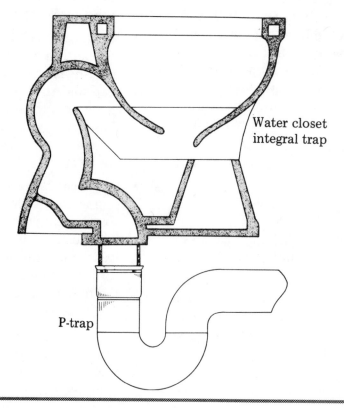

Figure 7–4

A trap must be level with its water seal

Figure 7–5

Fixtures with integral traps are not permitted to be double trapped

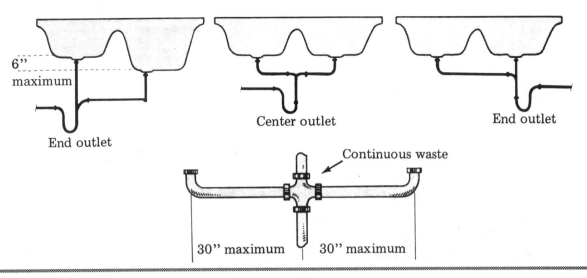

Figure 7–6
Two compartment sink and one trap approved

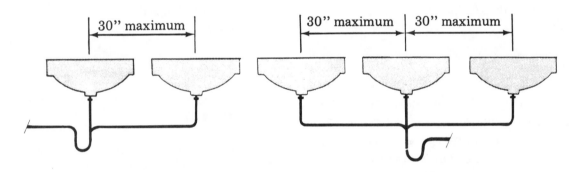

NOTE: Not approved in some codes. Check local code for approval.

Figure 7–7
Two or three lavatories on one trap approved

the fixture trap will be. The longer the fixture tailpiece, the greater the velocity of the waste water as it rushes out a fixture drain. Excess velocity can siphon the trap seal. For sinks, lavatories, showers, bathtubs and all similar fixtures, the vertical drop (tailpiece) can't exceed 24 inches (18 inches in some codes). See Figure 7–8.

• The vertical drop of the pipe serving floor-connected fixtures with integral traps, water closets and similar fixtures can't exceed 24 inches as shown in Figure 7–9. This is generally acceptable by most codes. Of course, a water closet is designed so that the contents can be

siphoned with each flush. When a water closet is flushed, the trap seal is lost, and it would remain so if the trap seal were not automatically restored by the refill tube in the flush tank or flush valve.

• Materials for concealed fixture traps, as for bathtubs, showers and bidets, must be cast iron, cast brass or lead. In a plastic system, plastic traps can be used. Concealed fixture traps must not be equipped with cleanouts.

• Materials for exposed fixture traps or otherwise accessible traps (except for fixtures with integral traps) must be cast iron, cast brass, lead, 20 or 17 gauge chrome brass or

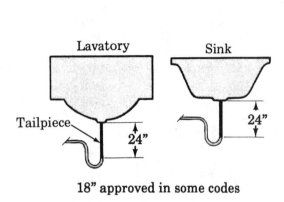

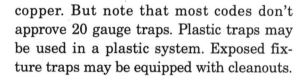

18" approved in some codes

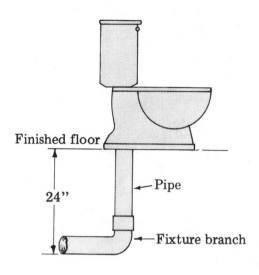

Figure 7–8
Maximum vertical drop (tailpiece) approved by most codes

Figure 7–9
Maximum vertical pipe drop for fixtures with integral traps

---

copper. But note that most codes don't approve 20 gauge traps. Plastic traps may be used in a plastic system. Exposed fixture traps may be equipped with cleanouts.

Drum Traps

Drum traps were commonly used in the installation of bathtubs and lavatories not too many years ago. Most experienced plumbers have seen drum traps installed in older buildings. A drum trap consists of a cylindrical metal shell with an inlet for the fixture near the bottom and a waste outlet near the top. The top has a removable screwed cover with a raised or countersunk head. This trap was installed in the fixture waste line with the top of the trap protruding through the finished floor for easy access. A scatter rug generally concealed the trap from view.

Today, use of drum traps is limited by most codes to special fixtures designed for drum traps. Drum traps must be approved by the local authority before installation and must have a minimum diameter of 4 inches. See Figure 7–10.

Building Traps

Many years ago, before we learned to use venting systems and fixture traps correctly, it was common for the water seal in traps to be destroyed by back pressure or siphonage, or both. Rats were able to travel freely from one building to another. Decomposing sewage generated gas and offensive odors which were released into buildings at fixture locations. Conditions like that gave indoor plumbing a bad reputation. Health department officials of that day recognized that conditions like this were a serious health menace, especially in larger cities and towns. To help solve problems like these, the code required a building or house trap (see Figure 7–11) in each building drainage line. These proved to be generally effective.

Building traps at that time provided a secondary line of defense against rats and sewer gas in the sewer system. The individual fixture traps provided the primary safeguard. The building trap was deemed a necessity until fairly recently. Today most model codes do not require (and some actually prohibit) a house trap in a building drainage line. Properly vented fixture

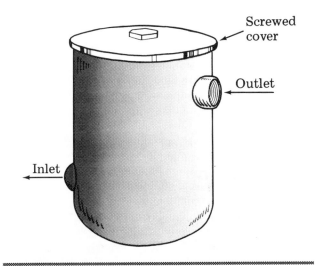

Figure 7–10
Drum trap

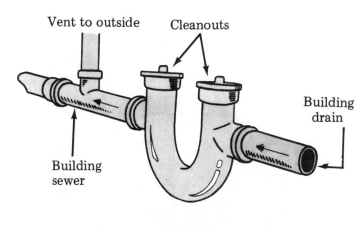

Figure 7–11
House (building) trap

traps in today's buildings no longer require the services of a building trap. Check your local code to see what's required in your community.

Broken Trap Seals

Trap seals may be broken five ways. We'll look at them one at a time.

Wind Effect

This is one of the least likely ways a trap seal may be broken. It can happen when the vent pipes are in strong upward or downward air currents. The pressure or suction created in the stack may cause the water to rise or fall within the trap. Each time it rises within the trap, a small amount may spill into the waste pipes and be lost, thus weakening the trap seal. The trap seal may then be vulnerable to lesser back pressure in a drainage system, allowing sewer gases to penetrate the weakened seal and enter the room at the fixture location.

Evaporation

Evaporation may reduce the depth of the water in a trap seal when a fixture is not used for a long time. When the trap seal is thus weakened, back pressure can break through and allow sewer gases to enter the room at the fixture location. Evaporation in the fixtures can occur when one or more bathrooms or fixtures is not used at least occasionally. Low humidity will tend to evaporate the water from the unused trap or traps faster than usual, weakening the seal or completely destroying it. Any plumbing fixture used at least once a week should retain its seal.

Any home that isn't used for part of the year (such as a vacation home) can lose the liquid seal in fixture traps. The water evaporates, especially in hot weather. If a home isn't going to be used for several months, consider pouring mineral oil in the trap seals. The oil retains the seal without evaporating and prevents vermin from using it as a pathway to enter the building. In winter, mineral oil trap seals prevent the trap from freezing.

Capillary Attraction

This happens rarely, but it can nevertheless cause the trap seal to be broken. Capillary attraction occurs when, for instance, a length of string enters the fixture drain and reaches down into the trap seal and on into the waste stack. The water from the trap seal will follow the string into the waste pipe and leak away, thus destroying the trap seal. See Figure 7–12.

Trap Siphonage

If negative pressure develops within the fixture drain, the fixture trap seal can be siphoned.

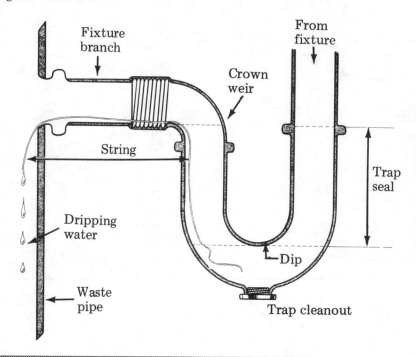

Figure 7–12
Capillary attraction

This may be caused by any of the following three conditions:

1. A fixture with a high unit rating is installed over a fixture with a low unit rating. For example, a water closet is installed on the second floor of a residence and a lavatory on a lower floor; both use the same waste pipe. When the water closet is flushed, a large amount of water is sent past the waste opening of the lavatory. A vacuum is formed, pulling the water from the lavatory trap. See Figure 7–13.

2. A fixture is installed creating a ¾ S-trap in the fixture drain. When the fixture is full and the waste water is released, the water rushing through the trap and drain pipe may carry the trap water with it. Thus, the trap will not retain enough water to properly reseal. This is a true siphon action, since the siphonage is caused by the differences in the weights of two columns of water. See Figure 7–13.

3. A fixture installed on a long unvented horizontal drain line can also siphon the water from the trap seal. This is not considered a true siphon, but it has a similar

effect on the fixture trap. In this installation, water rushing from the fixture into the trap and drain pipe builds up enough velocity to carry the trap water with it. Thus, the seal is broken. See Figure 7–14.

Back Pressure

Siphoning and water momentum tend to empty a fixture trap. But water may also be forced out of the trap in the opposite direction. This is known as *positive* or *back pressure*. This occurs when a higher pressure develops in a drainage system that is improperly designed and vented. This forces air into the fixture drain, through the trap water seal, and into the building. See Figure 7–15.

The Parts of a Trap

All plumbers should know and be able to identify the various parts and materials of the two most commonly used traps in the plumbing trade. Figure 7–16 illustrates the *one-piece trap*. This trap is generally constructed of the same material as the pipe and is installed in the building's drainage system. A cast iron system would use a cast iron trap, a copper system a

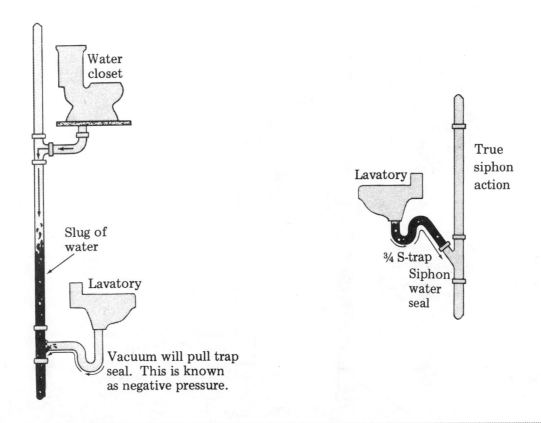

Figure 7–13
These installations not permitted by code

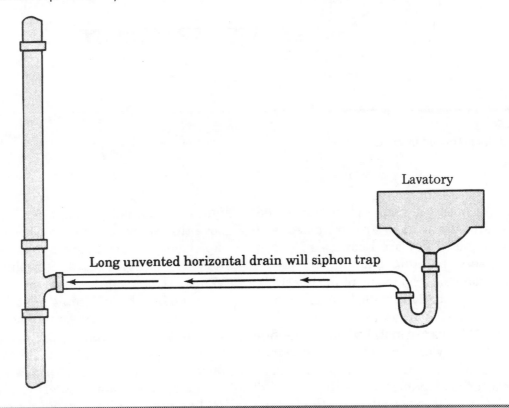

Figure 7–14
Installation not permitted by code

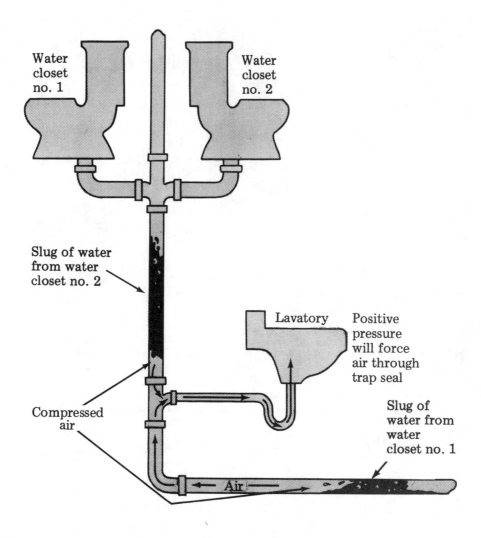

Figure 7–15
Installations not permitted by code

copper trap, a plastic system a plastic trap, and so on. Usually these traps are concealed and have no cleanout. They are known as P-traps and usually serve bathtubs, showers and bidets.

The most common trap in the plumbing trade is a P-trap. It's usually installed under lavatories, sinks, laundry trays and the like. See Figure 7–17. P-traps are installed above the finished floor and may or may not have individual cleanouts in their base. Chrome traps may be of a tubular construction and have slip joints and washers. A slip joint is permitted on the fixture inlet part of the trap and also on the outlet side, provided it is within the trap water seal. See

Figure 7–3. An old slip joint and washer located above the water seal on the discharge side of a trap can permit sewer gases to enter a building. Therefore, this type of trap is prohibited.

Plastic traps can be used for lavatories and sinks where the drainage system is plastic. Plastic pipe can resist heat damage at temperatures up to 180 degrees F. But use plastic traps and tailpieces for kitchen sinks with caution. Many plastic traps and tailpieces in this installation have deteriorated short of their intended life because residents pour boiling water directly into the sink. Codes may prohibit plastic traps and tailpieces in kitchens in the future.

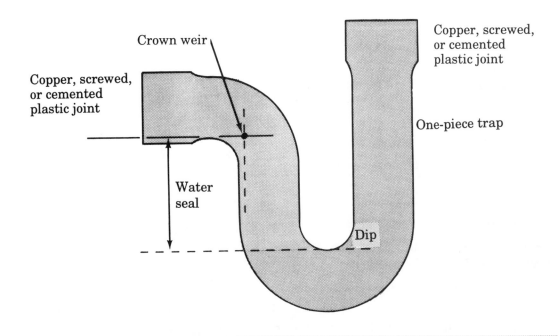

Figure 7–16
Parts of a one-piece trap

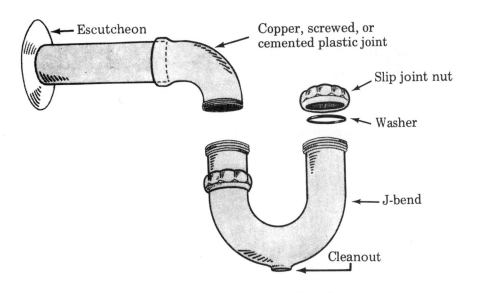

Figure 7–17
Parts of a two-piece P-trap

Questions

1. As shown in this chapter, all fixtures must be provided with what to meet code requirements?

2. How does a trap protect human health?

3. What is the most common trap used in the trade?

4. What is the term for a trap that does not collect and retain foreign substances?

5. Why does the code prohibit a 1½-inch trap connected to a 1¼-inch fixture drain?

6. Name three traps that are prohibited by code for use in new construction.

7. What is the minimum depth permitted by code of the water seal for a trap?

8. What is the maximum depth permitted by code of the water seal for a trap other than an interceptor type?

9. Why should all traps be installed level?

10. When two or more fixtures are permitted to use a single trap, what is the waste pipe for this installation called?

11. When two or more fixtures use a single trap, what is the maximum center-to-center measurement allowed?

12. When three individual fixtures are permitted by code on a single trap, on which fixture must the trap be located?

13. What is double-trapping and when is it permitted?

14. What is the maximum length of the vertical drop from a lavatory outlet to its trap water seal?

15. When a fixture tailpiece exceeds the maximum length set by code, what may happen?

16. What is the maximum length of the vertical drop from a water closet outlet to the building drain?

17. Water closets are designed so a certain trapping function takes place when they are operated. Can you describe this function?

18. In a tank-type water closet the trap seals are automatically restored by what action?

19. What is accomplished by using the shortest fixture tailpiece possible to connect it to its trap?

20. What is prohibited on concealed fixture traps?

21. Where can drum traps be used in today's plumbing?

22. All permitted drum traps must be equipped with what?

23. Why are building traps no longer used in plumbing systems?

24. Name five ways a trap seal may be broken.

25. Explain the difference between negative and positive pressures within a drainage system.

26. Why is it illegal for a trap to have a slip joint nut and washer on the discharge side above the water seal?

27. Why should caution be used in installing a plastic trap on a kitchen sink waste?

28. S-traps are available today for certain types of fixtures. How may S-traps be used?

29. What happens to a fixture trap when the fixture is installed on an unvented horizontal line exceeding the critical distance set by code?

30. Where is the crown weir located on a trap?

31. What portion of a trap is known as a J-bend?

32. What forms the dip of a trap?

33. What does a trap seal accomplish at fixture locations?

34. In a two-compartment sink with two different depths, what is the maximum depth of one compartment using a single trap?

Cleanouts

Before cleanouts were required on drainage piping, plumbers had to cut a hole in blocked drainage pipe to clean out obstructions. The plumber had to insert a cleaning cable to remove the obstruction and then patch the hole with a cement or some other material. These patch jobs often deteriorated and allowed raw sewage to seep out of the pipe and into the ground. This caused a health hazard for the building occupants and their neighbors. Cutting or drilling holes into drainage or vent pipe is now prohibited by code.

Current codes recognize the importance of properly-located and accessible cleanouts. They are now an essential part of the drainage system. Today's model codes specify the location, size, distance between cleanouts, and many other requirements. Every drainage pipe can get stopped up. Making cleanouts accessible saves the serviceman valuable time and saves the owner unnecessary expense.

Cleanouts or cleanout tees are required where a building sewer connects to the public sewer lateral at the property line. See Figure 8–1. A cleanout at this location serves a dual purpose. This is the point at which a test plug is inserted for performing a water test on a building sewer. See Figure 8–2. The test tee

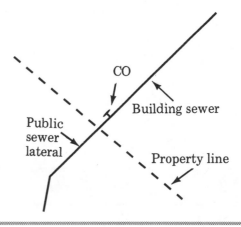

Figure 8–1
Properly installed cleanout at property line

also serves as a cleanout for cleaning any future stoppages that might occur in the public sewer lateral or building sewer. Some codes require that this cleanout be extended up to the finish grade, while others do not.

A full size cleanout *may* be located outside the building at the junction of the building drain and the building sewer, usually within 5 feet of the building line. This is not a requirement if other cleanouts are located upstream

57

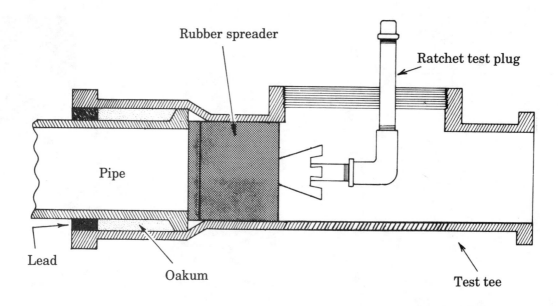

Figure 8–2
Test tee

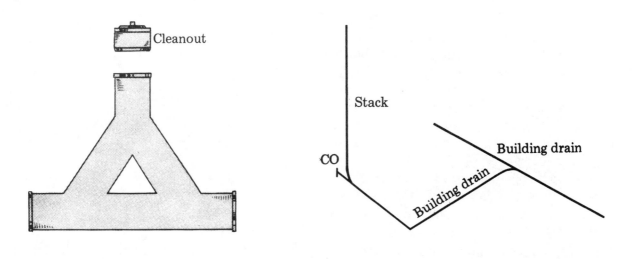

Figure 8–3
"No-hub" two-way cleanout fitting

Figure 8–4
90-degree change in direction requires cleanout

and if a 75-foot sewer cable will reach this area. If a cleanout is used at this location, it must permit upstream as well as downstream rodding. This means that a fitting known as a two-way cleanout must be used. Figure 8–3 shows a no-hub two-way cleanout fitting, which may be installed at the property line as illustrated in Figure 8–1. This fitting may or may not have to be brought to finish grade, depending on local code requirements.

Accessible cleanouts are required on all horizontal drainage piping. The distance between cleanouts can't exceed 75 feet (more or less in some codes). A cleanout should also be provided for each change of direction in a building drain greater than 45 degrees. For example, Figure 8–4 shows the change in direction on a 90-degree drain pipe. A cleanout would be required at the base of the stack or at the end of the drain pipe.

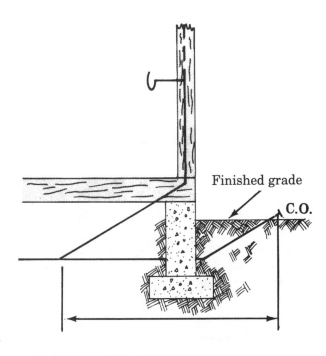

Finished grade

Figure 8–5
Dead end created with cleanout extension

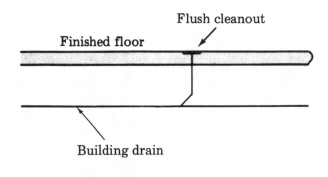

Finished floor · Flush cleanout · Building drain

Figure 8–6
Countersunk cleanout

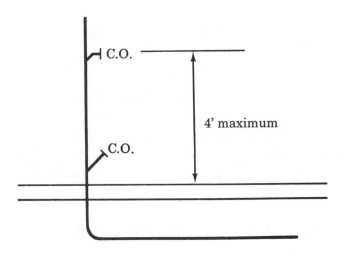

Figure 8–7
A cleanout is required at the base of all required stacks

Sometimes the extension of a cleanout to make it accessible to the outside of a building creates a "dead end." See Figure 8–5. This dead end is created when an extension has a developed length of 2 feet or more. Dead ends like

this are permitted only where there is no convenient way to get access from both directions.

When an underground piping cleanout terminates in a walkway, hallway or room, it must have a countersunk plug to prevent tripping or accidental injury. See Figure 8–6.

An exposed vertical stack may have a cleanout in the base or located within 4 feet of the finished floor, as shown in Figure 8–7. Either of the locations shown in the figure is acceptable.

If a cleanout can't be extended to the outside of a building and must be installed in a concealed vertical stack, a cleanout tee may be used. The cleanout plug must be accessible. Study Figure 8–8. A covering plate or access door must be provided within 4 feet of the finished floor to permit rodding. The covering plate may be held in place by a long screw. The raised head portion of the cleanout plug may be tapped for this purpose.

Cleanouts smaller than 3 inches must have a 12-inch clearance for rodding purposes. Cleanouts 3 inches and larger need an 18-inch clearance. See Figure 8–9.

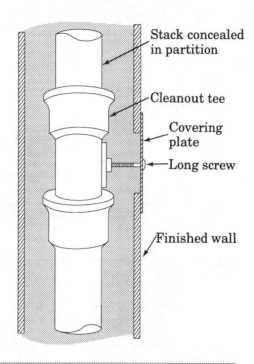

Figure 8–8
Cleanout tee in vertical stack

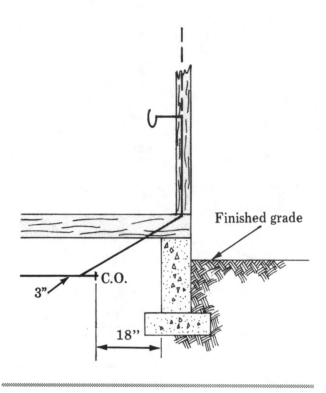

Figure 8–9
Minimum 18-inch clearance from wall

Cleanouts in one-story buildings can be omitted if certain code requirements are met. This section of the code is controversial, so check your local code before proceeding with the installation. Figure 8–10 may help you understand code requirements if this type of installation is permitted.

Vent stacks can sometimes serve as cleanouts. In certain types of installations a vent stack can be used both to supply and remove air and as a cleanout for inserting a cleaning snake. This type of vent must meet the following requirements to qualify for this dual role:

- The drainage system must not have more than one 90-degree change in direction.

- The vent stack must be vertical throughout (without any offsets) and must extend up through the roof.

- The vent stack must be of the same size as the waste pipe it serves.

- The vent pipe must not be reduced beyond the following minimums:

4-inch must not reduce to less than 3-inch

3-inch must not reduce to less than 2-inch

2-inch must not reduce to less than 1½-inch

Look at Figure 8–10. A two-way cleanout in the building drain permits rodding upstream to the base of Stack A. Installing a cleanout in the base of Stack A isn't necessary.

Stack B requires a cleanout at its base. The stack is properly sized, but is not vertical throughout as required by the code. Stack C has the same diameter as the waste pipe it serves (2 inches) and is vertical up and through the roof. A cleanout isn't required in its base.

Figure 8–11 shows a countersunk cleanout plug. Figure 8–12 shows a raised hex cleanout plug. A raised hex cleanout plug with cleanout body installed in the hub of a cast-iron pipe is illustrated in Figure 8–13. The size of these cleanout bodies, up to 4 inches, must be the same nominal size as the pipe to which the cleanout is joined. A waste pipe 2 inches in diameter would require a 2-inch cleanout; a 3-inch waste pipe would require a 3-inch cleanout, and so on. Building sewers up to 8 inches in diameter may be served by 4-inch cleanouts.

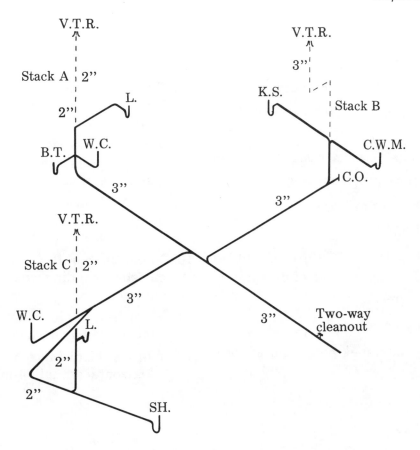

Figure 8–10
When vents are or are not permitted as cleanouts

Figure 8–11
Countersunk cleanout plug

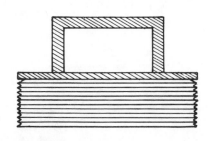

Figure 8–12
Raised hex cleanout plug

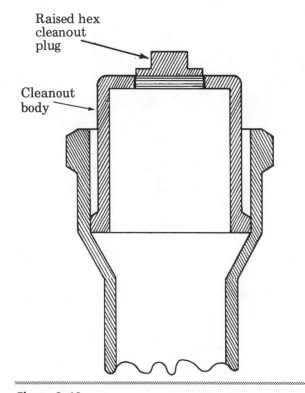

Figure 8–13
Cast iron pipe with hub

Questions

1. Why does the code prohibit cutting a hole in a drainage or vent pipe to remove a blockage?

2. What is the dual role of a cleanout tee when installed at the end of a building sewer?

3. What is the advantage of a two-way cleanout fitting?

4. A cleanout is required in a horizontal drain pipe for each change of direction of how many degrees?

5. What creates a dead end?

6. When is a countersunk plug required for a cleanout?

7. What is the maximum height from a finished floor a cleanout may be installed in a vertical stack?

8. What are the two possible procedures in making a concealed cleanout in a vertical stack accessible?

9. What is the necessary clearance from walls or other obstructions, required by code, for a 3-inch cleanout for rodding purposes?

10. Name three of the four requirements in a one-story building where cleanouts may be omitted.

11. What is the minimum size of a cleanout which serves a 4-inch horizontal drain pipe?

12. What is the distance between cleanouts in a horizontal straight 4-inch drain line?

Floor Plans and Layouts

This chapter will describe how to lay out the plumbing system in a home. I'll use isometric drawings and floor plans to show how the plumbing system will be installed.

I suspect that plumbing design has changed more in the last 50 years than any other part of the home. Many years ago bathrooms were purely utilitarian. The plumbing systems were simple installations and maintenance problems were minimal. While they were clean and antiseptic, you'd seldom find much style expressed in an older bathroom.

Modern home builders have learned that prospective owners want bathrooms to be both useful and attractive. It's easier to add style and a decorative flare if there's more space to work with. Maybe that's why better homes usually have at least one bathroom that measures more than the standard 5 feet by 8 feet. New bathtub, shower, lavatory and water closet designs usually require extra floor space.

Many builders allow more space for bathrooms because prospective owners expect the master bathroom to be a showplace. Bathroom design may be almost as important to the overall appeal of the house as design of the kitchen or dining room. Home buyers look for more than functionality in their plumbing. That

offers a challenge for plumbers that didn't exist a generation ago.

Beginning Your Layout

Every plumbing job begins with a floor plan or isometric layout. This layout identifies the location of fixtures and piping to be provided by the plumbing contractor. The floor plan will also show floor dimensions and the location of walls, doors and windows. That's important information to the carpentry contractor, of course. But it's also important information for the plumbing contractor. You don't want to be caught trying to install waste, vent or water pipes through door or window openings, for example. Plan the layout carefully so it's coordinated with the work of other trades.

▓ Use the Isometric to Plan the Installation

Draw the isometric to illustrate the type of installation best suited for the job, either a stack system, a flat system, or a vertical or horizontal wet vent system. Your isometric drawing will make it easy to figure the number, size and type of fittings to be used. Once the isometric is complete, begin a list of the materials needed. Scale the pipe footage off the plan (rather than the isometric) to reduce waste and leftover materials.

Be sure to space the fixtures properly on the isometric. Supply and DWV piping must be installed to the correct location so plumbing fixtures will fit and function properly. Nearly all the fixture manufacturers furnish booklets that show proper spacing for supply and DWV piping. Specially designed fixtures usually come with fixture dimensions and roughing-in measurements. Chapter 18 includes dimensions and measurements for the more common plumbing fixtures.

There are many acceptable ways to lay out the rough plumbing for most residential jobs. The code doesn't require any particular layout, so you're free to use your own ingenuity.

▒ Sample Bathroom Floor Plans and Isometrics

Seven bathroom floor plans are shown in Figure 9–1. Figure 9–2 shows complete floor plans for four residential units. Room dimensions have been omitted as they are not relevant here. The building drain or sewer location is given so that you have the same starting point for all of these isometric drawings. There are two illustrations for each bathroom floor plan in Figure 9-1. One assumes a flat installation and one is a stacked installation. Figure 9-2 shows only one type of layout for each floor plan. These isometric drawings should meet the requirements of any plumbing code in the U.S.

Use these isometric drawings as samples when preparing your own plumbing layouts.

Questions

1. Why are plumbing maintenance problems easier in older homes?

2. Why have builders allotted more space for bathrooms in today's homes?

3. Why is it necessary to draw an isometric layout of the plumbing before beginning work?

4. Name at least three items that must be included in professionally-prepared floor plans.

5. What should an isometric drawing show? Be specific.

6. What are some of the other important steps that should be considered before beginning the installation of rough plumbing?

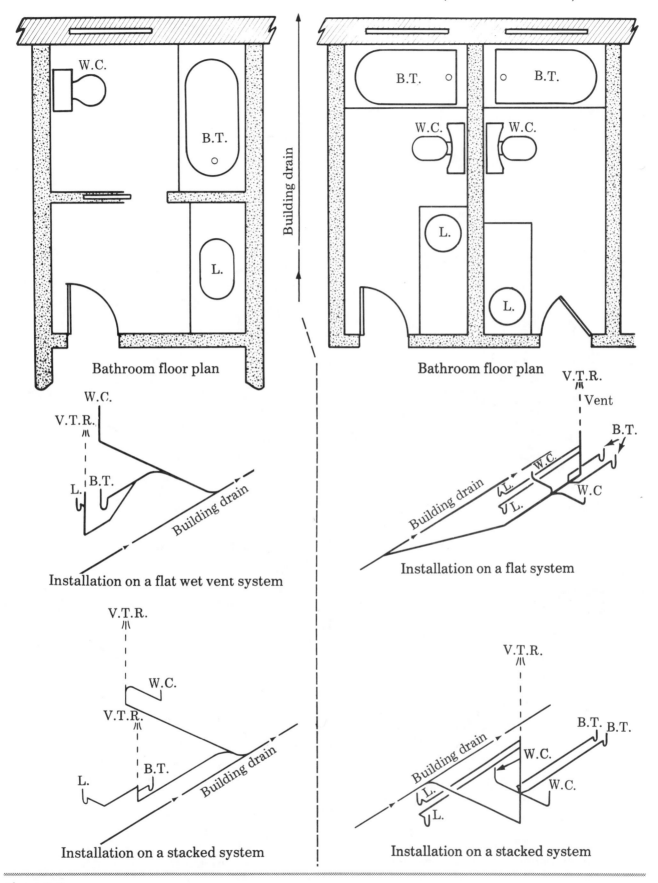

Figure 9–1
Typical bathroom installations

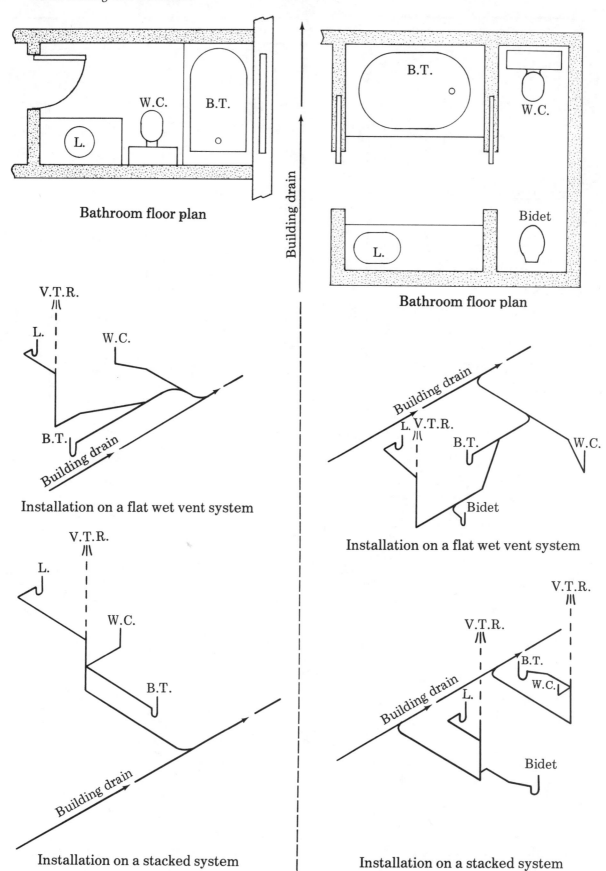

Figure 9–1 (continued)
Typical bathroom installations

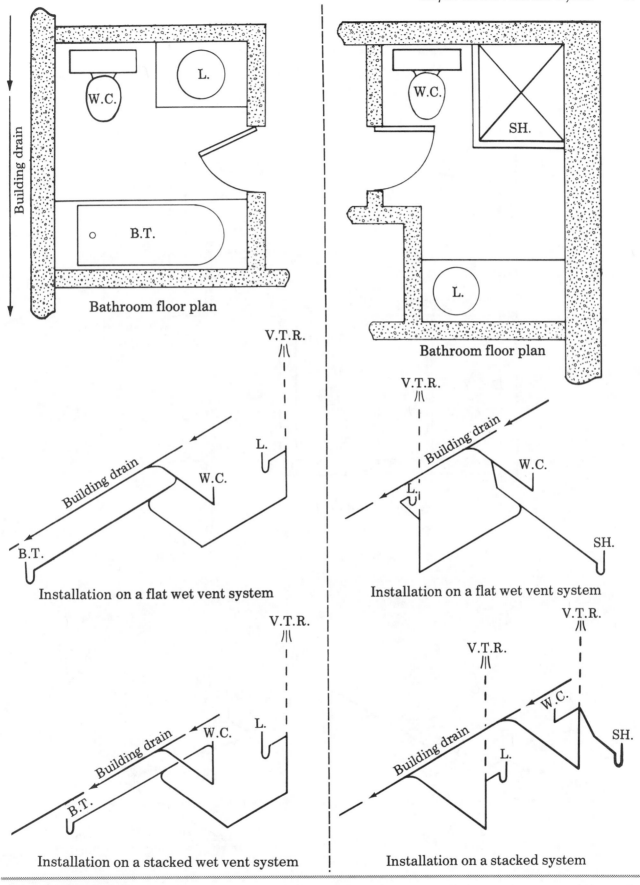

Bathroom floor plan

Bathroom floor plan

Installation on a flat wet vent system

Installation on a flat wet vent system

Installation on a stacked wet vent system

Installation on a stacked system

Figure 9–1 (continued)
Typical bathroom installations

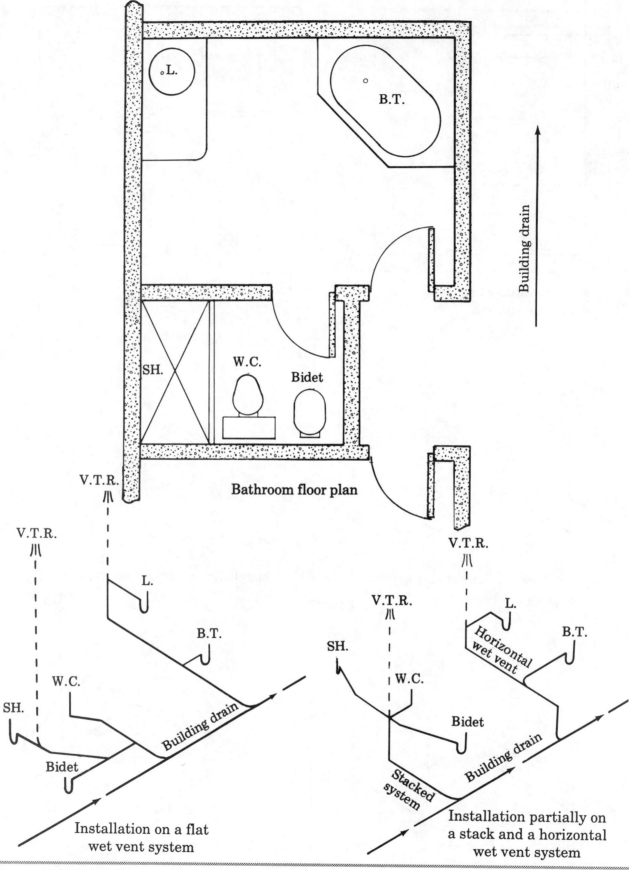

Bathroom floor plan

Installation on a flat wet vent system

Installation partially on a stack and a horizontal wet vent system

Figure 9–1 (continued)
Typical bathroom installations

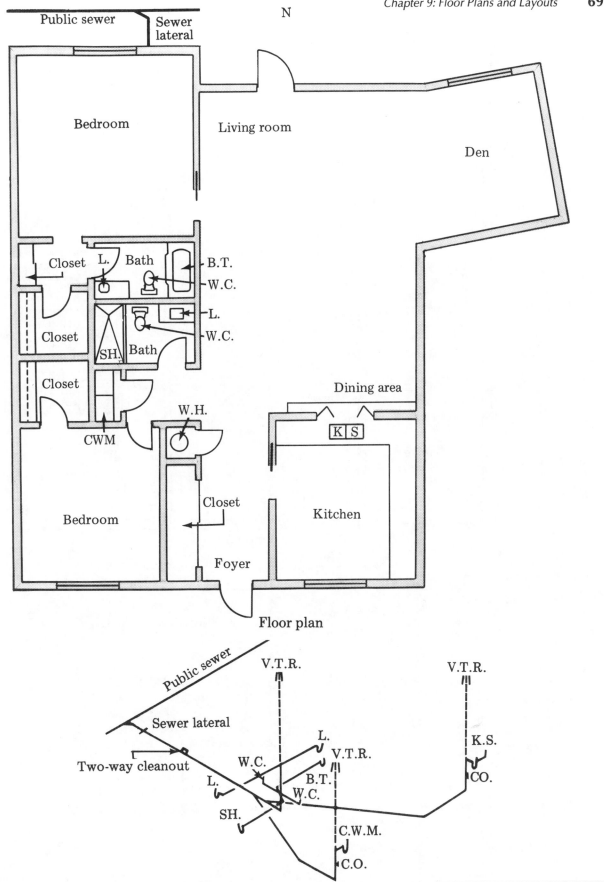

Figure 9–2
Typical residential installations

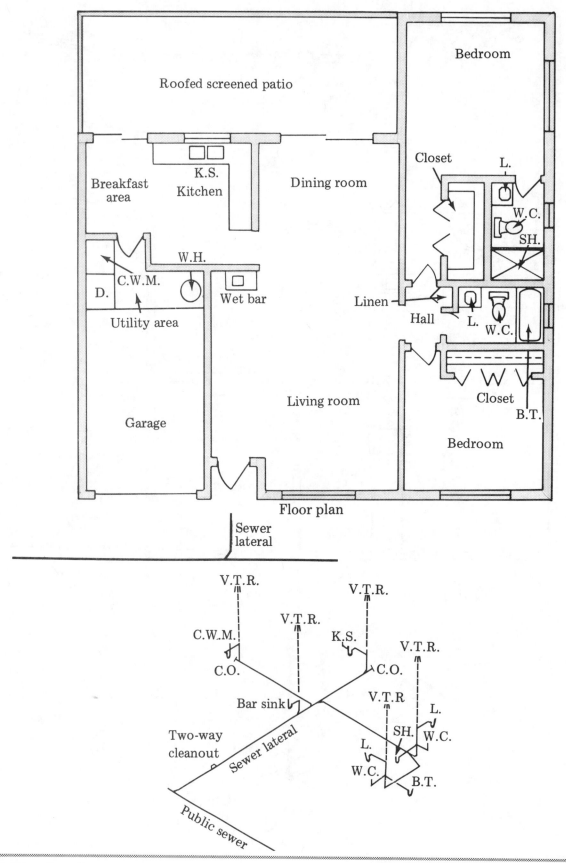

Figure 9–2 (continued)
Typical residential installations

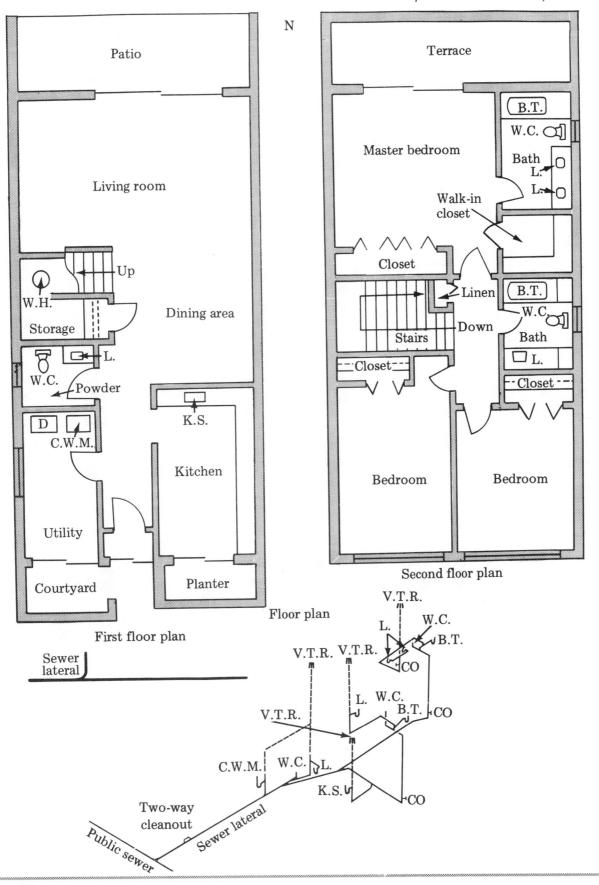

Figure 9–2 (continued)
Typical residential installations

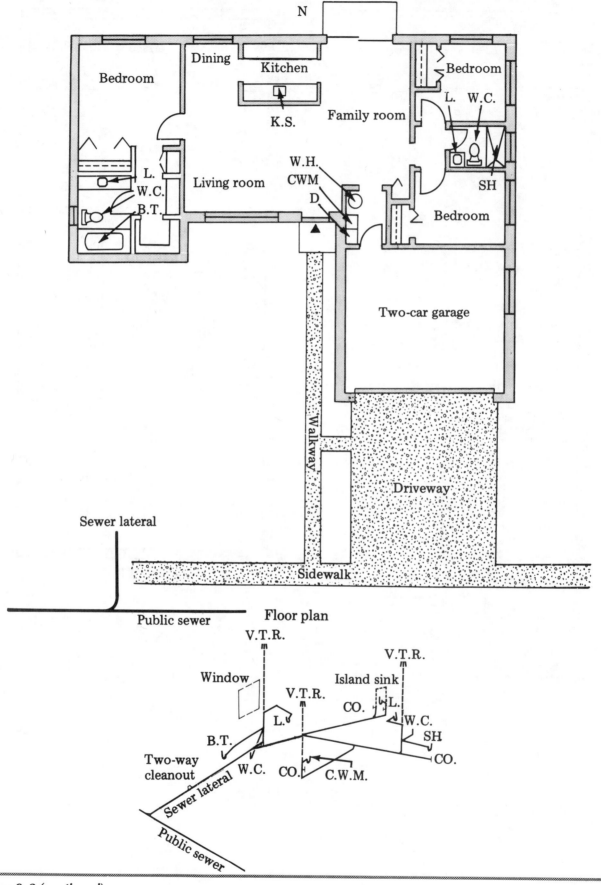

Figure 9–2 (continued)

Typical residential installations

DWV Materials and Installation

The purpose of the plumbing code is to protect the health, welfare, and safety of the public. While codes vary on design, installation, and maintenance details, the same basic principles of sanitation and safety are followed by all codes.

The Principles of DWV Systems

This chapter describes how to install drain, waste, and vent systems in residential and light commercial buildings. The principles outlined here do not cover every plumbing situation, but they should make the intent of the code clear and understandable. The principles in this chapter are so important that anyone violating them may assume that the inspector will reject the completed work.

- Drainage systems should be designed to prevent fouling or the deposit of solids along the pipe walls. That's why you should use the minimum pipe sizes established by the code.

- The drainage system must be properly vented to provide free circulation of air. Vent pipes must be sized and arranged to relieve pressure that builds up in the drainage system.

- Correct sizing prevents back pressure or siphoning action from destroying the fixture trap seal. But stoppages can occur in even the best-designed system. Adequate cleanouts must be installed so that all portions of the drainage system are accessible by cleaning equipment. (See Chapter 8.)

▓ Direction Changes

Every DWV system includes wyes and bends. The code regulates the use of fittings that change the direction of DWV piping. Single and double sanitary tees, quarter bends, and one-fifth bends may be used in vertical sections of drainage lines only where the change of direction in the flow is from horizontal to vertical. Proper use of fittings for changes in direction is illustrated in Figures 10–1, 10–2 and 10–3.

Double hub fittings (as in Figure 10–4) are prohibited on drainage lines. Fittings with a drainage pattern in the direction opposite to flow (Figure 10–5) are prohibited. An inexperienced plumber is likely to make this mistake when using no-hub or plastic drainage fittings.

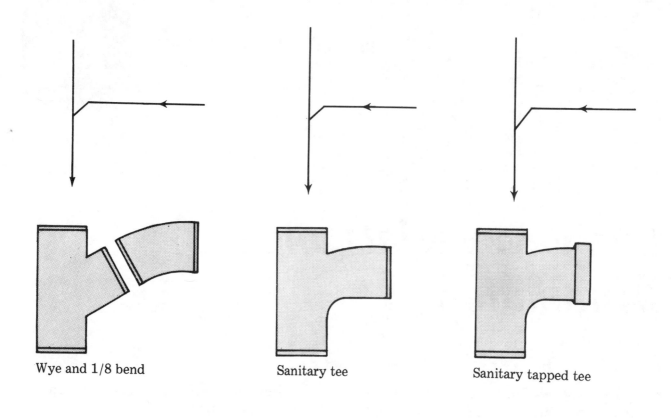

Wye and 1/8 bend   Sanitary tee   Sanitary tapped tee

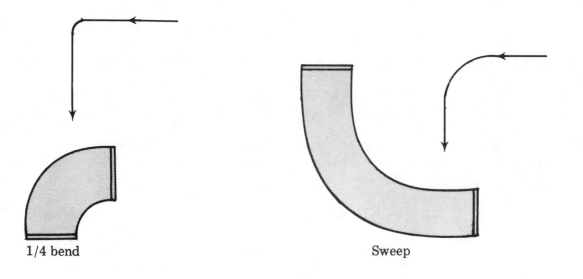

1/4 bend   Sweep

Figure 10–1
Horizontal to vertical change of direction

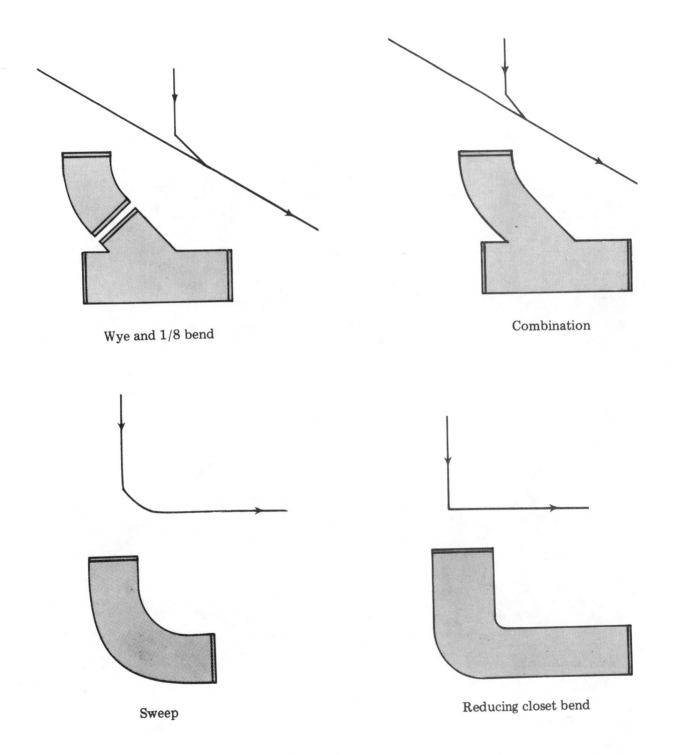

Wye and 1/8 bend

Combination

Sweep

Reducing closet bend

Figure 10–2
Vertical to horizontal change of direction

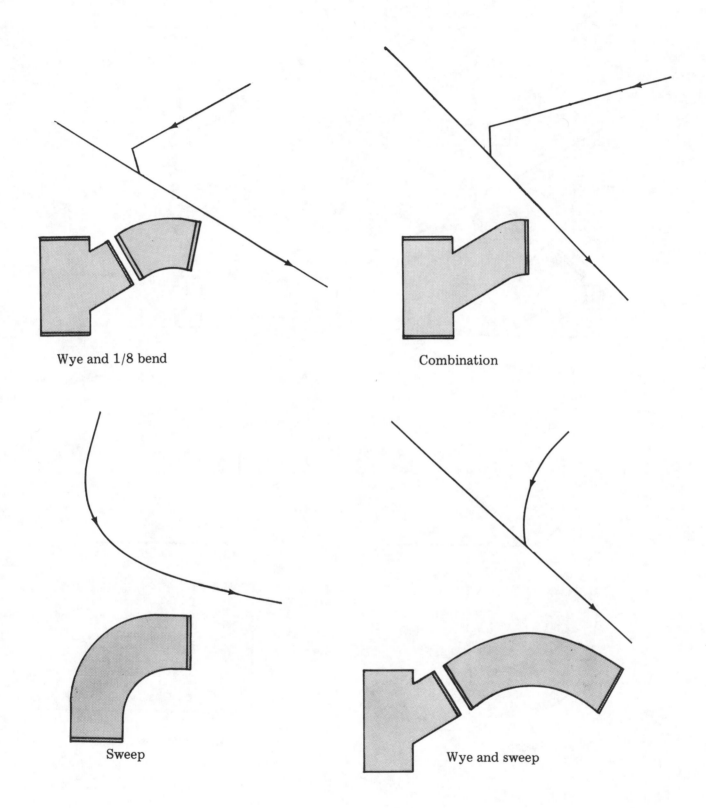

Wye and 1/8 bend

Combination

Sweep

Wye and sweep

Figure 10–3
Horizontal to horizontal change of direction

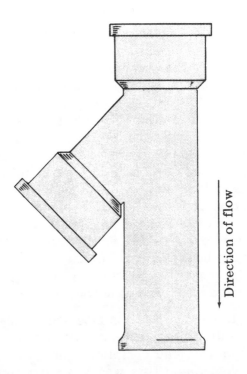

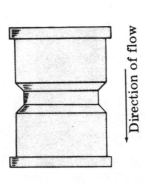

Figure 10–4
Double hub fitting with hub opposite to flow
(prohibited)

Figure 10–5
Inverted Y fitting with branch drainage pattern opposite
to flow (prohibited)

When you install these fittings, be sure to consider the direction of flow.

Tapping into Lines

The code does not permit drilling or tapping any drainage or vent piping. Drilling or tapping can create fractures that will allow raw sewage to seep out of the pipe. In the case of a vent pipe, sewer gases could escape into a building.

Laying Pipe in Trenches

Plumbing drainage pipes must be installed in open trenches and must remain open until the piping has been inspected, tested, and accepted by the plumbing inspector. This means that you are not allowed to tunnel under driveways or other permanent structures to lay drainage piping. There is no way to properly support drainage piping installed in this manner, so it would be likely to sag and cause stoppages.

Testing

The code requires that all drainage system piping be water tested with at least a 5-foot head. With all other openings plugged, at least one vent pipe in the system must be extended a minimum of 5 feet higher than the horizontal drainage piping and filled with water. This is to make sure that the joints of the system don't leak. The plumbing inspector may require the removal of cleanout plugs or caps to check whether the water has reached all parts of the system.

Materials

Types of materials and installation methods used in drain, waste and vent systems are regulated by the plumbing code. A wide variety of piping materials will meet code requirements. The materials included in this chapter are commonly used in residential or simple commercial jobs. Other types of materials and their use in commercial and special waste systems are included in a more advanced book by this author, *Plumber's Handbook Revised*.

Code standards and specifications for plumbing materials change every year. For example, no-hub cast iron pipe and plastic pipe and fittings are among the newer piping materials. They are accepted in codes being used in most parts of the country. But some counties and cities still prohibit the use of plastic pipe or no-hub cast iron pipe. Your local plumbing inspector or an experienced plumber will be able to answer your questions on the type of materials that can be used in your area.

The most frequently used materials for drainage, waste and vent systems are:

1. Tar-coated cast iron pipe and fittings with hub and spigot ends. Oakum with hot poured lead completes the joint. Many years ago, this was the most common DWV material.

2. Tar-coated cast iron pipe and fittings of the no-hub type. Elastomeric sealing sleeves and stainless steel clamps complete the joint. The cast iron pipe and fittings may be extra heavy or centrifugally spun service weight. Extra heavy and service weight pipe and fittings should not be mixed within the same system. Where permitted by code, most plumbers prefer service weight pipe and fittings because they're easier to work with and cut.

3. Schedule 40 ABS or PVC plastic pipe and fittings with cemented joints. The code prohibits mixing ABS and PVC plastic pipe and fittings in the same installation.

4. Copper type DWV or M pipe and galvanized steel pipe with recessed drainage fittings are commonly used in cast iron systems for fixture drains. Some codes now permit using plastic pipe and fittings for fixture drains for cast iron installations.

Vent extensions through the building roof from a cast iron system are frequently M copper, galvanized steel or plastic pipe with proper conversion adapters. These materials are lighter in weight than cast iron and come in long sections so few joints are required. This saves labor and material.

Fittings used in all portions of the drainage system must conform to the material and type of piping used. In other words, a drainage system installed with PVC plastic pipe cannot use ABS plastic fittings, or vice versa. A drainage system installed with cast iron pipe can't use plastic fittings, even with approved conversion adapters. Any fittings you use on screwed, copper or plastic pipe within a sanitary drainage system should be of the recessed drainage pattern type. *Fittings designed for use on water pipe should not be used in drainage systems.*

Cast Iron Soil Pipe and Fittings

Cast iron pipe was imported from England and Scotland in 1813 to replace deteriorated wooden water mains in Bethlehem, Pennsylvania. As the demand for cast iron pressure pipe increased, foundries of New Jersey and Pennsylvania supplied most of the nation's needs for pressure pipe used in water mains. In the 1880s cast iron pressure and soil pipe for building construction was manufactured to meet the requirements of the first plumbing code, published in Washington, DC in 1881.

After that, many cities began installing water works and sewage systems. There were 64 cast iron foundries producing pressure pipe and soil pipe in 1894. By 1915, the center of the soil pipe industry had shifted from the Northeast to the South. Approximately 70 percent of total soil pipe production in the United States was in southern foundries, most of them in Alabama, in 1940.

Cast iron soil pipe and fittings are used primarily in buildings for sewers, drains, soil and vent piping. It's available in two weights: centrifugally spun service weight and extra heavy cast iron. The service weight, unless prohibited by code, is recommended for use in all residential construction.

Cast iron soil pipe can have a hub and spigot joint using lead and oakum to make a watertight, gastight joint. The waterproofing characteristics of oakum fiber have long been recognized by the plumbing trades. To get an idea of how well oakum can seal a joint, try the following. When a job has been completed and the lead joints are caulked, fill the system with

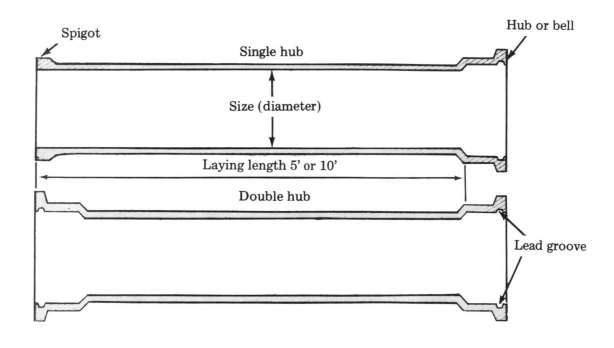

Figure 10–6
Cast iron soil pipe

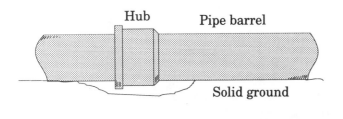

Figure 10–7
Excavate for hub

water. Disregard any leaks. Let the water stand overnight before calling for inspection. The wet oakum will expand overnight and fill each crevice, making joints watertight and gastight. This technique will save the time it would take to recaulk each seeping joint.

The strength, durability and good resistance to trench loads make cast iron soil pipe the preferred material for sewers.

Cast iron soil pipe is manufactured in sizes of 2 inches (inside diameter) and up. It comes in 5-foot and 10–foot lengths. When a piece of pipe shorter than 5 feet is needed, it's usually cut from a double hub pipe to avoid waste. See

Figure 10–6. This procedure will leave two usable lengths of pipe, each with a hub. With no-hub pipe, any lengths cut off can probably be used elsewhere, thus eliminating most waste.

▓ Installation of Cast Iron Pipe

When installing cast iron soil pipe, be sure to keep the pipe barrel in firm contact with solid ground. Always excavate for the hub (bell) or for the steel shield on no-hub pipe. The weight of the pipe should be evenly distributed along the full length of the pipe barrel. See Figure 10–7.

Because of cast iron pipe's high resistance to trench loads, depth is not too important. However, don't backfill with large boulders, rocks, cinder fill or other materials that could damage or corrode the pipe.

Most cast iron soil pipe installed is either no-hub or hub and spigot joint. The no-hub joint uses a one-piece neoprene gasket, a stainless steel shield and retaining clamps. These joints are an advantage in limited-access areas (such as close to ceilings or in corners) because the joints are easy to make up. Since the pipe and fittings don't have hubs, they can be used in thinner partitions. Installation is fast and

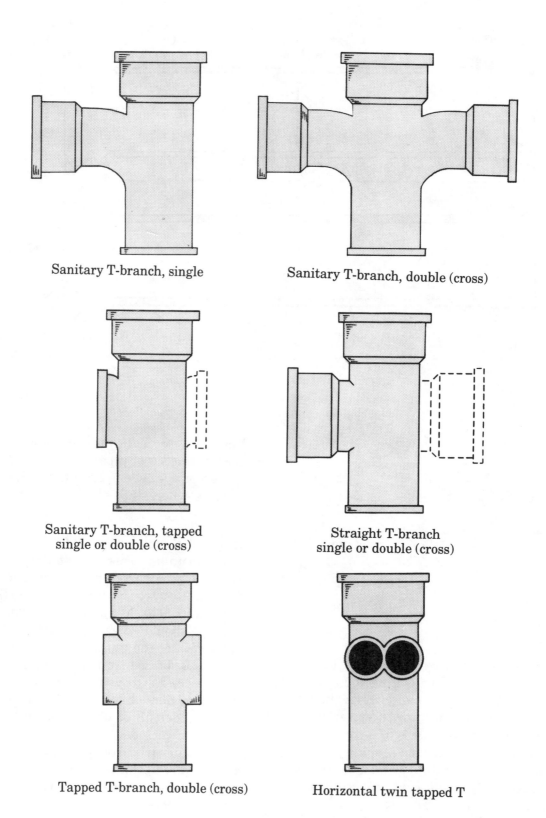

Sanitary T-branch, single

Sanitary T-branch, double (cross)

Sanitary T-branch, tapped
single or double (cross)

Straight T-branch
single or double (cross)

Tapped T-branch, double (cross)

Horizontal twin tapped T

Figure 10–8
T's or T-branches

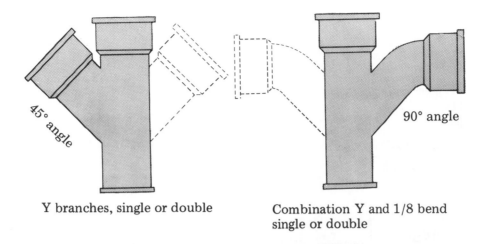

45° angle

90° angle

Y branches, single or double

Combination Y and 1/8 bend
single or double

Figure 10–9
Wyes (Y-branches)

efficient. No-hub joint pipe can be used in combination with a lead and oakum joint pipe to meet the needs and specific requirements of any particular job. Properly installed, lead and oakum joint pipe and no-hub pipe last as long as cast iron soil pipe. Generally, lead and oakum joint pipe and no-hub pipe can be installed and forgotten.

■ Soil Pipe Fittings

Figures 10–8, 10–9, 10–10, and 10–11 illustrate the more common hub and spigot soil pipe fittings. No-hub fittings for the same uses will have the same designs (as in Figures 4-4A, B and C back in Chapter 4), but without the hub and spigot ends.

Tees

Tees (sometimes called T's or T-branches) are used to make 90-degree turns. See Figure 10–8. The main pipe must be vertical and may be either a soil, waste or vent stack. This and other types of horizontal to vertical directional change fittings are illustrated in Figure 10–1.

Many types of cast iron tees are in general use. Sanitary tees are designed to carry waste substances. Straight tees are designed to carry air only and are used in a vent system. Tees may be tapped, cemented or may have a hub and can be used to connect threaded or unthreaded branch drains or vent lines to the stack vent.

The side takeoff of a sanitary tee enters the straight through section of the fitting on a downward curve. This is known as a *drainage pattern*. This changes the direction of flow smoothly from the horizontal to the vertical, which helps to prevent stoppages at this location. Tees are available as singles, as double horizontals, as crosses and as angles.

The *test tee* is a fitting with a screw plug. It is generally used for testing a system for leaks and can be used as a permanent cleanout after the test. See Figures 8-2 and 8-8.

Wyes

Wyes, or Y-branches, are used to make a 45-degree change in direction (see Figure 10–9). They are used mainly to connect horizontal waste pipes and branch drains to the building main drain. Their design allows a smoother change of flow direction than either straight or sanitary tees, which are prohibited in horizontal installations. Some of the more commonly-used fittings and combinations of fittings for horizontal to horizontal and vertical to horizontal changes of direction are shown in Figures 10–2 and 10–3.

A 45-degree Y-branch, also called simply a "Y-branch," has the side takeoff entering the straight through section of the fitting at a 45-degree angle. If the side takeoff is the same size as the through section (3 inches, for example), it is called simply a "3-inch wye," as all three

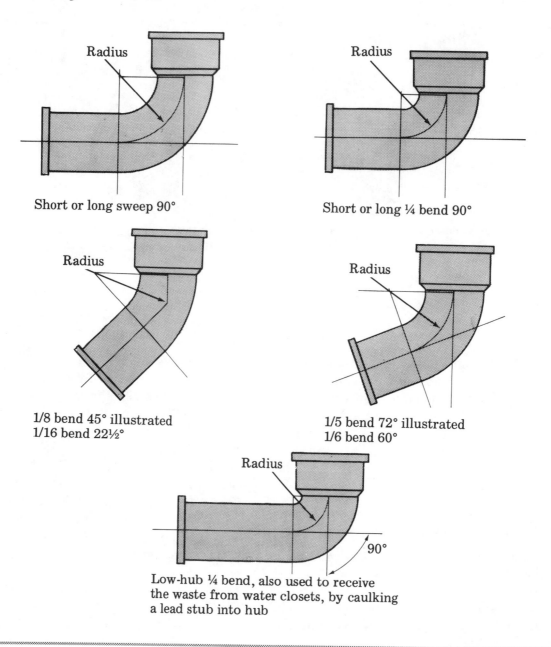

Short or long sweep 90°

Short or long ¼ bend 90°

1/8 bend 45° illustrated
1/16 bend 22½°

1/5 bend 72° illustrated
1/6 bend 60°

Low-hub ¼ bend, also used to receive
the waste from water closets, by caulking
a lead stub into hub

Figure 10–10
Common bends

openings are the same size. The side takeoff may be smaller (never larger) than the through section. Then the fitting is called a "reducing Y-branch" (3 × 2, for example). Wyes are available in single as well as double design.

A 90-degree Y-branch is also known as a "combination Y and ⅛ bend" (usually called a *combination* in the trade.) These are manufactured as one-piece units. See Figure 10–9. Where this particular fitting is required but not available, a Y and a ⅛ bend of the proper

size can be quickly assembled into a combination wye using only an extra joint.

Combinations are made in various shapes and sizes. They may be single, double or reducing. As in the 45-degree wye, the side takeoff may be the same size as the through section. If the side takeoff is 3 inches, for example, the fitting would simply be called a "3-inch combination," as all three openings are the same size. The side takeoff may be smaller (2 inches, for example), but never larger than

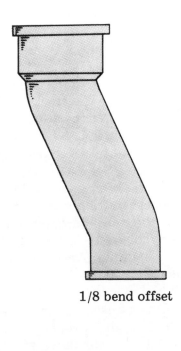

1/8 bend offset

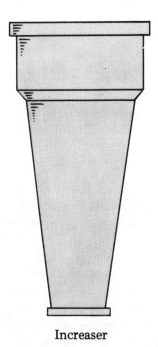

Increaser

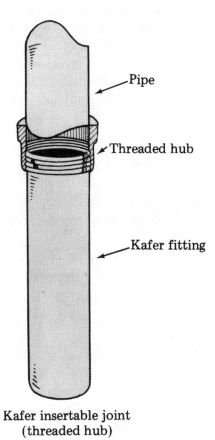

Pipe

Threaded hub

Kafer fitting

Kafer insertable joint
(threaded hub)

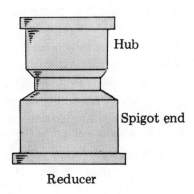

Hub

Spigot end

Reducer

Figure 10–11
Other fittings

the through section. In such a case, the fitting is called a "reducing combination," known correctly as a 3 × 3 × 2 reducing combination or simply a 3 × 2 combination.

You always size the through section first, reading from the spigot end to the hub end, and then size the branch. This is the correct method for sizing all plumbing fittings, whatever their use.

Bends

Bends are classified by the degree of turn and the radius of the curve. Bends are used to change the direction of either horizontal or vertical drainage or vent lines. Figure 10–10 shows a ⅛ bend, a ⅕ bend, two ¼ bends, and a short sweep. The ¹⁄₁₆ bend turns at 22½ degrees, the ⅛ bend turns at 45 degrees, the ⅙ bend turns at 60 degrees, the ⅕ bend turns at 72 degrees, the ¼ bend turns at 90 degrees. The sweep also turns at 90 degrees. A 4-inch ¼ bend has a turning radius of 4 inches, while a short sweep has a turning radius of 6 inches. A long sweep has a turning radius of 9 inches.

Bends are classified according to the radius of the curve as regular, short or long. As the radius increases, the change in flow direction becomes smoother.

The *closet bend* illustrated in Figure 10–2 is a special fitting to receive the waste from water closets. Closet bends are made in different styles to fit different types of construction and local code requirements. One type has a scored end (marked with lines) which fits into the closet floor flange. See Figure 4-4B in Chapter 4. The scoring makes it easier to cut the bend to the desired length for a given connection.

Other Fittings

Other types of common and seldom-used soil pipe fittings are shown in Figure 10–11 and explained below.

A *⅛ bend offset* is used to carry the soil or waste lines past an obstruction such as a tie beam. This bend is manufactured in different sizes and lengths. Two ⅛ bends will do the same job with an extra joint, but the one-piece ⅛ bend offset gives a smoother transition than the two-piece unit.

A *cross* (T-branch) joins two lines that are in the same plane and perpendicular to each other. The two types of cross, the straight cross and the sanitary cross, are identical except for the takeoffs. The *straight cross* is used to connect vent lines to the stack vent. The *sanitary cross* is used to connect sanitary branches to the soil or waste stack. Both types of cross are permitted only in a vertical pipe, never on a horizontal line. See Figure 10–8.

An *increaser* is used to increase the size (diameter) of a straight through line. It is often used for the vent stack terminal in very cold climates. This is to prevent frost from closing the vent opening. *Never use an increaser to increase the size and flow within a sanitary installation.* See Figure 10–11.

A *reducer* is used to reduce the size (diameter) of a straight through line. The reducing end may be one or more pipe sizes smaller than the pipe or fitting it is connected to. A reducer may be used to decrease the size and flow within a sanitary installation.

A *kafer fitting* is commonly used when it is necessary to replace a section of soil pipe line or install a fitting into an existing soil pipe line. The hub is threaded onto the kafer fitting body and can be unscrewed. In times past this fitting made it easier to make cuts into existing cast iron lines. The kafer fitting is still used today. But no-hub pipe and fittings now serve the same purpose. Thus the demand for kafer fittings has dropped considerably.

Plastic Pipe and Fittings

Rigid vinyl plastic pipe and fittings have been used in Europe since 1937 and in the United States only since 1955. The Navy used PVC DWV pipe and fittings for a 500-home installation in Key West, Florida in 1960. Plastic was preferred because pipe had to be laid in corrosive salty soil dredged from the ocean.

The use of plastic DWV pipe and fittings for residential construction has grown rapidly since then. By 1968 there were over 650,000 plastic DWV installations. Nearly all manufactured housing uses plastic DWV pipe and fittings. The smooth inner surface and superior resistance to deposit formation makes plastic drain, waste and vent piping ideal for residential sanitary systems.

The manufacturers of plastic pipe and fittings claim that their products allow a consistently higher flow rate than pipe of brass, copper, cast iron, or steel. Moreover, plastic pipe and fittings maintain these excellent flow characteristics indefinitely.

The heat transfer characteristic of plastic drainage pipe is a distinct advantage. Hot liquids discharged into plastic sewer line retain more heat than liquids draining in metal pipes. This means that water containing greasy waste material is less likely to solidify in the pipe and cause a stoppage.

Rigid vinyl plastic pipe is also known for its excellent chemical resistance. It is used throughout the country for chemical process piping and acid drain waste systems. Vinyl pipe doesn't corrode like metal pipe and is immune to galvanic corrosion.

ABS is the common designation for acrylonitrile-butadiene-styrene, a family of thermoplastics. PVC is an inert *thermoplastic* polyvinyl chloride, and is commonly used for drain, waste and vent pipe and fittings.

Not every plumbing code approves use of plastic pipe. However, nearly all communities on the west coast and southwest permit use of plastic pipe. It is also used widely in southern and southeastern states. Before bidding a job or buying plastic pipe, check with your local authority.

▓ DWV Plastic Pipe Fittings

Figures 10–12 and 10–13 show several DWV plastic fittings. These are manufactured to meet all code requirements. In domestic sanitary system installations, solvent welded (cemented) fittings can be used throughout. If you have to connect ABS or PVC pipe to steel, cast iron or other types of pipe, use a specially-made conversion adapter.

Plastic drainage fittings are nearly identical in design to similar cast iron screwed drainage fittings. The only real difference is the socket-type hub. Rough dimensions and sweeps are nearly identical. Adequate pitch per foot is provided in all branch sockets. The roughing-in dimensions of plastic fittings (except in hub lengths) meet the specifications given in

American Standards Association for cast iron screwed drainage fittings.

Molded fittings designed for solvent-welded (cemented) joints have a slight taper. The ID (inside diameter) at the opening should be slightly larger than the maximum OD (outside diameter) of the pipe. The ID at the seat should be less than the minimum OD of the pipe. When the pipe is inserted into the fitting, you should feel a good interference fit. Some fittings don't make a good joint with certain types of pipe. It's best to use the fittings recommended by the manufacturer of the plastic pipe you're installing.

As with no-hub pipe, plastic pipe and fittings are a good choice where access is limited (such as near ceilings or in corners). The solvent-welded or cemented joints can be made with ease. The fitting sockets are much thinner than cast iron hubs. That makes plastic pipe ideal for use in thinner partitions. Installation is fast and efficient.

Another advantage to using plastic pipe is its ease of handling. A complete bathroom assembly can be easily lifted into place by one person. Properly made solvent-welded joints are as strong and last as long as the pipe itself.

Use *Schedule 40* plastic pipe and fittings for DWV systems. This pipe is made in 20-foot lengths and is ideal as a labor saver for long runs such as sewer installations.

Plastic pipe is more fragile than cast iron pipe and requires special handling when used for building sewers. The entire bottom quarter of this pipe should be continuously and uniformly supported by the trench bottom. Support the pipe with 4 inches of fine sand that will pass through a ¼-inch screen. Some codes now permit use of up to ¾-inch screen material. Hub and coupling projections should be excavated as shown in Figure 10–7 so that no part of the pipe load is supported by the hub or coupling. The supporting material should extend 4 inches on each side of the pipe.

Backfill should be firmly compacted with selected material which will pass through a ¼-inch screen. Where permitted, this material may be ¾-inch screen. The backfill should extend from the trench bottom to a point 6 inches over the top of the pipe. The minimum

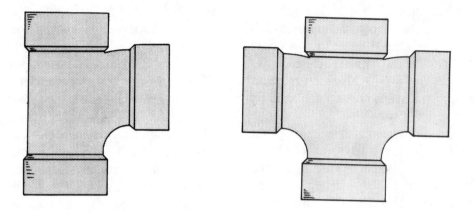

Sanitary tee, single and double

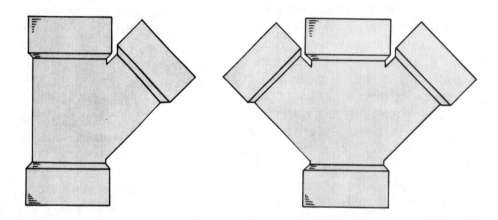

45° wye (Y), single and double

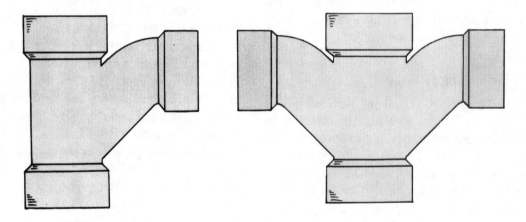

Combination wye 1/8 bend, single and double

Figure 10–12
Plastic fittings

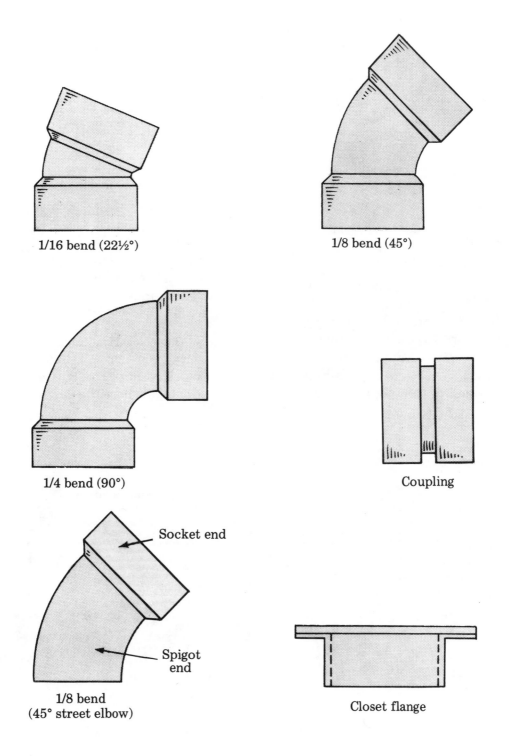

1/16 bend (22½°)

1/8 bend (45°)

1/4 bend (90°)

Coupling

Socket end

Spigot end

1/8 bend
(45° street elbow)

Closet flange

Figure 10–13
Plastic fittings

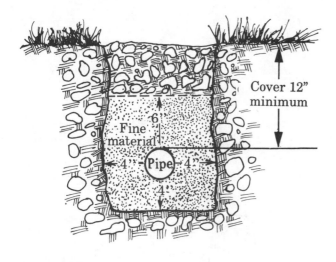

use plastic pipe. The pipe comes in 20-foot lengths, eliminating most waste.

Although solvent welded joints can be handled two or three minutes after joining, they should be permitted to dry 30 minutes before the system is water-tested for inspection.

General Installation Methods

▓ Cutting Cast Iron Pipe

Uninstalled cast iron pipe is easy to cut. But pipe that's been connected is more difficult to cut and generally requires a special tool. The most common way of cutting free or uninstalled cast iron pipe, especially on small jobs, is the hammer and chisel method shown in Figure 10–15.

1. Make the cut square. Unless you have a very good sense of direction, draw a chalk mark around the entire circumference of the pipe where the cut is to be made.

Figure 10–14
Protection for plastic drain pipe

pipe depth below ground level must be 12 inches. See Figure 10–14.

Plastic DWV pipe is manufactured in sizes as small as 1¼ inches inside diameter. This means that the complete plumbing system can

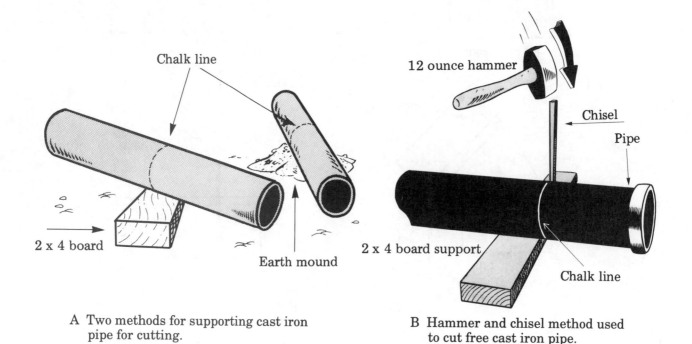

A Two methods for supporting cast iron
   pipe for cutting.

B Hammer and chisel method used
   to cut free cast iron pipe.

Figure 10–15
Cutting cast iron pipe

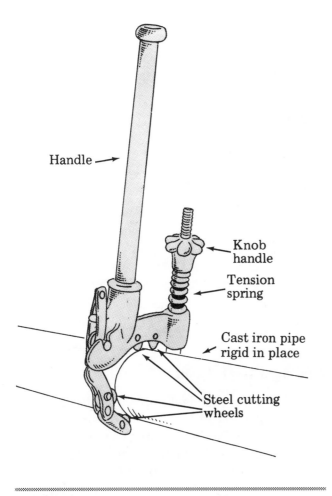

Figure 10–16
Soil pipe cutter, cutting a sewer pipe in trench

to wear protective eye shields. This method often causes bits of metal to fly off the blade.

The best tool to cut cast iron pipe that is rigidly in place is a *soil pipe cutter*. See Figure 10–16. This simple tool can be rented inexpensively for short-term jobs. It consists of a short steel handle attached to a slightly curved base. The base has two steel cutter wheels. A chain with additional steel cutter wheels can be hooked into slots at the base to assist the cutting. The other end of the chain has a knob handle that works against a spring to provide the necessary cutting tension. Cut a groove into the pipe body by tightening the knob handle back and forth across the pipe. Gradually increase the pressure and the pipe snaps evenly.

■ Cutting Plastic Pipe

Plastic pipe should also be cut squarely, since the pipe and the fitting socket require a good, tight joint. Use a miter box for this purpose; if you don't already have one, construct one at the job site as in Figure 10–17. A fine-tooth hacksaw or a regular handsaw is the only tool needed. Plastic pipe cutters are also available. Remove all burrs on the cut end with a knife or file.

■ Cast Iron Hub and Spigot Joints

Joints in cast iron hub and spigot pipe and fittings are made by caulking with lead and oakum. See the caulked joint in Figure 10–18. Make the joints in a vertical position whenever possible to make centering easier. Caulking in both vertical and horizontal positions is explained in detail below.

The usual method is to place the spigot end of the pipe or fitting in the hub of the following length of pipe. Pack the inner part of the gap between the hub and the spigot with oakum (usually twisted jute fiber). The oakum may be either plain or oiled. Pack the oakum well into the gap, leaving 1 inch at the outer end of the hub to receive the molten lead. Use 1 pound of lead for each inch of pipe circumference. For example, a 2-inch joint should take 2 pounds of lead, a 3-inch joint should take 3 pounds, and so on. Figure 10–19 shows the tools required to make a lead joint.

2. Lay the pipe over a 2 × 4 board, or use a mound of earth to support the pipe while it is turned.

3. Score the pipe barrel around the chalk mark with a ¾-inch cold chisel which is not too sharp. Use a 12-ounce hammer to tap the chisel lightly, turning the pipe until it has been evenly scored all around.

4. Continue to turn the pipe and strike the cold chisel with increasingly heavy blows. Most often the pipe breaks evenly at the line of the cut on about the third turn.

For small diameters, cut free cast iron pipe with a hand-held circular saw equipped with a metal cutting blade. Use a chalk mark to produce a square cut. A second person should turn the pipe while the first makes the cut. Be sure

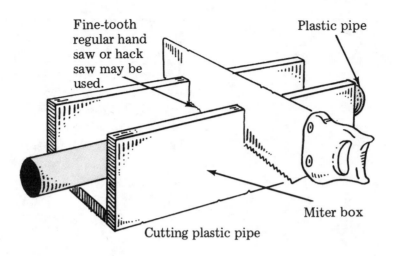

Fine-tooth regular hand saw or hack saw may be used.

Plastic pipe

Miter box

Cutting plastic pipe

Figure 10–17
Cutting plastic pipe

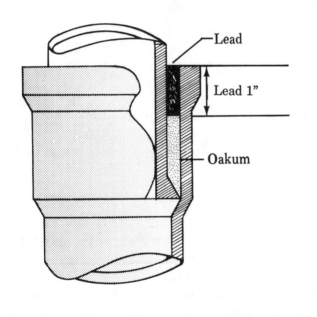

Lead

Lead 1"

Oakum

Figure 10–18
Lead and oakum joint

Plumber's ladles for pouring the hot lead come in several sizes. Use a ladle large enough to make the joint in one pour. Allow a minute or two for the lead to harden before attempting to caulk the joint as explained below. Heat the ladle over the edge of the melting furnace before dipping it into the lead. Take the same precaution when adding new lead to the pot. *Never dip a cold or damp ladle into a pot of hot lead. The lead will explode.*

Caulking a Vertical Joint

1. If the weather is damp, wipe the hub and spigot ends of the pipes to remove moisture and foreign matter. Water or dampness causes the melted lead to spatter and can result in serious burns.

2. Place the spigot end of the fitting into the hub of the pipe and align the joint so the spigot end is in the center of the hub as in Figure 10–20.

3. Using a yarning iron (Figure 10–21) firmly pack a strand of oakum into the hub completely around the joint. Then use a packing iron and an 8-ounce ball peen hammer to finish packing the joint, as in Figure 10–22. Align the joint so the fitting is straight and turned to face in the desired direction before pouring in the lead.

4. The joint is now ready to receive the molten lead. Pour the lead slowly into the joint until it rises slightly above the top rim of the hub.

5. Caulk the lead by striking an *outside* caulking iron gently but firmly against the lead with an 8-ounce hammer. Then repeat using an *inside* caulking iron. Caulking the lead too tightly by using a hammer that is too heavy may crack the hub. A cracked fitting or pipe must be replaced.

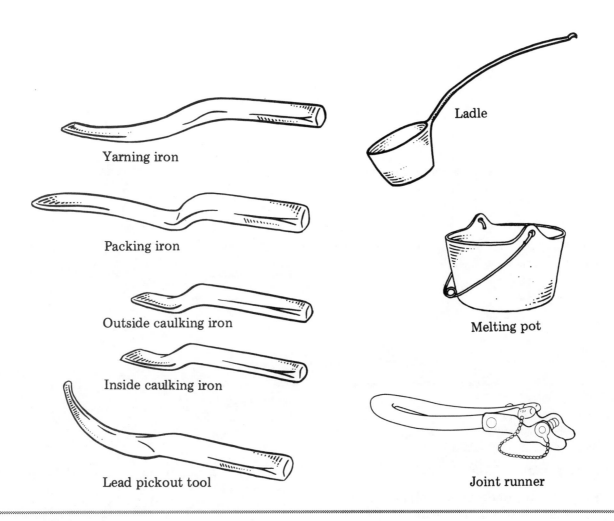

Yarning iron

Packing iron

Outside caulking iron

Inside caulking iron

Lead pickout tool

Ladle

Melting pot

Joint runner

Figure 10–19
Tools used to make lead joints

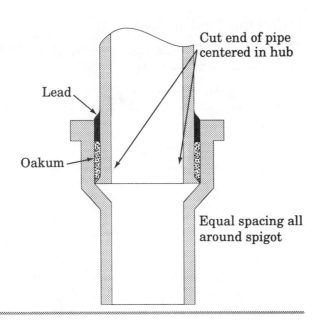

Cut end of pipe centered in hub

Lead

Oakum

Equal spacing all around spigot

Figure 10–20
Vertical lead and oakum joint

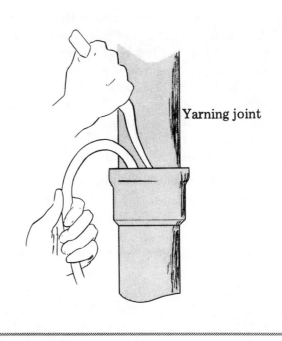

Yarning joint

Figure 10–21
Yarning an oakum joint

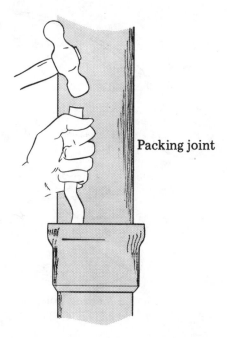

Figure 10–22
Packing an oakum joint

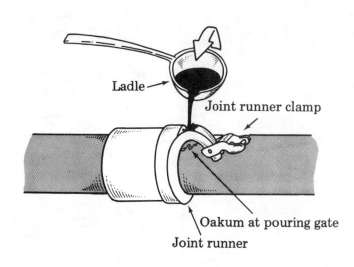

Figure 10–23
Using a joint runner clamp

Caulking a Horizontal Joint

1. Prepare the ends of the pipes and pack the joint with oakum as described above in 1 through 3.

2. Clamp a joint runner around the pipe. See Figure 10–23. The clamp leaves a pouring gate for the melted lead to flow through into the joint. Pack a small amount of oakum under the clamp to prevent wasting the lead. Pour the lead fast (faster than in a vertical joint) so it will not cool before filling the joint. After the lead cools, remove the joint runner.

3. Use a cold chisel and hammer to trim off surplus lead left by the pouring gate.

4. Caulk the lead the same way as for a vertical joint, but begin by using the *inside* caulking iron first and the outside caulking iron last.

Lead Furnaces

There are several types of melting furnaces available. The most common type used today is a tank burning propane gas as a fuel, known

as an *insto tank*. It has a hood that receives the lead pot. The propane burner can be controlled easily. Regardless of the type of furnace used, always follow the manufacturer's instructions.

■ Cast Iron No-Hub Joints

The no-hub joint is one of the easiest joints to make in a plumbing installation.

1. To join two pieces of pipe together, place the one-piece neoprene gasket onto the end of one pipe.

2. Slide the stainless steel shield and retaining clamps over the end of the second pipe. See Figure 10–24.

3. In the center of the gasket is a separator ring (see Figure 10–25) which fits firmly over the end of the pipe. Position the heads of the retaining clamps where they are accessible for tightening. Push the end of the second pipe into the neoprene gasket tight against the separator ring. Slide the stainless steel shield over the neoprene gasket until it is covered com-

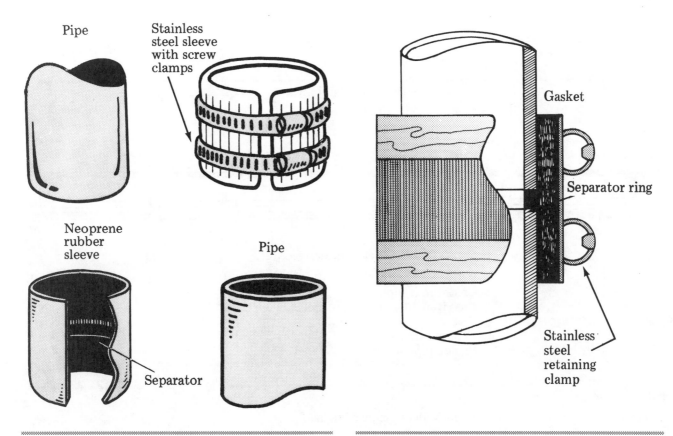

Figure 10–24
No-hub retaining clamp parts

Figure 10–25
No-hub joint

pletely. Push both pipes against the separator ring and tighten with a special torque wrench.

The torque wrench is the only tool needed to assemble no-hub pipe and fittings. It is preset to give 60 inch-pounds of torque. This eliminates guesswork and prevents undertightening or overtightening the retaining clamp.

Plastic Pipe Joints

Plastic pipe is popular because it's light, cheap and easy to assemble. But there are disadvantages. Rigid plastic pipe is harder to install correctly in cold weather. When the air temperature is below 40 degrees F, the solvent-cement gets stiff and may produce a poor bond.

Plastic pipe is more fragile than cast iron pipe. Be sure to align ends correctly so no bending is necessary to make up joints. Always install plastic pipe and fittings so that identifying marks are visible to the inspector.

Follow these steps to obtain welded plastic joints with maximum strength:

1. Wipe or brush off all dirt from the end of the pipe, both inside and outside. Wipe the fitting socket clean. Then use one of the following methods to prepare the pipe and the inside of the fitting socket for the application of solvent-cement:

   • Apply a cleaner to the spigot and socket contact areas using a natural bristle brush about half the diameter of the pipe in width. Immediately wipe these surfaces dry with a clean cotton cloth to ensure that these surfaces are clean and that surface gloss has been removed.

   • Remove the gloss with emery cloth and wipe away the dust.

2. Using a natural bristle brush (not the same brush used for the cleaner), apply a

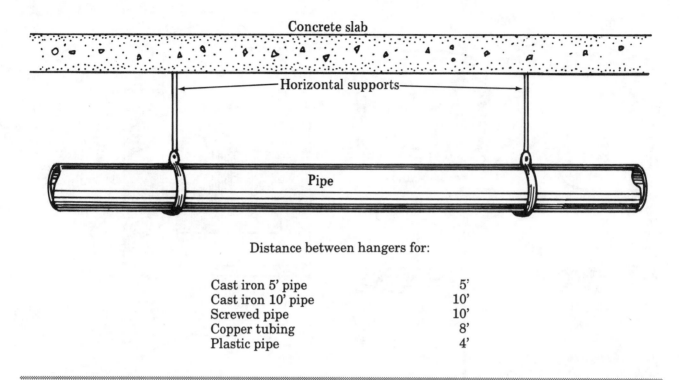

Distance between hangers for:

Cast iron 5' pipe	5'
Cast iron 10' pipe	10'
Screwed pipe	10'
Copper tubing	8'
Plastic pipe	4'

Figure 10–26
Distance between horizontal supports for various piping materials

light coat of solvent cement to the fitting socket and a heavy coat to the pipe spigot.

Immediately insert the pipe to the full socket depth, rotating it one-quarter turn in the fitting to get even distribution of the cement. A bead of cement around the fitting socket shows that you used enough cement. No bead means not enough cement was used.

3. Within two minutes after joining, the assembly will have started to set and can be handled if you work carefully. Setting time is longer at temperatures below 75 degrees F. Check the fitting for correct angle and direction immediately. Trying to adjust the pipe or fitting later may crack the joint. The way to fix a cracked joint is to replace the fitting.

4. Let all the joints dry 30 minutes before filling the system with water for inspection.

5. Plastic pipe may be joined to any other type of pipe by using the proper conversion adapters. Use threaded adapters to connect plastic pipe to threaded metal pipe and fittings.

Use only approved tape or lubricant when connecting plastic pipe with a threaded joint. Teflon tape or petroleum jelly are recommended for this purpose. *Do not use conventional pipe joint compounds.* Be careful not to over-tighten threaded plastic pipe joints. After hand tightening the joint, make just one more full turn with a strap wrench.

6. Any conventional pipe clamps, hangers or brackets with a bearing width of ¾-inch or more can be used to support plastic pipe. Provide supports for horizontal runs of pipe 1½ inches or less at least every 3 feet. Supports for pipe of larger diameters can have a maximum spacing of 4 feet. See Figure 10–26. Trap arms should be supported at the trap discharge. Vertical pipe should be supported to maintain alignment. Where expansion and contraction of the piping can be expected, supports should permit the pipe to move without binding. Where pipe expansion will be absorbed at an elbow or tee, be sure clamps

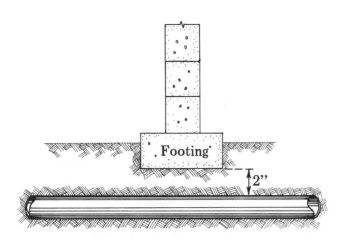

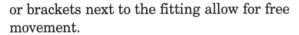

Figure 10–27
Code-required separation

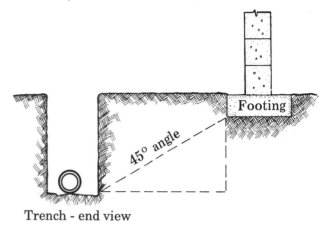

Trench - end view

Figure 10–28
Code-approved trench excavation

or brackets next to the fitting allow for free movement.

7. Protect plastic traps in summer and vacation homes from freezing by filling them for the winter with one the following:

   • 4 quarts of water mixed with 5 quarts of glycerol

   • 2½ pounds of magnesium chloride dissolved in 1 gallon of water

   • 3 pounds of table salt dissolved in 1 gallon of water

The salt solutions are effective to approximately 10 degrees F. If lower temperatures are anticipated, the traps should be filled with the glycerol solution.

Be aware of the following restrictions placed by the code on plastic systems:

• Plastic pipe or fittings can't support the weight of any plumbing fixture. Be certain the weight of the fixture is supported by approved brackets or other means.

• Never mix ABS and PVC pipe or fittings within the same system.

Drainage, Waste, and Vent Piping Within a Building

Here's a summary of code requirements for installing DWV piping in or under buildings.

• Underground or horizontal DWV piping must be adequately supported. Approved hangers or masonry supports may be needed to keep the pipe in alignment and prevent sagging.

• Building drains passing beneath foundations must have a clearance from the top of the pipe to the bottom of the footings of at least 2 inches. See Figure 10–27.

• Drainage pipe passing through cast-in-place concrete should be sleeved to provide ½-inch free space around the entire circumference of the pipe. This prevents stress on the pipe when the building settles or shifts. It also protects pipe from the corrosive effects of concrete. Caulk the free space between the sleeve and the pipe with coal tar or asphaltum compound, lead, or other approved materials. This allows movement while keeping insects out of the building interior.

• Occasionally you'll have to excavate a pipe trench parallel to and deeper than the foundation of the building. You can't excavate for drainage piping within a 45-degree angle from the base of the footing without special approval by the building official. See Figure 10–28. Remember this any time you excavate a trench parallel to a foundation.

• Drainage piping must be protected by sleeves, coating, wrapping or other methods

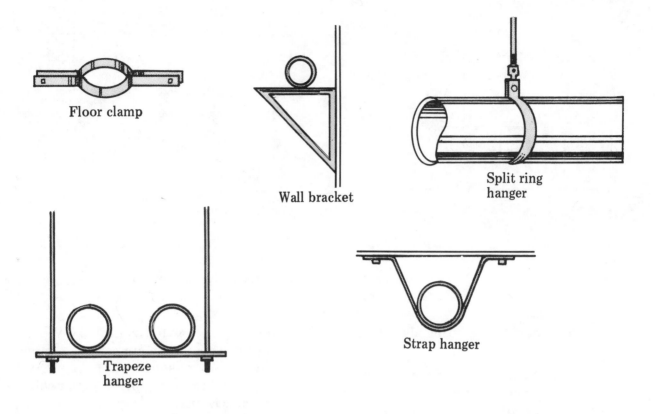

Figure 10–29
Five common horizontal and vertical pipe supports

approved by the local authority when pipe is installed in cinders or in corrosive soil.

■ Horizontal DWV Piping Supports

Support cast iron soil pipe with lead and oakum joints every 5 feet or less. Use approved hangers. If the pipe lengths are over 5 feet, supports can be spaced at not more than 10 feet. See Figure 10–26. Provide one support within 18 inches of the hub or joint.

No-hub cast iron joints must have supports adjacent to each joint or coupling if the developed length exceeds 4 feet. This is necessary because the joints aren't rigid and will sag.

Screwed pipe is stronger at the joint but still must be supported at approximately 10–foot intervals when it is used in a horizontal drainage and vent system. Copper pipe used as horizontal drainage or vent piping must be supported at 8-foot intervals. Plastic horizontal drainage or vent piping must be supported every 4 feet. Some of the more commonly used

horizontal pipe supports are illustrated in Figure 10–29.

■ Vertical DWV Piping Supports

In one-or two-story residential construction, vertical supports are needed more to keep the pipe in alignment than to carry any of the weight of the pipe. In two-story structures, the floor clamp shown in Figure 10–29 could be placed around the stack where it penetrates the second floor level.

■ Vent Terminals

Extensions of vent pipes should terminate at least 6 inches above the roof so gases are discharged well above the roof surface.

Where a portion of a two-story house is used as a sun deck, the vent should extend a minimum of 7 feet above the roof deck. Since cast iron pipe can be extended above a roof deck only 3 feet, use galvanized steel or copper pipe

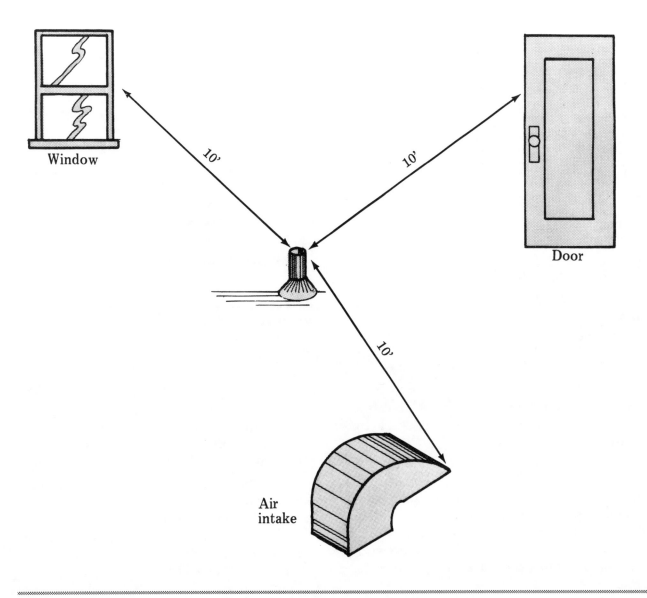

Figure 10–30
Proper separations from vent terminal

for the extension. Plastic pipe can be used if the extension is well protected and supported adequately.

In multi-level houses, a vent from a lower level can't terminate within 10 feet of any door, window or air intake openings for air conditioning on an upper level. See Figure 10–30. If there's no way to provide a 10–foot clearance, extend the vent to at least 3 feet above the height of the door, window, or opening on the upper level.

In very cold climates, moisture in the vent stack may freeze and obstruct the free flow of air. If ice seals off the vent stack, liquid in fix-

ture traps will probably be siphoned away. That makes fixtures drain much more slowly. The best way to eliminate this possibility is to increase the diameter of the vent extension through a roof to at least 3 inches. The change in diameter should start at least 12 inches inside the building. Figure 10–31 shows a long increaser used to increase the vent diameter. The long increaser may have either a spigot end for caulking into a hub or threads for joining to a galvanized threaded vent stack.

Install flashing where the vent stack passes through the roof. The two most common flashings are lead boot and galvanized steel.

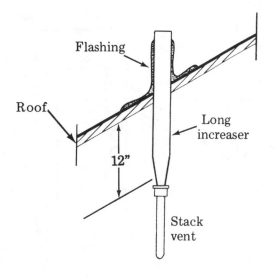

Figure 10–31
Cold climate vent closure prevention

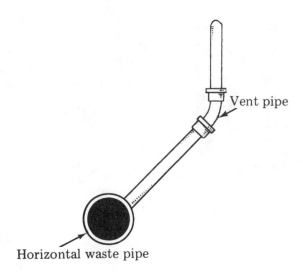

Figure 10–32
Proper connection of vent pipe to horizontal waste pipe

Galvanized steel has a neoprene collar that fits snugly on the exterior of the vent stack to make the joint watertight.

If lead boot flashing is used, it should extend above the vent opening and must be cut and turned down into the vent opening. Take care not to damage the lead as it is tapped into the end of the pipe. The base of the lead flashing has a flat plate. Coat the base of this plate with asphaltum or roofing cement to seal the boot to the roof.

Other important vent pipe installation factors you should know about are outlined below:

- Horizontal vent piping should be installed and sloped to drain to the vertical pipe section. Improper grading or sags in the pipe promote accumulation of condensation in low places. That restricts air circulation and reduces venting capacity. Vertical vent piping connected to a stack vent should be installed in an upward slope. That way, warmer air is free to circulate upward into the stack. Trapped moist air can accelerate pipe corrosion and greatly reduce the pipe's usable life.

- Vent pipes must not be connected to the side or bottom of horizontal soil or waste pipes. Instead, connect the vent pipe above

the center line of the soil or waste pipe as in Figure 10–32. The vent pipe should then rise vertically or at an angle not less than 45 degrees from the horizontal.

- Sometimes a fixture requires venting that runs horizontally to the nearest vent stack or stack vent. In that case, the horizontal vent pipe must be installed at least 6 inches above the flood-level rim of the highest fixture served by the vent. See Figure 10–33.

▓ Air Conditioning Condensate Drains

Many new homes have central air conditioning systems. Condensate from the cooler has to be piped to a legal drain. The following drainage method is acceptable by most authorities.

If permitted by your local code, use plastic pipe and fittings as indirect waste pipe for air conditioning drains. (Of course you may use other piping materials if you prefer.) The waste piping in a home constructed with a slab on fill should be installed 2 inches below the bottom of the floor slab. Install this waste piping after fill and compaction are completed to prevent sags in the pipe. Lay the piping on a firm base for its entire length and backfill with 2 inches of sand. All risers passing

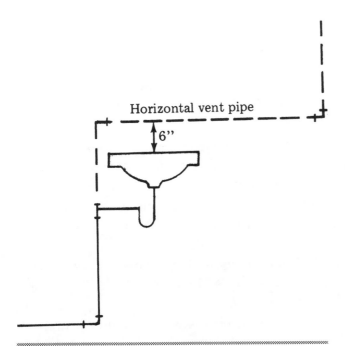

Figure 10–33
Horizontal vent pipe placement

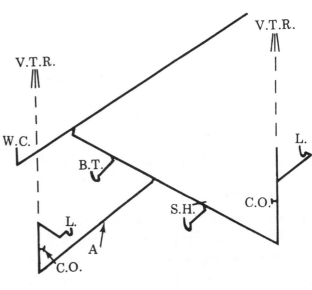

Figure 10–34
Faulty installation of fixture waste pipe (A)

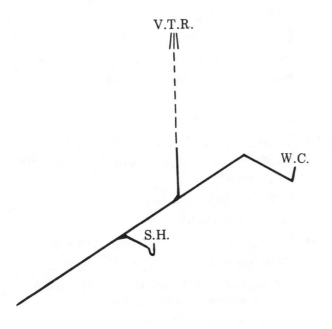

Figure 10–35
Faulty installation of major and minor fixtures

through the slab must be sleeved. Waste piping installed in a house with a wooden floor should be strapped securely to the floor joists to prevent sagging.

The minimum size for air conditioning condensate drain piping under a slab is 1¼ inches. It's not necessary to vent a condensate drainage system.

Air conditioning units not exceeding 5-ton capacity (most residential units have less) may discharge waste on a pervious area such as a lawn or bare soil. A dry well may also be used. Build a dry well from a length of 10–inch pipe that's 24 inches long. Leave both ends of the pipe open. Bury the pipe vertical in a hole 24 inches in diameter. Then fill the hole with ¾-inch rock. No cover is required for a dry well.

■ Common Venting Mistakes

The two faulty installations shown in Figures 10–34 and 10–35 are common mistakes made by novice plumbers. In Figure 10–34, fixture waste pipe A must not discharge into a horizontal wet vent. In Figure 10–35, the problem is that you may never install a major fixture (like a water closet) upstream from a minor fixture (like a shower). The inspector will reject these installations. Study the figures carefully to avoid these mistakes.

Questions

1. Why do we need a plumbing code?

2. Why should the drainage system be designed carefully?

3. For what purpose should vent pipes be sized and arranged?

4. Stoppages occur in the best-designed system. What provision does the code have to cope with this?

5. Name two fittings that can be used in vertical piping when the direction of flow is *from the horizontal.*

6. Name two fittings that can be used in horizontal piping when the direction of flow is *from the vertical.*

7. Name two fittings that can be used in horizontal piping when the direction of flow is *from the horizontal.*

8. Give two reasons why fittings with their drainage pattern opposite to flow *should not* be used.

9. Why is it against the code to drill or tap any drainage or vent piping?

10. Why is it against the code to install drainage pipes in trenches that are not open trenches?

11. What is the minimum head required to water test a drainage system?

12. Plumbing materials are subject to what standards and specifications?

13. Name the two pipes and fittings that are considered the newest materials for DWV installations.

14. Name the three most frequently used materials for DWV systems in residential buildings.

15. What are the two weights associated with cast iron pipe?

16. What is the schedule rating of plastic pipe and fittings used in a DWV system?

17. Can you name the two types of plastic pipe and fittings used in DWV systems?

18. Name any two piping materials commonly used for fixture drains in a cast iron system.

19. Plastic or copper pipe is frequently used for vent extensions in a cast iron system with what fitting?

20. Name the two materials the code prohibits to be intermingled in a plastic drainage system.

21. What is required by code for screwed, copper or plastic fittings used in a sanitary drainage system?

22. Cast iron pipe was first manufactured to serve what purpose?

23. Name the two things which DWV system joints must accomplish.

24. Name the characteristic of oakum in lead and oakum joints that is lacking in other types of joints.

25. How are no-hub joints made?

26. There are certain advantages to using no-hub pipe. Name two of these.

27. How long will a well-made lead and oakum joint last?

28. What characteristics of cast iron soil pipe make it superior as a building material for sewers?

29. How can the pipe barrel be kept in firm contact with solid ground?

30. What is the minimum size cast iron soil pipe manufactured for use in a DWV system?

31. When a cut is necessary in a hub and spigot system, from what type of pipe should the piece be cut in order to avoid waste?

32. Why is there less waste in a no-hub or plastic system?

33. Tees or tee branches must be installed only in soil, waste, or vent _____.

34. What is the term used for a tee that enters the straight through section of the fitting on a downward curve?

35. Straight tees are used in what section of a DWV system?

36. What are sanitary tees designed to carry?

37. What two purposes does a test tee commonly serve?

38. A wye or Y-branch is used to make changes of direction of how many degrees?

39. Combination Y and ⅛ bends are used together to make changes of direction of how many degrees?

40. Wyes and combinations are used in the building's main drain to serve what purpose?

41. What is the advantage of using a combination instead of a Y and ⅛ bend?

42. The side takeoff from a wye or combination may never be larger than what part of the fitting?

43. Describe the method used for sizing plumbing fittings.

44. How are bends classified?

45. Give the degree of turn for a ¹⁄₁₆, ⅛ and ⅙ bend.

46. When are ⅛ bend offsets generally used?

47. How does a straight cross differ from a sanitary cross?

48. How many vent lines may connect to a straight cross installed in a stack vent?

49. What two types of stacks may a sanitary cross use?

50. Where may an increaser be used?

51. What is the most frequent use of an increaser?

52. Where and for what purpose may a reducer fitting be used?

53. For what purpose are kafer fittings used?

54. What makes a kafer fitting unique?

55. In the United States, what city first used plastic pipe and fittings for sanitary installations?

56. What are two advantages to using DWV plastic pipe and fittings?

57. Describe the advantageous heat transfer characteristic of plastic drainage pipe over that of metal pipe.

58. Rigid vinyl plastic pipe is noted for its resistance to what substances?

59. What precaution must be taken before buying plastic DWV material in the United States?

60. What type of joint is used in a plastic DWV system?

61. Plastic fitting patterns are identical to what familiar type of fitting?

62. Plastic pipe and fittings can be used advantageously where access areas are limited. Why?

63. Why is it possible for only one person to work with plastic installations?

64. Why is it impossible to remove a plastic fitting once it is cemented to the pipe?

65. Rigid plastic pipe usually comes in lengths of how many feet?

66. When plastic pipe is used for building sewers, what precaution must be taken?

67. What is the minimum depth below ground a plastic building sewer can be installed?

68. What is the minimum size for plastic pipe and fittings used in a DWV system?

69. What is the most common method used on small jobs for cutting cast iron pipe?

70. To prevent cast iron pipe from breaking unevenly, what steps must be taken before the pipe is cut?

71. Why is a chalk mark necessary when the inexperienced plumber cuts pipe?

72. Why must a soil pipe cutter be used to cut cast iron pipe that is rigidly in place?

73. What device is used so that plastic pipe can be cut square?

74. What are the most common tools used to cut plastic pipe?

75. What is the depth of lead required by code in a lead and oakum joint?

76. Approximately how much lead is required to pour a 4-inch joint?

77. What may happen if a cold or damp ladle is plunged into a pot of hot lead?

78. What tool is used first in making a lead and oakum joint?

79. In caulking a vertical lead and oakum joint, which caulking iron should be used first?

80. What tool is necessary to retain hot lead poured into a horizontal joint?

81. In no-hub installations where access areas are limited, why is it important to position the heads of the retaining clamps in a certain direction?

82. What is the name of the tool used to assemble no-hub pipe and fittings?

83. Why is it best to work with plastic joints when the air temperature is above 40 degrees F?

84. What must be kept visible when installing plastic pipe and fittings?

85. How should plastic pipe and fittings be prepared for the application of solvent-cement?

86. What is indicated when a bead of cement shows around the plastic fitting?

87. What does it mean when no bead shows around the fitting?

88. What type of joint sealer should *not* be used on plastic threaded joints because of the injurious element it contains?

89. What must be adequately provided for when you install *horizontal* drainage, waste and vent piping?

90. What is the minimum clearance from the top of a horizontal pipe to the bottom of the footing of a building?

91. How much annular space must be provided for around the entire circumference of the pipe when the pipe must pass through cast-in-place concrete?

92. What procedure must be followed when excavating a trench parallel to and deeper than the foundation of the building?

93. What is the maximum distance between hangers for horizontal cast iron pipe in 10-foot lengths?

94. What is the maximum distance from a no-hub joint that a hanger should be placed?

95. What is the maximum distance between hangers for 3-inch plastic pipe installed in 20-foot lengths?

96. What common support is used to keep vertical pipe in alignment?

97. How high above the roof should vent pipes terminate?

98. What is the minimum distance from a window that a vent pipe opening may be installed?

99. What is used to make weathertight a vent stack where it penetrates the roof?

100. What can happen if vent pipes are improperly installed?

101. What happens when moist air is trapped in a horizontal vent pipe?

102. Where must vent pipe be connected to a horizontal soil or waste pipe?

Septic Tanks and Drainfields

Cesspools and outhouses were once the common means of disposal of human waste in both urban and rural areas. The inadequacy of cesspools and outhouses became clear when population densities increased. Now, local authorities who enforce model codes no longer accept outhouses or cesspools for human waste disposal. Some codes still permit cesspools as a means of sewage disposal for limited, minor, or temporary use, but only when they are first approved by the local authority.

Contamination from the sewage is a real possibility in rural areas where drinking water is taken from open or closed wells. Cross-connections allow untreated sewage to enter and contaminate the drinking water, spreading diseases like cholera and typhoid.

Where public sewers are not available, the most acceptable method for sewage disposal is the septic tank system. There is still some controversy over whether the septic tank is truly risk-free. The Septic Tank Association contends that there are no proven cases of septic tanks contaminating drinking water. But the Department of Environmental Resource Management (DERM) has questioned septic tank safety. The controversy and field testing continue. In many areas, DERM has established strict guidelines that have been adopted by most model code organizations.

Septic tanks and drainfields may vary in design, construction and installation methods. But the basic function of all septic tanks and the principles of sanitation and safety remain the same regardless of the code or geographical location.

Septic tank and drainfield installation are an important part of plumbing work, but few plumbing contractors and fewer homeowners attempt this type of work. A licensed septic tank contractor usually does the installation and maintenance for these systems. The plumbing professional is usually equipped to handle only the building drainage system to the inlet tee of the septic tank. But whether or not you intend to do this work, every professional plumber should know how to install the entire treatment system.

A properly designed and installed septic tank and drainfield needs no more maintenance than removal of the accumulated solids (sludge in the bottom and scum on the top) every few years. The drainfield requires no attention until it stops working, usually after about 15 or 20 years of service. When that happens, the old drain pipe and gravel have to be

103

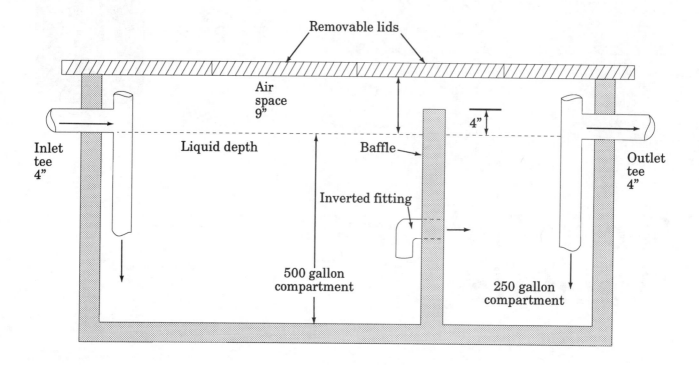

Figure 11–1
Cross section of typical 750 gallon septic tank with secondary compartment

How the Septic Tank Functions

removed and disposed of. A new drainfield with new drain pipe and gravel that comply with current code requirements must be installed.

A septic tank is a watertight receptacle which receives the sewage discharge of a drainage system through an inlet tee. See Figure 11–1. It's designed to separate solid from liquid wastes. The solids are usually about ¾ pound in each 100 gallons of water. The heavier portions settle to the bottom of the tank, while the lighter particles and grease rise to the top of the liquid. Anaerobic bacteria, which feed in the absence of air, decompose the solid matter, transforming it into gases and harmless liquids. The resulting gases agitate the tank contents and hasten the action of the bacteria. As new sewage enters the septic tank, gas is forced up and through the drainage vent pipes and into the atmosphere above the building roof. The tank is sized to have a capacity equal to approximately 24 hours of anticipated flow. This is necessary so that the bacteria have adequate time to fully digest the solids.

Digestion by bacteria creates a clear liquid, called *effluent*, which is forced through the outlet tee (see Figure 11–1) and into the drainfield. This is a subsurface system of open-joint or perforated piping installed on a bed of washed rock, gravel, slag, coarse cinders or other approved materials. The best-designed drainfields allow air to penetrate to the rock bed so aerobic bacteria have plenty of free oxygen available. The effluent seeps out between the joints or through the holes in the piping. Nitrification increases as the effluent oxidizes and seeps into the rock bed. The treated effluent is thus returned to the soil beneath the drainfield.

When you put in a drainfield, make sure that the ground surrounding the septic tank and drainfield can absorb the flow of effluent properly. Poor ground absorption will result in unpleasant odors.

Single Family Dwellings Number of Bedrooms	Minimum Liquid Capacity Required (gallons)
1 or 2	750
3	1,000
4	1,200
5 or 6	1,500

Figure 11–2
Capacity of septic tanks (Uniform Plumbing Code)

Single Family Dwellings Number of Bedrooms	Minimum Septic Tank Capacity (gallons)
1 or 2	750
3	1,000
4	1,200
5 or 6	1,500

Figure 11–3
Capacity of septic tanks (Standard Plumbing Code)

Locate drainfields in full sunlight, if possible, to promote good oxidation and evaporation of waste. Aerobic bacteria work more effectively when they are warm. In cold climates mound dirt over the drainfield as insulation.

Avoid planting trees or shrubbery on or near a drainfield. Trees and shrubs will thrive, but the roots will penetrate the drainfield and reduce its useful life.

Size and Location of Septic Tanks

The minimum capacity requirements for residential and duplex septic tanks vary slightly from code to code. Compare Figures 11–2 and 11–3 with what your code requires.

All codes base the minimum capacity of septic tanks on the number of bedrooms. The septic tank sizes in Figures 11–2 and 11–3 are designed to provide for sludge storage capacity as well as for domestic food waste disposal units. For each extra bedroom, add 150 gallons to the figures in the table. These figures allow enough capacity for bacteria to digest solids for 24 hours.

The code has placed a number of restrictions on the locations of septic tanks:

- The tank can't be located under or within 5 feet of any building.
- The tank can't be located within 5 feet of any water supply line (10 feet in some codes).
- The tank can't be located within 5 feet of property lines other than public streets, alleys, or sidewalks.

- The tank can't be located within 50 feet (25 feet in some codes) of the shore lines of open bodies of water.
- The tank can't be located within 50 feet (100 feet in some codes) of a private water supply well which provides water for human consumption, bathing or swimming.
- The excavation for a septic tank must not be made within a 45-degree angle of pressure as transferred from the base of an existing structure to the sides of an excavation. Look back to Figure 10-28 in the previous chapter.
- Circulation of air within the septic tank and drainfield must be through the inlet and outlet tees of the septic tank only. No other circulation is permitted.

Note that when a septic tank is to be abandoned, it must be pumped dry by a certified professional, the bottom of the tank broken in, and the tank filled with clean dirt.

Design and Construction of Septic Tanks

Blocks, brick, or sectional tanks are not permitted by most plumbing codes. Metal tanks are approved by local authorities in some areas. Fiberglass tanks are approved in most areas but have to be installed properly and protected from injury. Precast or cast-in-place concrete tanks are the most durable and meet the requirements of practically all plumbing codes. The interior walls of all septic tanks must be finished so that they are smooth and impervious. Voids, pits, or protuberances on these walls are prohibited by code.

Inlet and outlet tees for concrete septic tanks must be of terra cotta or concrete. They must have a wall thickness of at least 1 inch. The diameter of these tees must be no smaller than the diameter of the building sewer pipe. The outlet invert must be a minimum of 1 and a maximum of 3 inches lower than the inlet tee. Inlet and outlet tees must be installed at opposite ends of the septic tank and be a maximum of 4 inches above and 12 inches below the liquid level line. See Figure 11–1.

The connection of the building drain pipe to the inlet tee, as well as the tight-joint pipe used from the outlet tee, must be made watertight with a rich mixture of cement and sand. To protect metal tanks from corrosion, the inside and outside must have a bituminous coating applied. Some codes require that concrete septic tanks also have a bituminous coating applied to the inside.

Access to septic tanks for cleaning purposes must be provided by either manholes at least 20 inches in diameter (one located over each tee) or removable sectional lids. Septic tanks must be installed level.

Sizing Drainfields

Drainfields won't work if the soil surrounding a septic tank can't absorb the effluent properly. Codes vary on this requirement. Differences in soil absorption rates vary with:

1. topography (the lay of the land)
2. absorption capacities of the soil

Check your local code for any special requirements.

Topography

Where the land is relatively flat and the soil is sand or gravel, a conventional subsurface drainage system will work fine. See Figures 11–4, 11–5, and 11–6.

If the site is steep or if the soil is silt, clay or rock, a seepage pit or underdrain may be required. Otherwise the soil may not be able to absorb all the daily effluent. See Figures 11–7 and 11–8.

Underdrains are open-joint drain tiles buried about twice as deep as the regular subsurface drainage system. See Figure 11–8. Underdrains aren't connected to the distribution box or to the regular subsurface drainage system. They form a separate system and should be installed midway between the subsurface drainage lines, as in Figure 11–7. An underdrain filter bed will help prevent saturation of the drainage area due to poor percolation.

Rated Absorption

In sites with heavy clay, rock, or silt, the building official will probably require a percolation test before issuing a permit. Here's how to do a perc test.

Dig or drill at least three test holes to a depth equal to that of the planned drainage bed. Place a 2-inch layer of gravel in the bottom of each hole and fill the hole with water. If the soil is tight or has a heavy clay content, the test hole should stand overnight. If the soil is sandy and the water disappears rapidly, no soaking period is needed. Then pour water into the hole to a depth of 6 inches above the gravel. Measure the depth of the water every 10 minutes over a 30-minute period. The drop in water level during the *final* 10 minutes determines the percolation rate of the soil. If a test hole needs more than 15 minutes to absorb 1 inch of water, the soil is not suitable for a subsurface drainage system.

You'll use the average percolation rate for the three holes to figure the minimum effective drainfield absorption area in square feet per bedroom. Look at Figure 11–9, which shows drainfield absorption area per bedroom for both single family and duplex residences. At three minutes per inch of fall, you need 100 square feet of absorption area per bedroom. (Some codes require 150 square feet per bedroom as the minimum.) If your fall rate is more or less than 1 inch in three minutes, use Figure 11–9 to size the drainfield accordingly.

If the percolation rate is five minutes for 1 inch of fall, you need 125 square feet of percolation area per bedroom. If you're sizing a drainfield for a four-bedroom residence, multiply 4 times 125. The required drainfield size

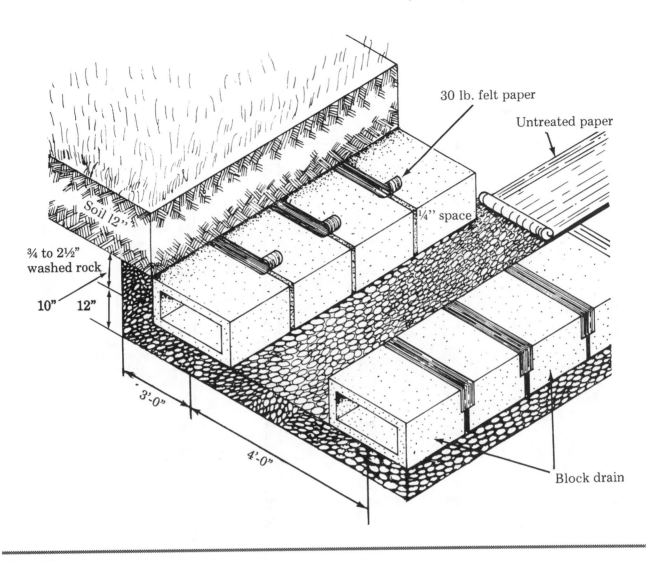

Figure 11–4
Cross section of reservoir-type drainfield

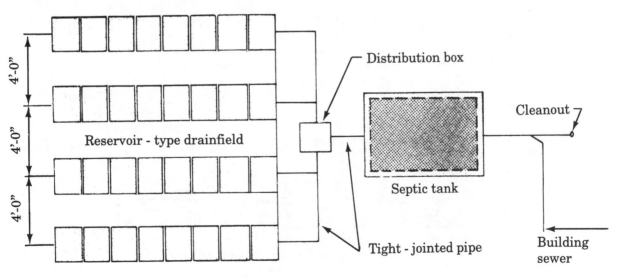

Figure 11–5
Detail of septic tank distribution box and drainfield

From
septic tank

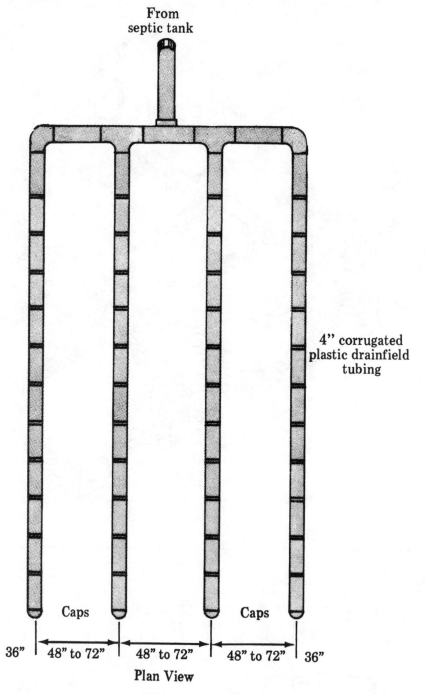

4" corrugated
plastic drainfield
tubing

Caps   Caps

36" | 48" to 72" | 48" to 72" | 48" to 72" | 36"

Plan View

Backfill

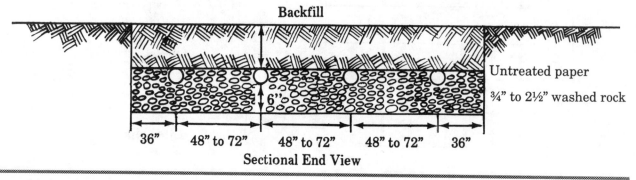

Untreated paper

¾" to 2½" washed rock

6"

36" | 48" to 72" | 48" to 72" | 48" to 72" | 36"

Sectional End View

Figure 11–6

Conventional drainfield with plastic tubing

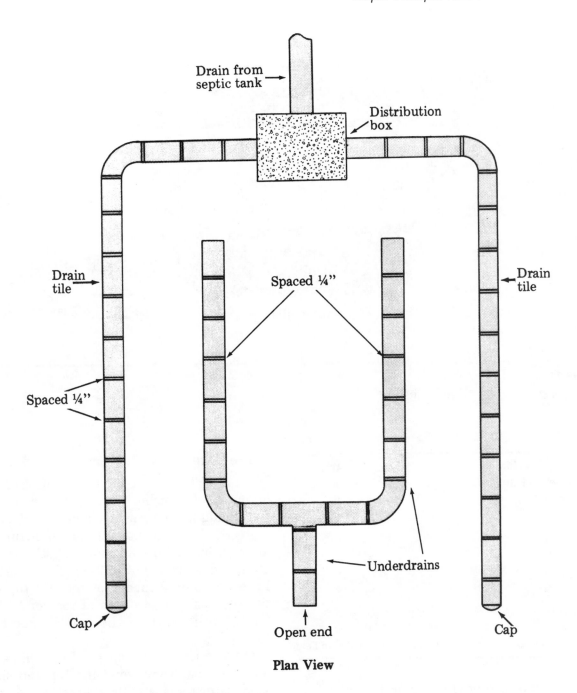

Figure 11–7
Underdrains used with drain tile (see end view Figure 11–8)

would be 500 square feet of *trench bottom*. Either individual trench bottoms or the overall square footage of a single drainfield's bed (length times width) may make up the total 500 square feet required for the installation.

Drainfield Construction and Installation

Several materials can be used for drainfield piping. The most common are open-jointed or perforated clay tile, perforated bituminous

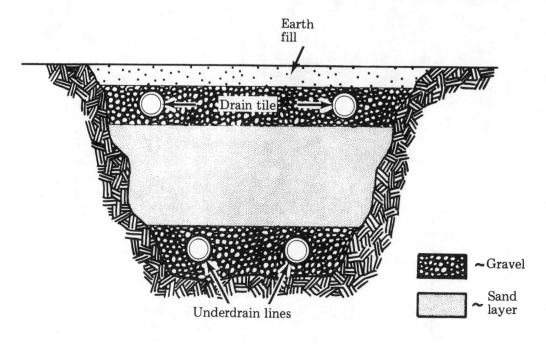

Figure 11–8
Sectional end view of underdrain installation

Percolation Time for Water to Fall 1" (minutes)	Absorption Area per Bedroon (square feet)
1 or less	70
2	85
3	100
4	115
5	125
10	165
15	190
Over 15	Unsuitable for absorption field

Figure 11–9
Drainfield absorption area for single family and duplex residences

fiber pipe, block or cradle-type drain units, and the newer corrugated plastic perforated tubing. Each type of piping must be able to carry and distribute the effluent evenly throughout the drainfield bed. The end of each distribution line must be sealed by capping or by cementing a block to it, as shown in Figures 11–6 and 11–7.

Tile Drainfields

Drainfield tile must have a minimum inside diameter of 4 inches. The open-joint drainage tile should have a gentle downward slope of no more than ½-inch per 10 feet to ensure even distribution of the effluent. Lay the tile on a 12-inch deep bed of ¾ to 2½-inch washed rock. Level another 2 inches of rock on top of the tile. This is a total depth of rock of 18 inches for the full width of the trench. See Figure 11–10.

Lay the tile with a space of ¼-inch between the tile ends. Then cover the top and sides of each ¼-inch space with a 4-inch wide strip of 30-pound bituminous-saturated paper. This prevents sand or other small particles from filtering into the openings. Then cover the entire length and width of the trench with untreated paper. This keeps sand and other small particles from filtering down and through the washed rock when the trench is backfilled.

Regardless of the width of the trench, consider each linear foot of drainfield tile to be 1 square foot. Each trench must have a minimum width of 18 inches and a maximum width of 36 inches. See Figure 11–10.

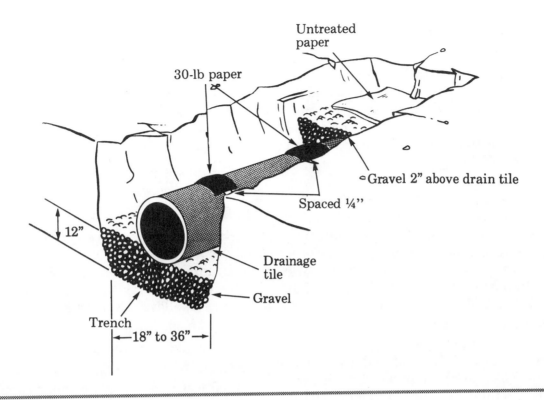

Figure 11–10
Tile drainage trench installation

The maximum length of a single tile drainfield trench is 100 feet. Where more than one tile drainfield trench is required, the trenches must be spaced at least 6 feet apart from center to center. Where two or more drain lines are used, they should be the same or nearly the same length and must be connected by a distribution box.

▓ Reservoir-Type Drainfields

Figure 11–4 shows a reservoir-type drainfield using concrete block. Block or cradle drain units can be used. The absorption area is based on the overall length and width of the reservoir rather than the length of trench, as in a drainfield.

Lay these drain units on a 12-inch deep bed of ¾- to 2½-inch washed rock. Then cover the drain with more of the same rock, leveling the top layer at 10 inches above the bottom of the units. This yields a total of rock depth of 22 inches for the full width and length of the drainfield. The block or cradle drain units

should also be installed on a gentle downward slope of not more than ½ inch per 10 feet. This ensures even distribution of the effluent over the entire drainfield.

Drain units that have a fixed opening to provide seepage may be butted tight against each other. But drain units without a fixed opening must be laid out with a ¼-inch space between the ends. Cover the top side of this opening with a strip of 4-inch by 16-inch 30-pound bituminous-saturated paper to protect the seam at the top and sides. The paper keeps sand from filtering into the openings. Then cover the entire reservoir area with untreated paper. See Figure 11–4.

Separate the drain units at 4 feet on center. See Figure 11–5. Distribution lines must not exceed 100 feet in length. Where two or more lines are used, they should be as near the same length as possible and must be connected to a distribution box. A tight-jointed pipe (a pipe with no open joints or fixed openings) must connect the distribution box to the septic tank. See Figure 11–5.

▨ Corrugated Plastic Perforated Tubing Drainfields

Plastic tubing can be used in reservoir drainfields where the absorption area is based on the overall square foot rule (length times width) rather than the length of the individual trenches. Plastic tubing is now widely accepted for use in drainfields.

Plastic tubing drainfields must also have a gentle downward slope of not more than ½ inch in each 10 feet. Lay the tubing on a bed of 12-inch deep ¾- to 2½-inch washed rock. Dump more rock over the tubing to a level 6 inches above the tubing bottom. This rock depth of 18 inches should extend for the full width and length of the drainfield.

Plastic tubing is made with small perforations and all joints are tight. No extra precautions are needed to keep sand and gravel out of the tube. The plastic is lightweight and easy to install. The tubing can be cut to length with a knife or fine-toothed saw. Note in Figure 11–6 that plastic fittings replace the distribution box. Cover the entire area of drainfield (all the washed rock) with untreated paper.

The distance between centers of plastic drain lines must not exceed 72 inches. The outside drain lines should be a minimum of 36 inches from any excavated wall. The length of each corrugated plastic drain line should not exceed 100 feet. Where two or more lines are used, they should be as near the same length as is practical. See the plan view in Figure 11–6.

▨ Drainfield Replacement

In time, most drainfields fail. The soil can no longer absorb the effluent properly, resulting in offensive odors. "Dead" drainfields can't be repaired, in most cases, but must be replaced. All drain lines and used filter rock must be removed and replaced with new materials. At one time, most codes permitted replacing an existing drainfield with a new drainfield of the same type and size as the original. Now, however, any replacement drainfield must be sized according to *current* plumbing code (as in Figure 11–9). *Check with your local building department before replacing an existing drainfield.*

▨ Basic Restrictions on Drainfields

- Drainfields must have a minimum separation distance of 8 feet from basement walls or terraced areas.
- Drainfields must have a minimum earth cover of 12 inches and a maximum of 18 inches.
- Drainfield must not be located under any building or within 8 feet of any building.
- Drainfields must not be located within 5 feet of water supply pipe lines (10 feet in some codes).
- Drainfields must not be located within 5 feet of property lines other than public streets, alleys or sidewalks.
- Drainfields must not be located within 50 feet of the shoreline of open bodies of water.
- Drainfields must not be located within 100 feet of any private water supply well which provides water for human consumption, bathing or swimming.

Care of Septic Tanks and Drainfields

As with any plumbing system, septic tanks and drainfields should be treated with care. If not abused, they'll give years of trouble-free service. Solid materials such as rags, sanitary napkins, heavy paper, cooking fats, coffee grounds, and food leftovers should not be discharged into the waste system. These items can't be decomposed by anaerobic bacteria. As a result, sludge and scum form faster than normal, meaning that the tank must be cleaned more frequently.

Even in normal service, the tank has to be cleaned occasionally to remove undigested particles and accumulated sludge and scum. Most codes require that a tank be cleaned by a certified professional who has the equipment to dispose of the waste properly. This waste is a health hazard and should not be buried or used as fertilizer on private property.

Paint thinners, chemical cleaners, or photographic chemicals should never be dumped in a plumbing system that connects to a septic tank and drainfield. In a home served by a septic tank, *chemical drain cleaner should never be used to unclog drain pipes.* These chemicals can kill bacteria in the septic tank, causing the tank to stop digesting solids. If undigested particles find their way through the outlet tee to the drainfield, it too will stop working. The result will be a serious plumbing problem and unnecessary expense.

If building drains are flowing too slowly, it's probably because the septic tank isn't draining properly. If you can't find a blockage in the line, the trouble might be too little bacteria in the septic system. You can increase the population of these microscopic creatures by flushing a yeast cake through the water closet once a week or so. Some commercial bacteria stimulants are also available.

Odor in the drainfield can be corrected in several ways:

1. Reducing the flow of water into the septic tank may help solve the problem. Disconnect the clothes washing machine and reconnect its waste to a separate drainfield, if this is permitted by your code.

2. The percolation rate of the soil may not be as high as it was when the drainfield was originally installed. Dig a seepage pit at the far end of the drain pipes to receive the effluent that has not yet seeped into the soil.

3. Check whether trees or shrubbery have been planted too close to the drainfield. Their roots could be clogging the drain pipes, making them useless.

4. Check for excess shade over the drainfield. Keep the drainfield area in full sunlight, if possible.

Questions

1. Under what circumstances may a cesspool be used?

2. What is the term used for the condition that allows drinking water to become contaminated?

3. Name two diseases that can be caused by lack of proper sanitary facilities.

4. What is the most acceptable method for sewage disposal where public sewers are not available?

5. The plumbing professional or homeowner usually connects the building sewer or drain to what portion of a septic tank?

6. What two substances must a septic tank be kept free of in order to maintain it in good working order?

7. What must be done to correct a drainfield that stops working?

8. What is a septic tank designed to do?

9. Anaerobic bacteria live and feed within a septic tank because of the absence of what element?

10. Into what form or substance do bacteria living in a septic tank transform solids?

11. What takes place within a septic tank when new sewage is discharged into it?

12. Approximately how many hours of anticipated sewage flow must a septic tank be sized for?

13. What is the liquid called after it has been processed?

14. The liquid enters a subsurface system containing what type of pipe?

15. Name the substance that is permitted to penetrate to the bedrock of a properly-designed drainfield.

16. Why should a drainfield be located in full sunlight?

17. What happens if trees or shrubbery are planted over or near a drainfield?

18. What determines the size of a septic tank?

19. What determines the size of a drainfield?

20. How close can a septic tank be located to a building?

21. How many feet must a septic tank be located from a private (potable) well?

22. An excavation for a septic tank must be made at no less than a _____ degree angle from the base of an existing structure.

23. How is air circulation within a septic tank and drainfield accomplished?

24. What must be done when a septic tank is abandoned?

25. Who is authorized by most codes to clean a septic tank?

26. What types of material are not permitted by most codes in the building of a septic tank?

27. What is the material that is most often used in septic tank construction because it meets most model codes requirements?

28. What is the minimum distance the outlet tee invert must be below the inlet tee?

29. Where must the inlet and outlet tees be placed in a septic tank in relation to each other?

30. What must be done to protect metal septic tanks from corrosion?

31. Give the two ways the lid of a septic tank can be arranged to permit access for cleaning purposes.

32. What is the essential characteristic of soil suitable for a septic tank installation?

33. Name two factors necessary for sizing drainfields.

34. What is the purpose of underdrains?

35. What is required where soil porosity appears to be less than usual?

36. When do you need more than 100 SF (150 in some codes) of absorption area per bathroom?

37. What must be done to the end of each distribution line in a drainfield to meet code requirements?

38. What is the minimum inside diameter of drainfield tile?

39. What is the most acceptable downward slope for all types of distribution lines?

40. Why should the entire width of filter material be covered with untreated paper?

41. What is the minimum center-to-center spacing for individual trenches?

42. On what rule is the sizing of reservoir-type drainfields based?

43. What is the minimum depth of rock required under drain units of a filter bed?

44. What must drain units be provided with to avoid spaces between the ends of each unit?

45. What is the maximum distance between centers of distribution lines for reservoir-type drain units?

46. What type of pipe must be used to connect the septic tank outlet tee to the distribution box?

47. Why is no seepage space needed when plastic tubing is used?

48. What is the maximum distance between the centers of drain lines in a plastic distribution system?

49. What is the maximum permitted length of a drain line?

50. What is the minimum depth of earth required to cover a drainfield?

51. What is the minimum distance of separation space between a drainfield and a basement wall?

52. What is the minimum separation space between a drainfield and a water supply line?

53. To help maintain a septic tank in good working condition, what solid waste materials should not be disposed of through a drainage system?

54. What kinds of chemicals should not be disposed of through a drainage system served by a septic tank?

55. What may be used to help the growth of bacteria in a septic tank?

Water Supply and Distribution System

Our drinking water comes from lakes, rivers and deep wells. Much of this raw water is not fit for human consumption until unpleasant taste, odors and impurities are removed at treatment plants. After treatment, the water is distributed through mains to our homes and offices.

The water main for a public water distribution system is usually located in the street, alley, or dedicated easement adjacent to each owner's parcel of land. This water main is installed, maintained and controlled by the local water authority.

The water main provides drinkable (potable) cold water at a pressure that meets certain standards. Pressures in the main usually range from 40 to 55 pounds per square inch (psi). If the pressure is expected to exceed 80 psi, a pressure-reducing valve should be installed in the water-service line where it enters the building.

Each property owner has to get a permit from the utility company to tap into the public water main. The utility installs a water meter at the property line to measure water usage. Figure 12–1 shows a typical water service connection detail. The building water service pipe is then connected at the outlet side of the water meter.

Private Water Supply

About one home in six in the United States is without a public water system. Almost 47 million people (that's approximately 18 percent of all households) get their water from lakes, streams or private wells. (Chapter 17 covers this subject.) These private sources must provide water that is safe to drink. In many cases homeowners have to provide their own treatment equipment.

Pumps are generally used to pump water from the source to a storage tank near the building. The pump also pressurizes water in the storage tank so the distributing pipes can supply the plumbing fixtures with enough water to function properly. The building water service pipe is then connected to the storage tank.

Note that local authorities usually require that a connection be made to a public water supply system when one becomes available.

Sizing the Water System

There's no way to determine exactly the maximum water demand in a building. The building code can't predict how many fixtures will

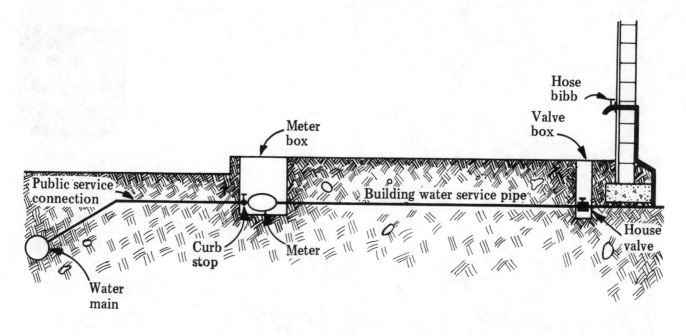

Figure 12–1
Service pipe

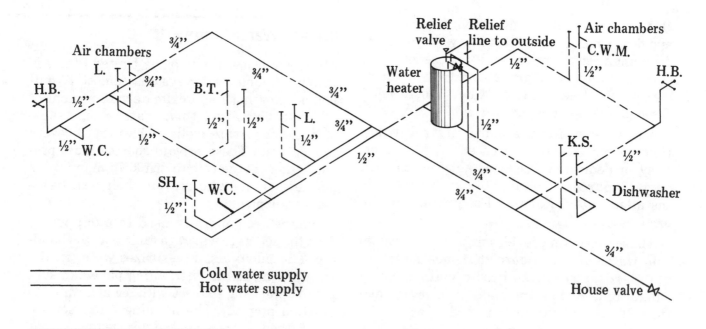

Figure 12–2
Water distributing piping diagram

be in use at the same time. But the code does a good job of estimating the maximum demand and requiring pipe sizes large enough so enough water will be available when it's needed. The objective is to avoid undersizing the pipe, but still not burden homeowners with unnecessary expense. Figure 12–2 is a water piping diagram for a small home.

The size of water supply pipes will depend on:

• The type of flush devices to be used on the fixtures

- The water pressure in pounds per square inch at the source
- The length of the pipe in the building
- The number and kinds of fixtures installed
- The number of fixtures expected to be used at any time

Sizing by Fixture Units

The size of piping in building water supply systems depends on the anticipated demand. The demand for water can be measured in water supply *fixture units*. For code purposes, one unit is equal to 7½ gallons (1 cubic foot) per minute. This method works well if there is enough pressure to supply the highest and most remote fixtures during peak demand. Sizing by the fixture unit takes into account two factors: velocity limitations recognized as good engineering practice and the recommendations of manufacturers of piping materials. The fixture unit values used for sizing supply lines are the same as the fixture units used when sizing the drainage system. Refer back to Figure 5-2 in Chapter 5.

Sizing from Tables

Most plumbing codes have a table that simplifies sizing procedures. Figures 12–3 and 12–4 are probably very similar to tables in your local plumbing code. You can use these tables for almost all one- and two-family dwellings if the minimum water pressure is at least 50 psi. In these types of buildings, 50 psi is usually more than enough to overcome the static head and ordinary pipe friction losses. You don't have to consider pipe friction when sizing pipe with Figures 12–3 and 12–4.

Figure 12–4 shows pipe sizes that are considered adequate by plumbing authorities in most areas. This table takes into account the major physical properties that reduce or restrict the flow of water. For most one- and two-family homes, I recommend using tables in your code like Figures 12–3 and 12–4.

Fixture Type	Pipe Size (inches)
Water closet	1/2
Shower	1/2
Bathtub	1/2
Lavatory	1/2
Kitchen sink	1/2
Clothes washer	1/2
Hot water heater	3/4
Hose bibb	1/2 or 3/4

NOTE: For fixtures not listed, the minimum supply branch may be sized as for a comparable fixture.

Figure 12–3
Size of fixture supply pipe

Number of Bathrooms and Kitchens (Based on Tank-type Water Closets)		Diameter of Water Service Pipe (inches)	Recommended Meter Size (inches)	Approximate Pressure Loss (Meter and 100' of Pipe) (psi)
Copper	Galvanized			
1 - 2	—	3/4 or 1	5/8	27
—	1 - 2	3/4 or 1	5/8	40
3 - 4	—	1	1	22
—	3 - 4	1	1	24
5 - 9	—	1 1/4	1	28
—	5 - 8	1 1/4	1	32

NOTE: For residential use not exceeding two stories in height

Figure 12–4
Minimum water service pipe size for one- and two-story buildings

Examples that follow explain how these tables are used. More complete tables for larger installations can be found in *Plumbers Handbook Revised* by this author. There's an order form for this and several other plumbing and construction reference books bound into the back of this manual.

▓ Sizing the Water Service Pipe

The water service pipe is the heart of the water supply system and must be sized first. See Figure 12–4. If this pipe isn't sized correctly, there may not be enough water available, no matter how the remainder of the system is designed. The minimum practical size for the water service line is ¾ inch, and that size can be used only in certain instances.

The size of the water service pipe must not change when it enters the building to become the building water main, whether the pipe is horizontal only or horizontal and vertical, as in a two-story dwelling. After water-distributing branch pipes are connected to the building main, the main may be reduced in size with proper fittings. As the main progresses through the building, the demand likely to be placed on the line decreases.

Figures 12–3 and 12–4 show the correct pipe sizes for residential buildings. The tables have certain sizing restrictions which are not difficult to follow when sizing water supply pipes. *These tables are adequate if the following restrictions are kept in mind:*

- Figure 12–4 applies only where water main pressure does not fall below 50 psi at any time.

- In single-family residential buildings more than two stories high, use the next larger pipe size.

- Buildings located where the water main pressure falls below 50 psi should use the next larger pipe size.

- Figure 12–3 shows the minimum sizes for fixture supply pipe from the main or from the riser to the wall openings. See Figure 12–2.

- For fixtures not listed in Figure 12–3, size the minimum supply branch for a similar fixture.

- Only two fixtures can connect to a ½-inch cold water supply branch.

Note in Figure 12–4 that pressure loss is greater using galvanized steel pipe than when using copper pipe. That's why more fixtures can be installed on a copper system than on a galvanized steel system of the same size.

Let's test your understanding of Figures 12–3 and 12–4 with several examples.

Assume that you're sizing the cold water service piping in the one-bath dwelling shown in Figure 12–5. Ignore the hot water piping when sizing the water service and distribution lines because it does not increase the water demand. The horizontal and vertical piping in Figure 12–5 serves one kitchen sink, one clothes washing machine, one hot water heater, one hose bibb, one water closet, one lavatory and one bathtub.

Study the next few paragraphs carefully to see how the 14 sections of pipe (A through N in Figure 12–5) are sized using the tables. Assume that the water pressure at street level is 53 psi and galvanized pipe is selected for this job. Begin with Figure 12–3. Without hesitation, you should be able to size pipe sections A, B, F, I, K and L as ½-inch pipe and section D as ¾-inch pipe.

The minimum pipe sizes for individual fixtures are shown in Figure 12–3. To size pipe sections C and J, you must recall that no more than two fixtures can connect to a ½-inch water supply branch. Since only two fixtures are connected to sections C and J, they may also be ½ inch. Pipe sections E and G must be ¾ inch to supply pipe section D, which supplies the hot water heater. Pipe section H would also be sized at ¾ inch, as this water supply branch supplies water to more than two fixtures (a water closet, bathtub and lavatory). Pipe section M would have to be sized at ¾ inch to the house valve, as no water supply pipe can be reduced between the source of supply and demand.

To size pipe section N, refer to Figure 12–4. Under "Number of Bathrooms and Kitchens," column 2, "Galvanized," note that a ¾-inch

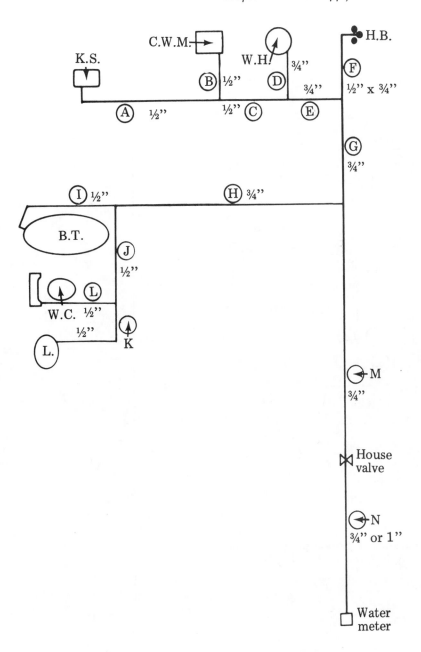

Figure 12–5
Sizing cold water piping, single-bath residence

water service pipe (1 inch in some codes) is sufficient for a one- or two-bathroom dwelling.

Let's look at two more examples. Refer to Figures 12–6 and 12–7. Compute the demand for these two dwellings. Calculate the size of each section of pipe by using Figures 12–3 and 12–4, the same way we did for the home in Figure 12–5. Assume that the street level water pressure is 50 psi and that copper pipe will be used.

Maintenance of Water Systems

Some types of pipe can't be used when hydrogen sulfide gas or other substances are present in the soil. Corrosion actually breaks down the walls of metal pipe, causing a thinning or pitting of the metal. The life of galvanized or copper water service pipe is very short when installed in sandy soil below the high tide level.

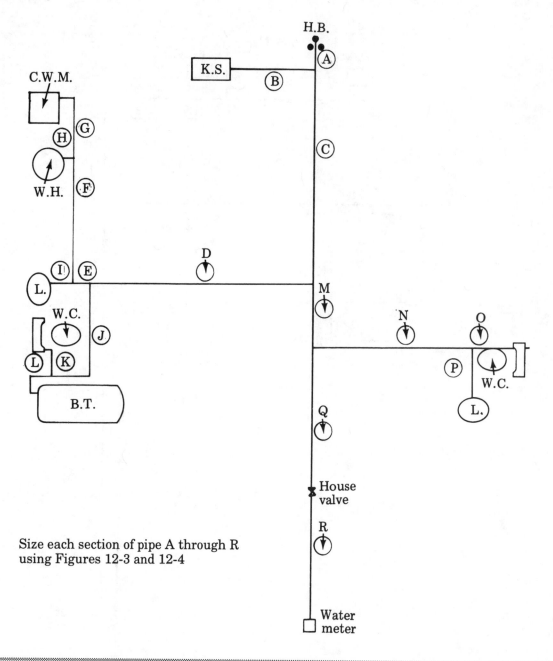

Size each section of pipe A through R
using Figures 12-3 and 12-4

Figure 12–6
Sizing cold water piping, two-bath residence

Under some conditions, metal pipe has to be protected by sleeving it in polyethylene plastic tubing, painting it with asphaltum paint, or using other approved protection methods.

▨ Types of Corrosion and Their Effects on Piping

Corrosion caused by electrolytic action (electrolysis) is very common. This is sometimes called *galvanic* corrosion. It occurs where two types of metal pipe (such as galvanized steel pipe and copper pipe) are joined. The two metals conduct electricity at different rates. When they are covered by an electrolytic solution such as water, a weak electric current begins to flow. By a process known as ionization, the metal from one pipe passes into the electrolyte and is plated onto the other metal.

Another example of galvanic corrosion occurs when a water heater with a galvanized steel tank is installed in a system with all copper

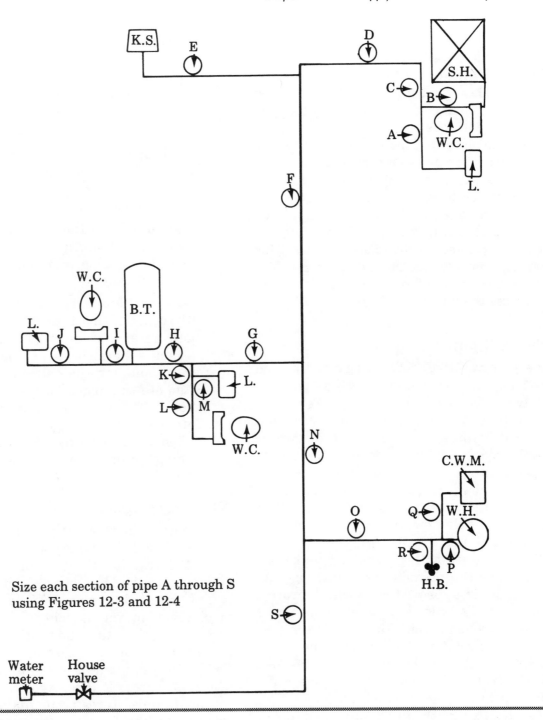

Size each section of pipe A through S
using Figures 12-3 and 12-4

Figure 12–7
Sizing cold water piping, three-bath residence

supply piping. In this system, weak electric currents generated from the tank begin to pass into the water and plate the copper pipe. This reduces the thickness of the tank walls and, if the process continues, the entire thickness of the metal could be corroded away in some spots.

Water with a high acid content is another cause of corrosion. Over a period of time, acidic water will corrode the inside surface of galvanized steel pipes and tanks, resulting in rust formation. In this type of corrosion, the iron in the pipes forms a compound with the oxygen dissolved in the water. This compound coats the metallic surface.

Rust forms faster as the water temperature rises. Thus, hot water piping is affected by rust more readily than cold water piping. Corrosion also varies with the velocity of water moving through the pipes. The more rapid the flow, the greater the opportunity for corrosion.

Scale is still another type of corrosion. Hard water contains large amounts of calcium and magnesium compounds. These compounds keep soap from lathering and form a scum which is deposited on the inside of pipes. This slows the flow of water through the pipes. The deposits harden and form scale that can eventually close the interior of the pipe.

How to Reduce Corrosion

There are several ways of slowing corrosion and prolonging the life of a water supply system.

Galvanic corrosion of water tanks can be controlled by placing dielectric unions or other approved fittings in the cold and hot water takeoffs from the tank. These dielectric unions or special fittings have a fiber washer or special material which insulates the tank from the rest of the water piping installation. That stops the weak flow of electric current from the tank to the piping.

Water heaters with galvanized tanks are equipped by the manufacturer with magnesium or other approved rods which protect against rust and corrosion. These rods act as electrolytic cells in which the magnesium particles go into solution, flow through the water, and are deposited on the metal to be protected. The electrolytic action dissolves the rods, which must be replaced periodically.

Metal water-distributing pipe should not come in contact with ferrous pipes, electrical conduit, building steel or reinforcing steel. In short, metal pipe can't come into physical contact with any good conductor of electricity. Where contact is unavoidable, a piece of felt or other insulation material should be securely fastened around the pipe to separate the two metals.

You can help control corrosion caused by excess velocity by installing a reducing valve in the line where the pipe enters the building.

Using the next larger pipe size will also reduce water velocity. Small pipes corrode more quickly than large pipes when carrying the same quantity of water under high pressure.

Here are some ways to reduce scale formation caused by hard water:

- Use copper pipe, CPVC or polybutylene plastic where permitted. The smooth interior of copper and plastic pipes tends to resist scale formation.

- Regularly flush the hot water heater of sediments (calcium and magnesium deposits) to keep them out of hot water distributing lines.

- Use a water softener. Water softeners usually contain the mineral zeolite. When hard water containing calcium and magnesium sulfates passes through a zeolite water softener, the calcium and magnesium are exchanged for the sodium in the zeolite. The softener water is then piped into the distributing system. Zeolite water softeners must be recharged regularly by adding sodium chloride (table salt).

Protection Against Freezing

In climates where water piping is subject to freezing, install the water service supply pipe below the frost line. If the water service supply pipe enters a building above ground or where it isn't protected from the cold, provide enough insulation to prevent freezing.

Remember also that the heating system may not always be on during the winter. Temperatures inside the building may drop below 32 degrees F. Pipe exposed to continued freezing temperatures accumulates ice. Since ice takes up about one-twelfth more space than the same amount of water, frozen pipes may burst. The ice transmits pressure along the cold water line to the weakest point. The point of rupture is not necessarily the point where the ice formed.

Insulation

The insulation materials in Figure 12–8 and 12–9 are recommended to prevent freezing. Apply these over pipe in crawl spaces, in

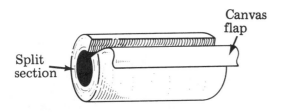

Figure 12–8
Woolfelt pipe covering

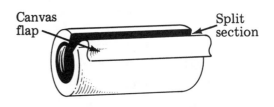

Figure 12–9
Frostproof insulation

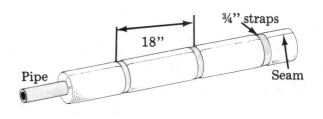

Figure 12–10
Pipe insulation with metal straps in place

unheated basements, and at the point where pipe enters a building above ground.

Woolfelt (Figure 12–8) is made of matted wool fibers or wool and fur or hair, pressure-rolled into a compact material. Woolfelt comes in a canvas jacket in thicknesses of ½ to 1 inch and is available in 3-foot lengths. It fits most pipe sizes.

Frostproof insulation (Figure 12–9) is made for cold water service lines located outside or in unheated areas. Frostproof insulation usually has five layers of felt. Three of these are woolfelt and two are asphalt-soaked asbestos felt. It is sold in 3-foot lengths with a canvas cover and is 1¼ inches thick. It fits most pipe sizes.

The coverings described are easy to install because each section is split in half and the canvas has a flap for quick sealing. Joint collars should be used to cover joint seams on piping exposed to outside temperatures. Cheesecloth can be used instead of canvas, but it must be glued in place. A thick paste of flour and water is an ideal glue for the purpose, but it must be applied in an area protected from the weather. Metal straps are usually supplied with the insulation and should be used to hold the insulation firmly in place. See Figure 12–10. These straps should be at least ¾ inch wide and spaced no more than 18 inches apart. If the insulation is exposed to view, a special type of white enamel paint can be used on the insulated pipe.

Properly-installed insulation requires little maintenance. Insulation exposed to weather or subject to damage from sharp objects should be inspected occasionally. If the canvas cover is torn or punctured, it should be patched with another piece of canvas. Use only a waterproof paste when installing or repairing outside insulation.

Occasionally you will find loose straps or loose insulation around valves and fittings. The straps should be retightened and the loose insulation replaced or pasted down. Make inspections before cold weather arrives. This could save the inconvenience and expense of frozen water lines.

Procedures for thawing frozen lines are described and illustrated in Chapter 20.

Questions

1. What are the sources of water for public water supply systems?

2. What is potable water?

3. What is the approximate percentage of households in the United States that depend on a private water supply system?

4. Most plumbing fixtures operate satisfactorily when the water main pressure is __ psi.

5. When should a pressure-reducing valve be installed on a water service line?

6. What does the law require owners of private water systems to do when public water is available?

7. Name three factors affecting the sizing of water supply piping.

8. Each water supply fixture unit represents how many gallons per minute flow?

9. What part of a water supply system must be sized first?

10. What is the minimum practical size for water service line?

11. When may a building main be reduced in size?

12. What rule of thumb is recommended when you are sizing the water piping in a building with a water main pressure of less than 50 psi?

13. How many fixtures may connect to a ½-inch cold water supply branch?

14. Why is hot water piping not considered when sizing cold water supply piping?

15. What is the minimum cold water pipe size to supply a hot water heater?

16. What preventive measures should be taken to protect metallic water pipes where hydrogen sulfide gas or other injurious elements exist?

17. Does corrosion affect only the *outside* of metal water pipe?

18. What causes galvanic corrosion?

19. What type of corrosion occurs as a result of weak electric currents?

20. Oxygen and iron in the pipe will cause what type of corrosion to occur inside a water line?

21. Name two other causes of corrosion that are common within a water supply system.

22. What is the most common way of reducing corrosion in a plumbing system?

23. Why is it necessary to isolate ferrous water pipes from other metals within a building?

24. Where water velocity is high, what should be done to protect a water system from corrosion?

25. How does water containing large amounts of calcium and magnesium affect a water system?

26. Name two ways of reducing the formation of scale within pipes where water is unusually hard.

27. In climates where water service supply pipe is subject to freezing, how deep should the trench be?

28. What maintenance precautions should be taken where water pipes are not protected from the cold?

29. When water freezes within a pipe, what causes the pipe to rupture or burst?

30. When water freezes within a pipe, where is the pipe likely to rupture or burst?

31. Name two common insulation materials used to protect pipe against freezing.

Valves and Faucets

Valves and faucets regulate the flow of water in a plumbing system. Valves up to 2 inches in diameter are usually made of brass. There are many types of valves, but those you install in residential jobs have an opening that can be closed with a disc or washer. Closing the valve presses the washer or disc tightly against a seating surface in the body of the valve.

Valves

Figure 13–1 shows five common valves: gate, globe, stop and waste, angle, and swing check. We'll discuss each in some detail here.

The *gate valve* has a wedge-shaped or tapered metal disc which fits into a smooth-ground surface or seat with the same shape. When the valve is open, the disc gate clears the opening formed by the seat and allows a straight line flow. Turning the valve clockwise moves the disc downward onto the seat. Gate valves are a good choice on lines which are to remain either completely open or closed most of the time. This limits wear on the metal gate and seat. The valve shown is a rising stem, single disc gate valve. Other common types are the double disc and the nonrising stem.

The advantage of a gate value is that it offers minimum obstruction to the flow of water. Globe and angle valves obstruct the flow more than a gate valve when completely open. Gate valves are generally required in water service pipes (in Chapter 12) and as the control valve for hot water heaters (See Chapter 14).

A *globe valve* looks like a gate valve from the outside but is very different on the inside. The operating principle is similar to that of a compression faucet. See Figure 13–2. A partition between the inlet and the outlet blocks the flow through the valve. The seat, which is at a right angle to the direction of flow, changes the direction of flow passing through it. The horizontal seat is a removable ground joint. A replaceable composition or fiber washer or metal disc at the end of the stem makes a tight fit against this joint.

Use a globe valve where fine adjustments are needed in the rate of flow, and where a tight shut-down is essential. Globe valves are suitable for all kinds of service except where full flow is required. Install the globe valve in the line so that the inlet side will carry the pressure when the valve is closed. An arrow on the valve body usually shows the correct direction of flow.

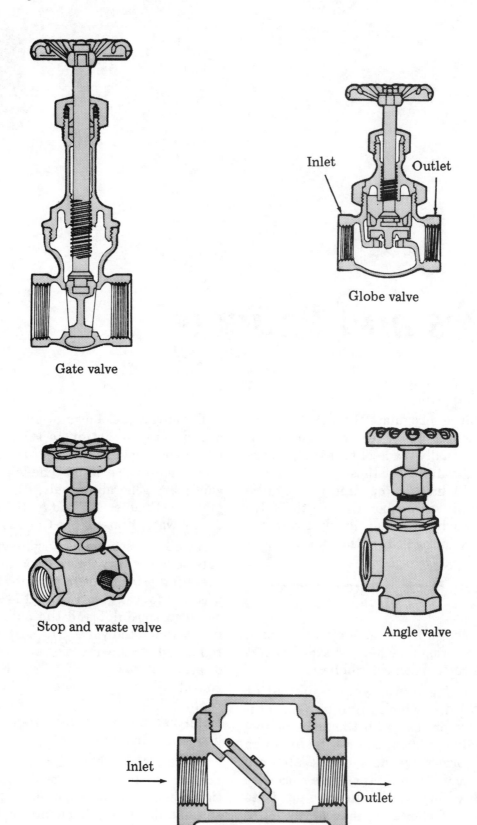

Gate valve

Globe valve

Inlet   Outlet

Stop and waste valve

Angle valve

Inlet

Outlet

Swing check valve

Figure 13–1
Types of valves

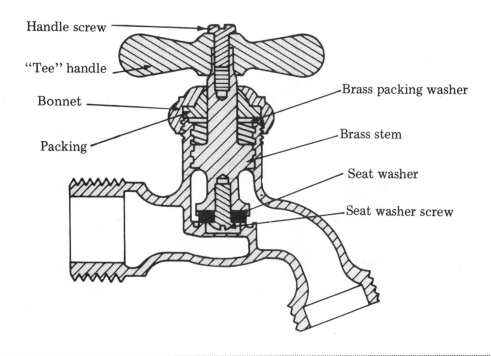

Figure 13–2
Typical compression faucet

The *stop and waste valve* is similar to the globe valve except that it has a small drain opening on the outlet side of the valve body. This small opening lets you drain water from the building supply system when the main valve is closed. This protects piping which may be exposed to freezing and expansion in cold weather. Most codes used in colder climates require that a stop and waste valve be installed near where the water service pipe enters the building. *Never install a stop and waste valve in a trench with water service pipe.*

An *angle valve* is a globe valve with the inlet and outlet at right angles to each other rather than in line. This valve controls the flow and changes the direction of a line.

A *check valve* prevents reversed flow in a plumbing system. It permits free flow in one direction but automatically closes if the liquid begins to flow the other way. The most common type in use is the horizontal swing check valve shown in Figure 13–1. Use this type of valve on domestic or irrigation wells. Install the valve on the suction line between the well and the pump. Other types of check valves are the disc check, the ball check, the adjustable check, and the spring check.

Faucets

A *faucet* is a valve installed at the end of a water supply line. The faucet most generally used for water outlets is the *compression type.* See Figure 13–2. It can be made of chrome-plated or unplated brass, depending on the location in the piping system. The spout usually curves down and may be threaded or plain. Most compression faucets have a brass stem with an enlarged threaded portion. This threaded portion fits threads in the body so that the stem moves up or down as the handle is turned. At the lower end of the stem is a fiber or composition washer held in place by a bibb screw. When the handle is turned to shut off the flow of water, the bibb washer is forced down against the ground seat. A compression faucet bibb washer may be flat or beveled. Always use the correct replacement washer to fit the seat configuration of the faucet you are servicing.

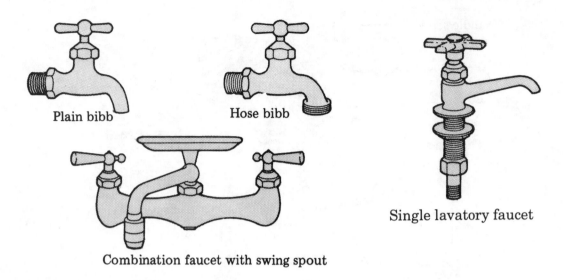

Plain bibb

Hose bibb

Single lavatory faucet

Combination faucet with swing spout

Figure 13–3
Some types of faucets

Figure 13–4
Compression deck-type lavatory faucet

Figure 13–5
Mixing single lever deck-type sink faucet

The stem threads are not watertight. Thus a packing washer or ring fits snugly around the stem, held in place by a packing nut which screws down onto the body of the faucet. As the packing washer wears away, water can begin to come up through the threads and out near the handle when the faucet is opened. These packing washers must also be replaced periodically.

Your plumbing supply dealer offers a wide variety of compression faucets. Figures 13–3 shows a plain bibb, a hose bibb faucet, a typi-

cal lavatory faucet and a combination faucet. Here are the descriptions:

A *plain bibb faucet* will show up occasionally in the kitchen of an older home. They are still sometimes installed in commercial-type kitchen sinks and fixtures such as pot sinks, scullery and service or slop sinks.

The *hose bibb faucet* has a threaded spout end that's useful for attaching a garden hose or a supply hose for a clothes washing machine.

The *single lavatory faucet* isn't used in residential lavatories any more. But the self-closing type is still used in lavatories in public toilets to conserve water.

The *combination faucet* usually has a swing spout. The advantage is that water passing through each of the faucets mixes in a single water outlet. Water temperature and flow are more easily controlled. The mechanics for servicing combination faucets are the same as for individual compression faucets. Notice that the combination compression sink faucet in Figure 13–3 is wall mounted. The lavatory faucet in Figure 13–4 is deck mounted. A single lever or knob mixing faucet is common in modern sinks and lavatories.

The *mixing faucet* shown in Figure 13–5 has largely replaced the faucets in Figures 13–3 and 13–4 for most applications. This type of faucet is usually the best choice for kitchen sinks, bathtubs, showers and lavatories. Flow and temperature are easily controlled without having to grind the washer against a metal seat. Because of the simplicity of design, this faucet can give many years of use before repairs become necessary.

Valve and faucet repairs are discussed in Chapter 21.

Questions

1. What is the function of valves and faucets?
2. Where are gate valves generally required?
3. Why should a gate valve be installed where little use of a valve is expected?
4. How does the flow control of a gate valve differ from the flow control of a globe valve?
5. Does a gate valve restrict the flow of liquids through it?
6. Which direction is the wheel turned to close any valve?
7. On the outside, a globe valve looks similar to what other valve?
8. Is the direction of the seat in a gate valve horizontal?
9. Is the direction of the seat in a globe valve horizontal?
10. In which direction does the seat of a globe valve change the flow?
11. Globe valves must usually be installed where?
12. Explain how a globe valve must be installed in the line it serves.
13. What is the main reason for using a stop and waste valve?
14. What makes a stop and waste valve different from other valves?
15. Under what conditions are stop and waste valves generally used?
16. Is it permissible to install a stop and waste valve in the same trench as the water service pipe?
17. Name the two main purposes an angle valve serves in a water line.
18. What is the working principle of a check valve?
19. (a) Where is a check valve most generally used in residential construction? (b) Where is it located?
20. What type of faucet is most commonly used for water outlets?
21. What advantage does a mixing faucet have over a compression faucet?
22. What is the working principle of a compression faucet?
23. What prevents water from leaking out around the stem of a compression faucet?

Hot Water Systems

The hot water system heats water and distributes it through pipes to fixtures as needed. The materials used in hot water systems are the same as or similar to those used in cold water supply systems. Galvanized steel pipe was once the most popular material, but in recent years copper hot water systems have become more common. Copper is especially good for hot water distribution systems because of its ability to resist corrosion, a problem worse in hot water systems than in cold, as the rate of corrosion increases as the temperature of the water increases. Maintenance and repair of hot water systems are similar to maintenance and repair of cold water supply systems.

The newer CPVC plastic and polybutylene (PB) plastic pipe are now accepted by most codes in both hot and cold water systems.

Most codes don't *require* a hot water system for most buildings. But you won't see many homes built with only cold water piping. All codes require hot water piping for certain types of buildings designed for special occupancy. In any case, if you install a hot water system, the code is quite specific on the safety devices required for that system. Safety equipment is

necessary to prevent damage to property and injury to persons using the plumbing fixtures.

The code does not specify design requirements for hot water distribution systems. It leaves the design of the systems to the installer. The code simply states: "The sizing of the hot water distribution piping shall conform to good engineering practice." Most plumbers aren't engineers. Fortunately, you don't have to be an engineer to design a good hot water system. This chapter has all the information you need to design safe and functional hot water systems for residential and light commercial buildings.

The Hot Water System Components

Hot water pipe carries water from the storage tank to the plumbing fixtures. Installation begins with a supply line connected to a heating tank. Water for each fixture in the building is taken from the main hot water supply line as needed. You can use Figures 12–3 and 12–4 in Chapter 12 to size the hot water main supply line and the fixture supply lines.

Follow the same sizing procedures as for cold water systems, with one exception. The hot water main supply and branch lines may be one pipe size smaller, to a minimum of ½ inch. For example, the cold water main supply to a bathroom is ¾ inch; the hot water main supply to this same bathroom may be ½ inch. Figure 12–2 shows a typical-size hot and cold water distributing piping diagram.

Two principal design objectives must be met in any good hot water system. First, the system must satisfy the expected hot water demand. Second, safety features must be built into the system to prevent excessive pressure and temperature.

The Hot Water Heater

The design temperature for hot water in most plumbing fixtures is between 130 and 140 degrees F. A water heater thermostat setting of 150 degrees F is recommended where a dishwasher is used. This isn't hot enough to scald the skin of someone working at the sink. The water heater thermostat is generally preset within the correct range at the factory and does not need additional adjustment. All modern water heaters have an energy transfer rate that's appropriate for the quantity of water stored in the tank. Water heaters are also made with insulation adequate to prevent excessive heat loss from the water stored.

There are three categories of hot water heaters in residential buildings.

- Electrical storage heaters
- Fuel-burning storage heaters
- Solar energy storage heaters

The most commonly-used water heaters today are the electric and the fuel-burning storage tank types. Electric water heaters are clean and attractive and can be installed nearly anywhere within a building. Water heaters designed to burn gas or oil must be located in a well-ventilated area and must have flues to carry away combustion gases. See Figure 14–1. The confined space must be provided with two permanent openings, one beginning within 12 inches of the top and one starting 12 inches from the bottom. They must

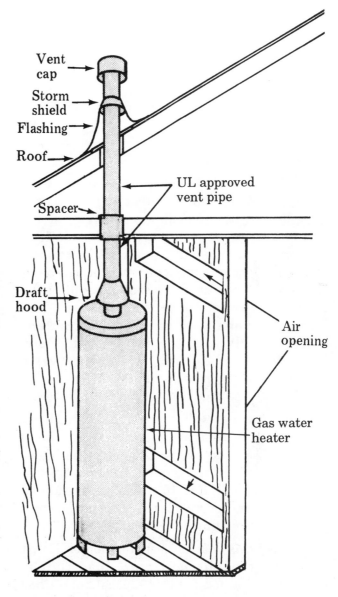

Figure 14–1
Installation of fuel-burning water heaters

provide adequate air for combustion, ventilation and dilution of flue gases.

Tank locations for solar water heaters are also limited by the size of the tank and by the type of system planned.

Constantly circulating hot water reduces water heating costs in larger systems. In one- and two-family homes, there's is no need for constant circulation of hot water. A small system has small pipes and short runs. The hot water in these pipes cools when water is not being used. But hot water is available again shortly after the faucet valve is opened.

Water expands slightly when it is heated, but is otherwise relatively incompressible. Expansion should not be a problem in most residential buildings with short piping runs.

▦ Heater Capacities

Electric and fuel-burning units have small storage tanks because the heating capacity is usually adequate to maintain the water at 130 to 150 degrees F even during the peak draw period. (The peak draw period is usually assumed to be one hour, although it can be as little as 20 minutes if several people take showers during those 20 minutes.) The amount of water used per person for bathing varies greatly with the habits of the user. Typically, hot water usage is spread over a number of peak times during the day.

When sizing hot water storage tanks, assume that only 75 percent of a tank's hot water supply is available during the peak draw period (one hour). Estimate the maximum hourly use and the number of hot water users. In general, 8 to 10 gallons is considered the maximum hourly use of hot water per person for dwelling units.

The *working load* of a water heater is the percentage of the *maximum* load that is expected to be used under normal conditions in any given hour. The working load factor is similar to the simultaneous use factor for cold water systems. The working load of single family residential buildings is approximately 35 percent.

The maximum load (storage tank capacity) and the working load (percentage expected to be used) determine the amount of hot water needed per hour. The recovery rate per hour and the storage tank capacity determine the amount of hot water that can be drawn from the tank per hour during peak demand periods.

For a peak draw period of one hour, a 20-gallon water heater should provide 15 gallons (75 percent of the tank's capacity) plus the recov-ery rate of at least 50 percent. This should provide 22.5 gallons per hour.

Assume that a one-bedroom residence has two occupants. If two people each use 8 gallons of hot water in one hour, a 20-gallon water heater will be adequate. Twenty gallons provide a backup of 6.5 extra gallons.

Notice that the size of both septic tanks and the capacity of the hot water storage tanks are determined by the number of bedrooms — *not* bathrooms. The number of bedrooms is the best indication of the quantity of water that will be used in the dwelling.

The individual tank storage capacities in Figure 14–2 are both economical and satisfactory for the average dwelling unit *when the peak draw period does not exceed one hour.* Special demand requirements which may last more than one hour might indicate that the next larger size should be used.

For each additional bedroom above four, use the next larger size hot water heater. You may want to split the system and install two water heaters, especially if long pipe runs are required. (Central hot water systems for multifamily dwellings are explained and illustrated in *Plumbers Handbook Revised* by the same author.)

Be sure that water heaters located in closets or under kitchen counters are installed so that the data plate showing working water pressure and other data is easily visible. The control and relief valves should also be readily accessible. Locate the heater where it can be easily serviced or replaced without removing any permanent part of the building. See Figure 14–3.

Safety Devices

A combination pressure and temperature relief valve must be installed on all hot water heater storage tanks. The temperature relief valve must be the reseating type and must be rated by its Btu capacity. *The Btu rating of the temperature relief valve must be greater than the Btu rating of the appliance it serves.*

In addition to the temperature and pressure relief valve, an *energy shutoff device* is now required on all automatically-controlled water

Number of Bedrooms	Hot Water Storage Tank Capacity
1	20 gallons
2	30 gallons
3	42 gallons
4	52 gallons

Figure 14–2
Recommended tank sizes for hot water

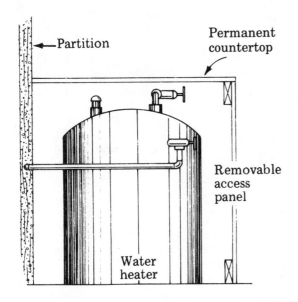

Figure 14–3
Installation of under-counter hot water heater

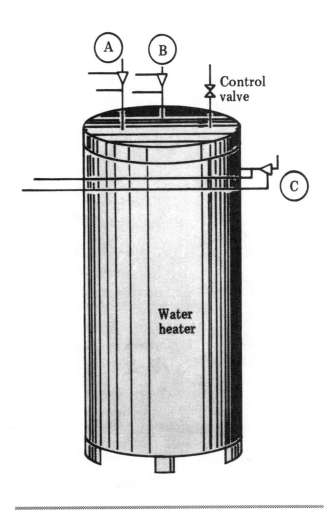

Figure 14–4
Relief valves are located in the top one-eighth of water heater. Positions illustrated at A, B, or C are acceptable

heaters. This device will cut off the supply of heat energy to the water tank before the temperature of the water in the tank exceeds 210 degrees F. Domestic hot water heaters commonly used in residential buildings don't require separate pressure and temperature relief valves. A combination pressure and temperature relief valve is acceptable.

Fuse-type relief valves are no longer acceptable on new construction and should be replaced on older heaters. In a fuse-type relief valve, the fuse melts and releases the excess pressure through the relief line. The water continues to run until the relief valve is removed and a new fuse is installed.

For relief valves to be most effective, they must be installed so that the temperature sensing element is immersed in the hottest water in the tank. This is always in the top one-eighth of the tank. See Figure 14–4.

A check valve or shutoff valve should never be installed between the relief valve and the hot water heater storage tank. If the check valve failed to operate or if someone accidentally closed the shutoff valve, the relief valve would then be useless. The tank could rupture or explode.

Relief valve drip pipes (popoff lines) should not be connected directly to any plumbing drainage or vent system. This could cause contamination of the potable water system. It could

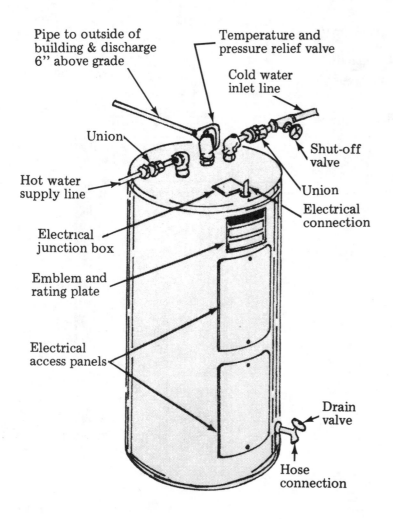

Pipe to outside of building & discharge 6" above grade

Temperature and pressure relief valve

Cold water inlet line

Union

Shut-off valve

Hot water supply line

Union

Electrical connection

Electrical junction box

Emblem and rating plate

Electrical access panels

Drain valve

Hose connection

Figure 14–5
Electric water heater piping detail

Figure 14–6
Combination temperature and pressure relief valve

also conceal from view any continuous discharge. Relief valve drip pipes should not terminate over any plumbing fixture or any other area where scalding might occur if the relief valve discharges.

The terminus of drip pipes should extend to an observable point outside the building. It should be unthreaded to prevent connections and should reach to within 6 inches of the ground. It should always be designed to drain dry.

Relief lines are sized by the Btu rating of the appliance. A ⅜- or ½-inch inside diameter pipe is adequate for sizing relief lines for single family residences. (All relief pipes are sized by their inside diameter.)

To install a relief line on a relief valve, first install a male flare or compression adapter (you can see one in Figure 15–6 in the next chapter) of the same size as the female threads in the relief valve. Spread pipe compound in the male threads and tighten the adapter into the female threads. Cut a length of the soft copper tubing to extend to the outside of the building. Allow enough tubing so the end can be turned down toward the ground. Shape the tube by hand to fit the wall line and secure it to the flare or compression nut on the relief valve. Strap the tubing securely in place. See Figure 14–5 (typical electric water heater piping detail) and Figure 14–6 (typical combination temperature and pressure relief valve.)

Hot water heaters must have the drain cock (valve) in an accessible location, both for flushing the tank of sediment and for emptying it for repairs or replacement.

The minimum size cold water supply pipe for a hot water heater, regardless of its capacity, is ¾-inch diameter. The cold water supply pipe must have a minimum ¾-inch shut-off valve in an accessible location. If a heater is located on a shelf in a utility room, for example, the shut-off valve should not be higher than 6 feet above the floor. The shut-off valve used must have a cross-sectional area equal to 80 percent of the nominal size of the pipe in which it's installed.

Hot water piping materials and installation methods are the same as those for other types of water distribution. See Chapters 15 and 16. The only difference is that hot water piping supports must permit pipe expansion.

Solar Water Heaters

Solar energy is a practical means of heating domestic hot water. Solar hot water heating can be added to existing systems as well as to new construction at a relatively modest cost.

The use of solar energy for heating domestic water is not new. In many areas of the world solar heating has been used successfully since before 1900. Solar water heaters have been commonly used in Florida and California for over 70 years. They can reduce the energy load of most homes and in most climates. A family of four using an electric water heater should save as much as $200 a year by adding a solar water heater.

▨ Solar Water Heater Components

There are at least three components in any solar heat collecting system:

1. the solar heat collector
2. the circulation system
3. the solar storage tank

A fourth component used in many systems is a backup heat source to ensure that hot water is available even during periods of peak water use and low solar radiation.

In a conventional fossil-fuel water heater, the heating element can work continuously to produce hot water. The storage tank can be rather small, 30 or 42 gallons. The tank in a solar heating system must be large enough to keep water warm through cloudy days and hours of darkness. The heating element of a solar water heater is known as a *solar collector*. See Figures 14–7 and 14–8.

The flat plate solar collector is the most practical and least expensive for residential use. It can produce temperatures up to 200 degrees F. The heat deck consists of a metal plate and tubing. The collector plate absorbs heat from the sun and transfers it to the liquid in the tubing. Heat deck materials can be copper, aluminum or steel. Thermally, these materials are the same. But both the tubing and the collector plate should be of the same metal so that they expand and contract at the same rates. Codes generally don't permit potable water to flow through aluminum tubing.

The heat collector box should be well insulated to shield the heat deck plate from the weather and to reduce heat loss. The heat deck plate and tubing are usually painted flat black to maximize absorption of the sun's energy. Light colors reflect the sun's rays.

A transparent cover on the collector box permits the sun's rays to strike the metal collector plate while reducing the loss of radiated heat back to the atmosphere. The glass used should have a low iron content to make it as transparent as possible to incoming rays. This glass admits solar radiation but is opaque to the long-wave energy trapped inside the box. This heat is transferred to the fluid in the tubing.

A clear plastic sheet is better than no cover at all, but glass is more efficient and more durable. Plastic tends to transmit both incoming and outgoing energy. And most plastics deteriorate fairly rapidly when exposed to heat and moisture. If a more heat-resistant plastic were available, it would be more desirable because glass breaks easily.

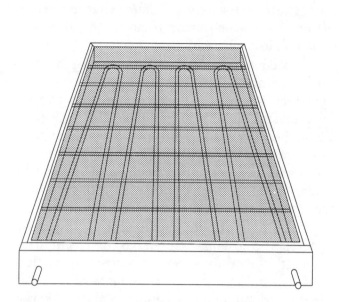

Figure 14–7
Flat plate collector

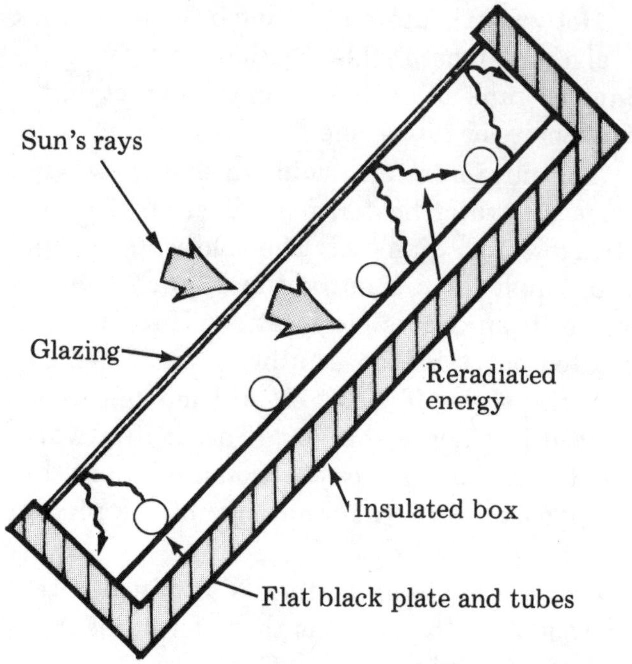

Sun's rays

Glazing

Reradiated energy

Insulated box

Flat black plate and tubes

Figure 14–8
Collector cross section

In cold climates, it's best to install a double layer of glass to prevent heat loss by convection when very cold air strikes the transparent surface.

A 4-foot by 12-foot solar heat collector heats approximately 80 gallons of water per day. This should be enough for the average family of four. Allow 12 square feet of heat collector surface per person to meet the typical hot water needs of most families at any time of the year in most sun belt areas of the U.S. The collector surface per person in northern states has to be larger, of course.

The average family of four requires a storage capacity for solar heated water of at least 80 gallons. But larger tanks of 100 or 120 gallons don't cost much more than an 80 gallon tank and store extra hot water for use when demand is unusually high.

Collectors are manufactured by many companies. Check for the approvals required by your local plumbing code before installing the unit.

▉ Installing the Solar Heat Collector

Only about 30 to 65 percent of the solar energy which strikes the glass surface of a collector during the day actually heats water circulating through the tubing. The rest is lost back to the atmosphere through the glass plate.

The most efficient solar collector would be perpendicular to the sun's rays at all times during the day and at all seasons of the year. But this is possible only if the collector turns and tilts to follow the sun's path. Motorized collectors are far too expensive for installation in most homes. The best compromise is a flat plate collector tilted in the general direction of the sun's path across the sky at an angle equal to the latitude plus 10 degrees. Fixed collectors should face south. However, collectors facing southeast or southwest work about 75 percent as well as those facing due south.

Locate the collector wherever it is most convenient and most attractive, as long as it's in full sun from two hours after sunrise to two hours before sunset. Shading before or after

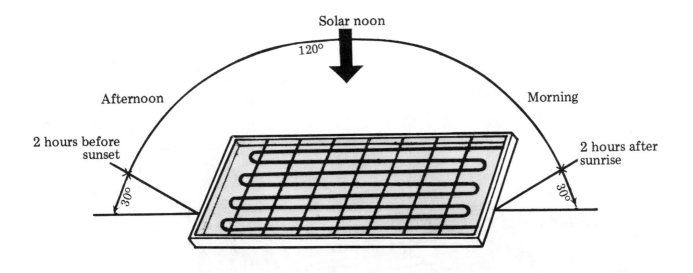

Figure 14–9
The optimum heat collection period

these periods has little effect. Late in the day and early in the day, when the sun is low on the horizon, most of the sun's energy is reflected back into the atmosphere. See Figure 14–9.

Mount the collector as close as possible to the storage tank to reduce heat losses and friction in the pipes. Make sure that all piping is well insulated. In northern states, shield the transparent cover to protect it from winds that would otherwise cool the surface.

Collectors that are to be installed as awnings or as fixed overhangs on a residence should be approved by your local authority before work begins. *This type of installation may not be acceptable under your code.*

Collectors can be mounted on the ground almost anywhere, but they must be securely anchored, as in Figure 14–10. Ground installations get damaged more often because they're more exposed to accidental breakage and vandalism.

In climates subject to freezing temperatures, all tubes should be installed so they drain dry. Draining the system is the best and cheapest way to keep the water from freezing and bursting the pipes.

▦ Mounting the Heat Collector

In many areas of the country, plumbers are responsible for mounting solar collectors. This is especially true in new construction where collec-

tor units are built into the roof as an integral part of the house. Built-in collectors look better because they can be designed to blend with the exterior. And they resist wind loads better, because they're usually flush or nearly flush with the roof surface. They can also be coordinated with roof surfaces to minimize leaks.

Roofing contractors make several recommendations for mounting collectors. They suggest that the collector have a structural frame. The frame should be securely bolted to the main roof structure. Flashing and a rain collar should be put around the pipes and the collector. The collector is then fitted into the frame and anchored securely to it.

Roof leaks can be a serious problem with roof-mounted collectors. Seal every hole drilled through the roof membrane. If a hole is too large or not properly sealed, a leak will occur. To prevent leaks, bore only small holes and caulk carefully around each. A pitch pan is a practical solution on built-up roofs. If a pitch pan is used, seal it carefully before securing the pan and filling it with pitch. Otherwise, the pitch will heat up under summer sun and create a pocket. Water can then get into the pocket and remain there. This moisture will eventually rust the top of the securing bolt. Advise the owner to check the roof annually to be sure expansion and contraction haven't

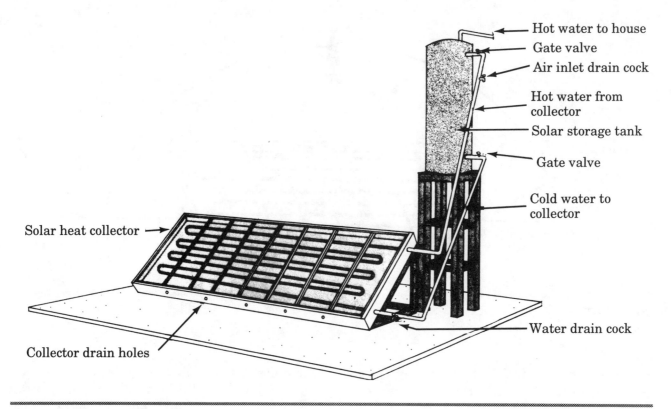

Hot water to house
Gate valve
Air inlet drain cock
Hot water from collector
Solar storage tank
Gate valve
Cold water to collector
Water drain cock
Solar heat collector
Collector drain holes

Figure 14–10
Ground-mounted thermosyphon solar water heating system

opened fissures where the bolts and pipes penetrate the roof membrane.

Install a bearing plate under the cliff angle to keep the collector from rocking. Screw the bolt all the way through the sheeting and then securely into the main roof structure. Lag bolts fastened only to the sheeting will pull out under the load of a collector vibrating in the wind.

On roofs with asphalt shingles, apply a layer of plastic cement on top of the shingles and beneath the cliff angle. Apply another layer of plastic cement on top of the cliff angle. Then securely fasten the cliff angle to the main roof structure with lag bolts.

▓ Water Circulation

Water circulates through the collector, moving hot water from where it's heated to where it's needed. A circulating pump can do this, but natural thermosyphon circulation moves the hot water through the system more simply. A thermosyphon requires no external energy source and no pumps, controls or other moving parts. The rate of movement of hot water from the collector to the storage tank, and cold water from the tank to the collector, is controlled by the intensity of sunshine. A thermosyphon results when hot water, which is lighter, rises to the storage tank and replaces heavier cold water drawn into the collector. See Figure 14–11.

The thermosyphon will not work properly unless the storage tank bottom is at least 2 feet above the top of the solar collector. Locating a large, heavy storage tank higher than the solar collector may mean putting the tank on the roof or in the attic. This presents problems of weight, construction and appearance. A leak in an attic-mounted tank can cause considerable water damage inside the house. For this reason, solar energy codes usually require that an adequate-size drain pan with a safe waste pipe extended to the exterior be placed under all hot water storage tanks located above the ground floor.

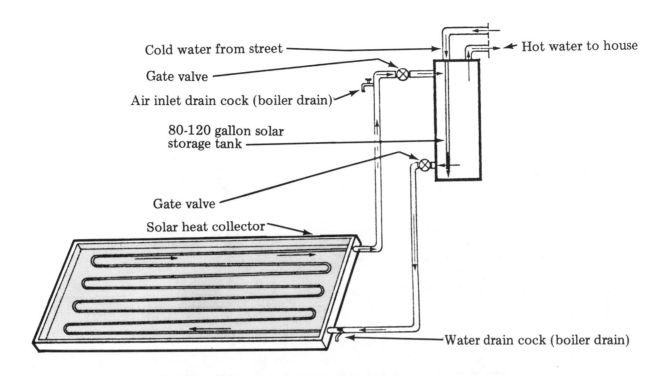

Cold water from street

Gate valve

Air inlet drain cock (boiler drain)

80-120 gallon solar storage tank

Gate valve

Solar heat collector

Hot water to house

Water drain cock (boiler drain)

Figure 14–11
Thermosyphon solar water heating system

In a thermosyphon system, use at least ¾-inch inside diameter piping both in the collector and in the circulation system. This reduces flow resistance. Be sure that the connecting pipe or tubing has a continuous fall with no sections that would permit the formation of an air pocket. An air pocket stops water circulation.

A pumped system uses the same basic piping as the thermosyphon system. But a pump forces hot water from the collector to the large storage tank. See Figure 14–12. In a pumped system, the heavy solar tank can be located in any convenient place—the garage or utility room, for example. This avoids the design and potential leakage problems of a tank mounted on the roof or in the attic.

You can use ½-inch copper tubing in a pumped system. The pump controls must be set so that water circulates through the heat collector only when water in the tank is cooler than water leaving the collector. The pump and controls add to the total cost, but this may be offset by the lower cost of installing the heavy storage tank at ground level.

Closed Solar Heating Systems

In a closed solar energy collection system, a fluid such as antifreeze is heated in the collector. A closed system may have a built-in heat exchanger within the storage tank body or on the outside of storage tank, as illustrated in Figure 14–13. The fluid circulates through the solar collector and transfers its heat to water in the storage tank through a heat exchanger. This system, although more complicated and expensive, eliminates the possibility of freezing. A hard freeze can destroy a heat collector if it's not properly protected. If you expect freezing temperatures, install a closed system or make sure that the thermosyphon system can be drained dry in the winter.

Materials and Installation

The rapid development of interest in solar energy has created a problem for many building code administrators. Building inspectors are faced with new installation problems not covered by some codes. Some codes now include

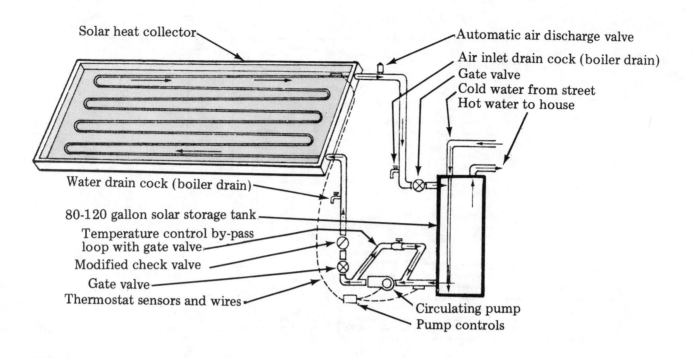

Figure 14–12
Pumped solar water heating system

specific provisions for solar heating. If your code doesn't cover solar heating, you still have to follow general provisions of the code when you do work on solar heating systems.

The rest of this chapter shows how existing codes and the *Uniform Solar Energy Code* apply to solar energy installations.

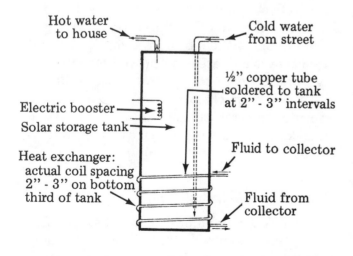

Figure 14–13
One type of heat exchanger

Pipe and fittings used for conveying fluids within a solar system must comply with the requirements for a potable water system. See Chapters 15 and 16.

Galvanized steel pipe and fittings and Type K or L copper pipe and fittings are the most common materials used in a solar circulation system. *Plastic pipe must not be used for conveying heated fluids within a solar system,* since plumbing standards limit the use of plastic pipe where temperatures could exceed 180 degrees F. *Aluminum tubing can't be used in a potable water system.* Its use in solar heating systems is limited to the closed-type system with a heat exchanger, and only if first approved by the local authority.

Built-in heat exchangers within the body of the tank must not be of lighter-weight pipe than Type L copper tubing. The heat exchanger must have no seams, joints, fittings or valves. It must be constructed of double-wall material designed to prevent leaks that could result in a cross-connection with the potable water supply.

Fittings used in a solar piping system should be of the same material as the pipe. An exception is made for valves and similar devices.

Where the use of dissimilar piping and fitting material can't be avoided, these materials should be electrically isolated using approved fittings which have been properly installed.

Valves installed in a solar piping system must be of brass or other approved materials. The fully-opened valve must have 80 percent of the cross-sectional area of the nominal size of the pipe to which the valve is connected. Control valves must be installed so that they can isolate the solar system from the potable water supply. All control valves must be readily accessible.

Where excessive water pressure is possible, an approved type of pressure regulator must be installed. Since the solar energy system is an integral part of the building's water supply system, the pressure regulator can be the regulator installed in the water service pipe. In this case, a second regulator would not be necessary in the solar supply system.

Temperature relief valves must be installed on all equipment used for the heating or storage of domestic hot water. A combination pressure and temperature relief valve is acceptable under most codes. The temperature sensing element must be installed in the hottest water within the top one-eighth of the tank. See Figure 14–4. *Temperature relief valves must be the reseating type.* Fuse-type relief valves are not acceptable.

Relief valves located inside a building must have a full-size discharge line. The line should discharge to the outside, turn down to within 6 inches of grade, and drain dry after use. Be sure that the line is securely strapped to the exterior of the building. The end of the discharge line must not be threaded. A discharge line can terminate at other locations if the installation is first approved by the controlling authority.

Some authorities may require a temperature relief valve located at the highest point of the solar piping system near the automatic air discharge valve. See Figure 14–12. Generally, codes don't require a separate discharge line from this relief valve to the ground but will accept the discharge of the solar system's relief valve onto the roof.

Automatic air discharge vents must be installed at all high points of a solar piping system. See Figure 14–12.

Storage Tanks

All storage tanks used for domestic hot water must meet the applicable A.S.M.E. requirements. Some codes require that any hot water storage tank located above the ground floor have an adequate-size pan and a minimum ¾-inch indirect (open) waste pipe which discharges to the outside of the building. This can eliminate interior building damage if the tank should leak.

All storage tanks must be equipped with an adequate and accessible drain cock. The tank and any devices attached to it must be accessible for repair or replacement.

Storage tanks must be permanently labeled with the maximum allowable working pressure and the hydrostatic test pressure which the tank is designed to withstand. These markings must be accessible to the inspector.

With special approval from the administrative authority, tanks designed and constructed to resist trench loads and corrosive soil effects can be buried underground. But no part of the tank can be covered or concealed prior to inspection and approval.

Installation Checklist for Solar Water Systems

- Only solar collectors approved by the local authority should be installed.

- Frames for securing solar collectors to the building must be of materials durable enough for outside use.

- Solar collectors must be anchored securely to withstand the expected dead, live and wind loads.

- Joints around pipes, ducts, bolts, or anything which penetrates the roof, must be made watertight.

- Collector panels that are not an integral part of the roof must be mounted a minimum of 3 inches above the roof surface.

- Collector panels and related piping must be installed so that they can drain dry.

- Collector boxes should have drainage holes at the low point for draining rain or other liquids that might collect in the box.

- Glass used in the collectors should have a low iron content. Only tempered glass should be used.

- All piping in a solar system must be installed to allow for expansion, contraction, and normal pipe movement.

- All piping which carries heated water, fluids or gases from a collector or heat exchanger to a storage tank must be insulated to minimize heat loss. The insulation must be thick enough to limit maximum heat loss to 50 Btu per hour per linear foot of pipe.

- Storage tanks must be insulated to limit heat loss to no more than 2 percent of the stored energy in a 12-hour period.

- Threaded, soldered and flare joints must be made as though for water piping. (See Chapter 15.)

- Piping in the solar heating system must be isolated so that gases, fluids or other substances can't enter any portion of the potable water system.

- The entire solar heating system should use only approved plumbing items and UL-approved electrical components.

- A permit is usually required to install, repair or alter any solar energy system.

- After installation, the solar system must be tested and proved tight under a water, fluid or air pressure test. The system must be able to withstand 125 psi for 15 minutes without leaking.

Questions

1. Define a hot water system.
2. Why is copper pipe and tubing the most popular material for hot water systems?
3. What type of plastic pipe may be used for hot water piping, under limited circumstances?
4. Why is safety equipment necessary for a hot water system?
5. What governs the sizing of hot water distribution piping?
6. With what does a hot water installation begin?
7. What two principal objectives must be met in designing a good hot water system?
8. What is the recommended temperature setting for the hot water heater thermostat when a dishwasher is used in a residence?
9. What are the two most common sources of energy for residential heating units?
10. Name another type of energy source for residential heating units.
11. Why are electric water heaters the most popular of all water heating units?
12. What two provisions must be made for gas- or oil-fired water heaters that are not necessary when installing other heating units?
13. Why are the possible locations of solar hot water storage tanks limited?
14. Why is it not necessary to circulate hot water lines for small jobs?
15. In designing a hot water piping system, what special consideration should be made for securing pipes?
16. Residential hot water heaters may be grouped into what three categories?
17. What are the two most common types of water heaters?
18. In most dwelling units, the peak draw period is usually assumed to be how long?
19. What percentage of a hot water storage tank capacity is used in sizing the tank for the peak draw period?

20. What is considered the maximum hourly use of hot water per person?

21. What is meant by the "maximum load" for a hot water heater?

22. Describe what is meant by the "working load" of a hot water heater.

23. For a peak draw period of one hour and a 50 percent recovery rate, how much hot water may a 20-gallon water heater be expected to supply?

24. What structural factor determines the quantity of hot water used in a residential unit?

25. When a special demand requirement is not considered, what is the recommended size of an individual storage tank capacity for a three-bedroom residence?

26. Water heaters must be installed where they are _____.

27. What determines the rating of a temperature relief valve?

28. In addition to the Btu capacity, what other special requirements must a combination pressure and temperature relief valve have?

29. Why is the older fuse-type relief valve not acceptable in new construction?

30. In what part of a water heater must a relief valve be installed to be the most effective?

31. Why is it against the code to install a check valve or shutoff valve between the relief valve and the hot water heater storage tank?

32. Why should relief valve drip pipes not be connected directly to any plumbing drainage system?

33. Where should a relief valve drip pipe terminate on a single-family residence?

34. Why should relief valve drip pipes not be threaded?

35. What determines the size of the relief lines?

36. What type of fitting should be used to secure a relief line to the female thread of a relief valve?

37. Where must the drain cock on a hot water heater be located?

38. What purposes does a hot water heater drain cock serve?

39. What is the acceptable minimum size cold water line to a hot water heater?

40. Why is solar energy for heating domestic water not considered new?

41. What are the three major components of a solar water heater system?

42. What are some of the obvious features that make the solar hot water system unique?

43. What type of solar collector is most practical for residential use?

44. What three materials may be used in building the heat deck for a solar collector?

45. What is the thermal difference between the three heat deck materials?

46. Aluminum tubing may be used in solar installation if it is limited to what type of system?

47. Why should the heat collector box be well insulated?

48. What provisions should be made to prevent heat loss in cold climates?

49. How many gallons of water per day will a 4-foot by 12-foot solar heat collector provide?

50. What is the minimum size recommended for a solar storage tank for a family of four?

51. A flat collector should be tilted in what general direction to be most efficient?

52. In climates subject to freezing, why is it important to install solar piping to drain dry?

53. What are some of the recommendations made by roofing contractors about mounting collectors?

54. Describe how a natural thermosyphon solar water heating system works.

55. What minimum size piping must be used in a thermosyphon circulation system?

56. Where may the hot water storage tank be located in a pumped solar water heating system?

57. What fluid is commonly used in a closed solar energy collection system?

58. Why is it not practical to use CPVC plastic pipe in a solar circulation system?

CHAPTER

15

Water Pipe and Fittings

The plumbing code regulates the materials, sizing and installation methods for water piping. The objective of the code is to provide a satisfactory supply of drinkable (potable) cold water to all plumbing fixtures so that they flush properly and remain clean and sanitary. The code establishes safeguards to avoid pollution of the water supply due to backflow or cross-connection. In this chapter, I'll cover the types of piping materials available for use in residential plumbing systems.

Consider the water supply in your area before selecting the material and size for water supply pipes, tubing or fittings. The water in many communities is filled with minerals which will corrode or leave deposits on the interior walls of some pipe. Some types of soil and fill can corrode the exterior of the pipe. (See "Maintenance of Water Systems" in Chapter 12.) Pipe that can leave toxic substances in the water supply must *not* be used for piping, tubing or fittings. Piping that has been used for anything other than a potable water supply system must not be reused in a water system that will be used for drinking. For example, pipe or fittings used in a gas system should *not* be reused in a potable water supply system. The reasons should be obvious.

Many water piping materials meet code requirements. Some materials are acceptable for use in underground installation; others can be used above ground only. Some materials are acceptable both above and below ground. The materials considered in this chapter include only those commonly used in residential or simple installations. Materials and installations methods for larger buildings are found in *Plumbers Handbook Revised* by this author, and available from the publisher of this book.

Galvanized Steel Pipe

Galvanized steel pipe (coated with zinc inside and outside) is usually sold in 21-foot lengths that are threaded at each end. It comes in nominal inside diameter sizes of as small as 1/8 inch. In most sizes it's available in three weights or strengths: standard, extra strong and double extra strong. Standard weight pipe is adequate in most installations and is the most common for plumbing jobs. Pipe of the heavier weights has the same outside diameter as the standard weight pipe but has thicker walls. Inside diameters of heavier pipe are reduced, thereby reducing the flow.

144

Galvanized steel pipe usually has a pipe coupling screwed to one end of each length of pipe. Use this coupling to join two pieces of pipe unless a special fitting is required.

Galvanized pipe has many advantages. It's used to distribute both hot and cold water. It's considered superior as a building material because of its strength, durability, and resistance to trench loads. It's especially desirable for outside use in water service supply piping. Because of its resistance to trench loads, the depth of placement is not critical except as protection against freezing. Galvanized steel pipe can be expected to give many years of service. Buried underground, it should last 15 to 30 years. Within a building it can last 50 years or longer.

Galvanized steel pipe has some serious disadvantages. It's more subject to corrosion than copper or plastic pipe. Water with a high acid content will rust the inside of the pipe. Hard water, which contains large amounts of calcium and magnesium compounds, forms scales or leaves deposits that harden in galvanized steel pipe more readily than in copper or plastic pipe. Where low water pressure is a problem, the rougher inside surface of galvanized steel pipe offers more friction against moving liquid than the smoother interior of copper or plastic.

In installations *not* subject to these conditions, galvanized steel pipe and fittings will provide a long service life.

■ Threaded Fittings

Galvanized malleable iron threaded fittings are usually found in water supply systems. They are pressure rated and used to connect lengths of pipe, change the direction of flow, or reduce the pipe diameter. The name of the fitting usually describes its shape and use. Some of the more commonly-used galvanized fittings are illustrated in Figure 15–1. The numbers in parentheses after the name of the fittings refer to the number of the fitting in Figure 15–1.

Tees (T's) are used to provide an opening to connect a branch pipe at right angles to the through pipe or main pipe run. A *straight tee* (1) has the same size opening at all three connections. For example, if the main pipe run is ¾ inch, and a ¾-inch side opening is needed for a branch pipe connection, use a straight ¾-inch tee.

When the branch opening is not the same size as the other two, the fitting is called a *reducing tee* (2). Specify a reducing tee by giving the straight-through (run) dimension followed by the side-opening (branch) dimension. Thus, a tee with run openings of ¾ inch and a branch opening of ½ inch is known as a ¾- by ½-inch tee. Assume you need a reducing tee with through (run) openings of 1 inch and ¾ inch and a branch opening of ½ inch. This tee would be called a 1 × ¾ × ½ inch reducing tee.

Tees may also have a run with male threads. These are known as *street* or *service tees*. A tee with two branch openings at right angles to each other is known as a *cross* (3). Crosses may also be straight or reducing.

Elbows (L's or ells) are used to change the direction of rigid pipe lines. Like tees, they are manufactured in many sizes and patterns. Specify elbows according to the angle in degrees between the two lengths of connected pipe. The more commonly-used elbows have female threads for connecting to threaded pipe. They may be 90 degrees or 45 degrees (4 and 5).

For special situations, use *street elbows* (6). These may be 90 degree (as illustrated) or 45 degree. *Reducing elbows* are available (7). A cross-sectional view of a typical 90-degree elbow connected to a piece of screw pipe is shown in Figure 15–1, number (8).

Unions are used where an additional branch line is installed in an existing screw pipe system or where certain types of appliances (a water heater, for example) must be replaced occasionally. A union permits removing a section of pipe to install the necessary fitting for the branch opening without disturbing any of the existing pipe or fittings.

The *ground joint union* (9) has a shoulder piece with female threads, a thread piece with female and male threads, and a ring or collar with both an inside flange (matching the shoulder of the shoulder piece) and a female thread which matches the male thread of the thread piece. The shoulder and thread pieces have a brass ground joint. The pipes to be connected are screwed to each end of the union

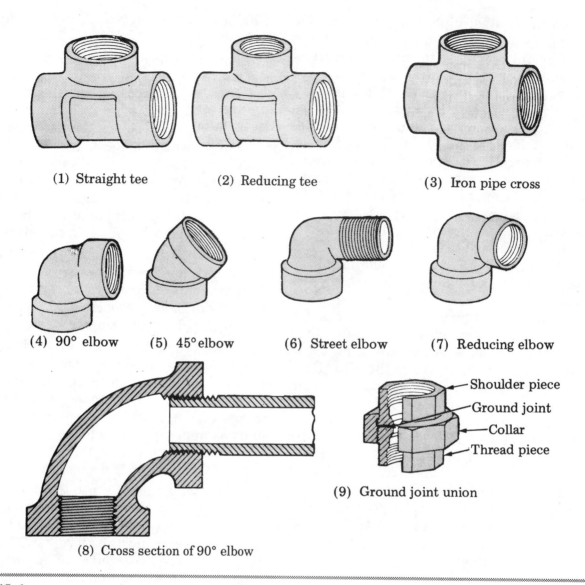

(1) Straight tee   (2) Reducing tee   (3) Iron pipe cross

(4) 90° elbow   (5) 45° elbow   (6) Street elbow   (7) Reducing elbow

Shoulder piece
Ground joint
Collar
Thread piece

(9) Ground joint union

(8) Cross section of 90° elbow

Figure 15–1
Threaded fittings for galvanized steel pipe

pieces. Both pieces are then drawn together by the collar, making a watertight joint.

Couplings are used to join two lengths of the same pipe size when making a straight run. See Figure 15–2. The coupling is a short fitting with female threads (1). It's not used to join two lengths of pipe already installed and fixed in place. You use a union in that case. The *reducing coupling* (2) is used to join pipes of different sizes. Another type of coupling commonly used in water supply systems is the *extension piece* (3), which has both male and female threads.

Among many other types of iron pipe fittings are the following:

Nipples (Figure 15–3) are used to make an extension from a fitting or to join two fittings. A nipple is considered to be a piece of pipe 6 inches or less in length, threaded on both ends. A *close nipple* (1), threaded its entire length, joins two fittings which must be very close to each other. A nipple threaded nearly its entire length with only a short unthreaded section in the center is called a *short* or *shoulder nipple* (2). When the unthreaded portion is longer than this, the nipple is called a *long nipple* (3 and 4).

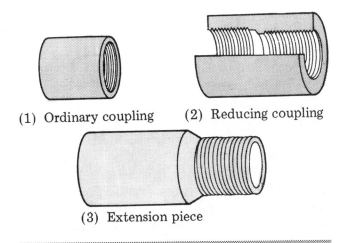

(1) Ordinary coupling   (2) Reducing coupling

(3) Extension piece

Figure 15–2
Iron pipe couplings

Long nipples are specified by lengths such as 3 inches, 3½ inches, 4 inches, up to 6 inches.

Miscellaneous iron pipe fittings are shown in Figure 15–4. A *plug* is a short, generally solid, length of galvanized metal, having a male thread and some means of being turned by a wrench. It's used to close openings in other fittings or to seal the end of a pipe. It has various types of heads. The *square head plug* (1) is the most widely used. The *slotted head plug* (2) is seldom used except in close spaces where the installer can't use a wrench.

A *cap* is a fitting with female threads. Used like a plug, it's screwed to the threads of a piece of pipe or nipple. It may be used for plugging up water outlets when testing the system or to create an air chamber to eliminate water hammer (3).

A *bushing* has a male thread on the outside and a female thread in the inside. It's usually used to reduce the opening in a fixed fitting to receive a smaller pipe. The ordinary bushing (4) has a hexagon nut at the female end for screwing the bushing into the fitting. The faced bushing, without a hexagon nut, used for close work, is prohibited by most codes.

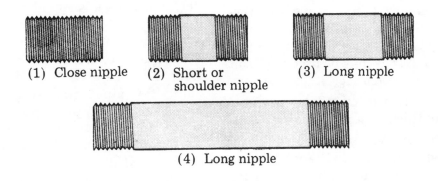

(1) Close nipple   (2) Short or shoulder nipple   (3) Long nipple

(4) Long nipple

Figure 15–3
Nipples

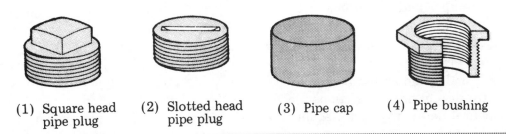

(1) Square head pipe plug   (2) Slotted head pipe plug   (3) Pipe cap   (4) Pipe bushing

Figure 15–4
Miscellaneous iron pipe fittings

Copper Pipe and Tubing

Copper pipe and tubing are used in both hot and cold water distribution systems. They can be used for both interior piping and underground installations, including the water service line. Copper is light, easily installed, and resists corrosion. There are two types of copper water line: pipe and tubing.

Copper *pipe* is rigid, hard tempered, and comes in 20-foot lengths. It does not bend easily unless annealed or softened by heating. Changes in direction are usually made with fittings similar to many galvanized iron fittings. Fittings for copper pipe are made of copper or brass. Also, with copper pipe, the joints are soldered. The code requires use of lead-free solder and flux in water distribution systems.

Copper *tubing* is flexible, soft tempered, and comes in coils of 50 to 100 feet, depending on size. It's easily bent by hand while cold to make changes in direction. Since the soft-tempered tubing may be easily bent, many fittings required for offsets with rigid pipe can be eliminated. Only 90-degree changes in direction, tees for branch line connections, and adapters must be used. Soldered fittings used for rigid copper pipe may be used on soft copper tubing. In addition, flare fittings may be used with soft copper tubing if the need arises.

Copper pipe and tubing come in three weights and wall thicknesses, as follows:

Type K has the heaviest or thickest wall and is generally used underground. It's available in hard and soft temper and in sizes useful for most piping installations.

Type L has a medium wall thickness and is most commonly used for water service and for general interior water piping. It's available in hard and soft temper and in various sizes for most piping installations.

Type M has a thin wall. Many codes permit its use in general water piping installations where Types K or L could be used. Check with your local authority for any restrictions placed on Type M before doing any installation. It's available only in hard temper and comes in 20-foot lengths and in sizes for most piping installations.

Copper Fittings

There are two types of copper fittings: solder type and flare type.

Solder type copper fittings are more widely used than flare fittings. Solder fittings can be used equally well with either rigid or soft copper pipe or tubing. They are made in the usual iron pipe patterns such as tees, elbows and couplings. Figure 15–5 illustrates some of the many solder type fittings. In addition to those shown, solder type fittings include reducing tees, reducing ells, tees with male or female iron pipe adapter openings, iron pipe adapter ells, and others.

Flare type fittings (Figure 15–6) are used only with flexible soft copper tubing. Their principal use is limited to certain specialized installations because they resist vibration better than solder type fittings. They should not be used for general plumbing where the flare fitting would be sealed in the wall.

Plastic Pipe

PVC plastic pipe is supplied in 20-foot lengths. It is rigid and comes in sizes suitable for most piping installations. Its recommended weight is Schedule 40. Where its use is permitted by code, it can be used only for outside installations. PVC pipe and fittings can be used for water service supply piping, swimming pool piping and other piping applications located outside a building wall. Its joints are usually solvent welded.

Two newer plastics that are now accepted by most codes for water distribution above or under the ground are chlorinated polyvinyl chloride (CPVC) or polybutylene (PB) plastic pipe or tubing.

CPVC pipe is rigid and comes in 20-foot lengths. It's designed for use in both hot and cold water systems. The fittings are solvent welded to the pipe. It comes in standard sizes suitable for most hot and cold water piping installations.

Polybutylene (PB) plastic comes in rigid 20-foot pipe lengths or in flexible coils of varying lengths. Polybutylene tubing (coil) has one distinct advantage over rigid plastics: it bends

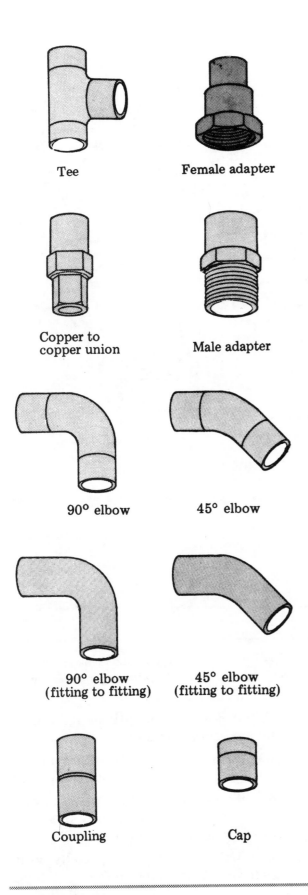

Figure 15–5
Solder type fittings

easily, which saves fittings. You can join polybutylene four ways:

1. compression fittings
2. flare fittings
3. crimp type fittings
4. heat fusion method

All plastic pipe and fittings must carry the ASTM (American Society for Testing Materials) identification numbers. All plastic pipe and fittings used in water distribution systems inside a building must have a minimum working pressure of 100 psi at 180 degrees F for both hot and cold piping material. Plastic piping installed outside a building must have a minimum working pressure of 160 psi at 73 degrees F.

Since plastic pipe is inert, it's resistant to rust, rot, scale and corrosion. Outside, plastic pipe may be installed without fear of exterior corrosion, even in soil where hydrogen sulfide gas is known to be present. It's not affected by electrolysis or by highly acidic water. Plastic pipe interiors are very smooth. This reduces frictional loss and prevents hard water scale buildup which would clog the lines.

Polybutylene pipe has a couple of additional advantages that some rigid plastics or metallic pipes do not enjoy. The pipe is not affected by freezing. It will expand to accommodate ice without splitting. Polybutylene's flexibility largely absorbs the effects of temperature change (pipe expansion or contracting). Its "give" reduces water hammer and virtually eliminates noise.

▓ Plastic Fittings

Plastic fittings for rigid plastic pipe are usually solvent welded to the pipe using a liquid cement recommended by the pipe manufacturer. (Polybutylene has other types of joining applications mentioned earlier.) Fittings are also made in iron pipe patterns such as tees, elbows and couplings. Plastic fittings of this type are similar in size and appearance to copper water pipe fittings. Care should be taken to use only *pressure rated plastic fittings* in a water piping system. Installation of plastic pipe and fittings is explained in the next chapter.

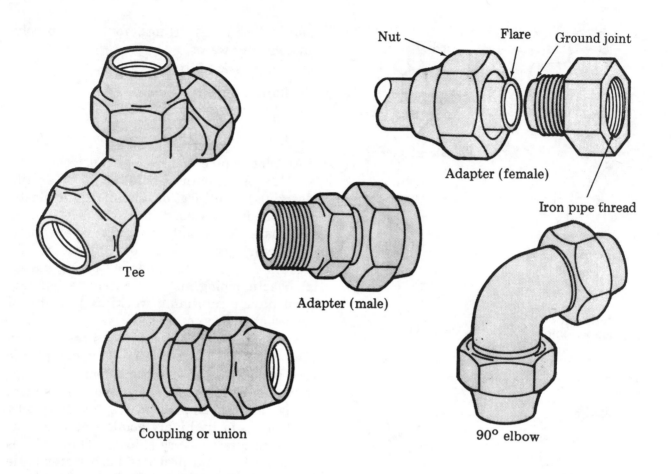

Figure 15–6
Flare type fittings

Questions

1. What are three aspects of potable water piping that are regulated by the plumbing code?

2. Why is it important that the code establish safeguards for public water supply systems?

3. How does the quality of water in the community affect the plumber's choice of piping materials?

4. How do various soils affect the choice of piping materials?

5. Why is it important not to reuse gas piping and fittings in a potable water supply system?

6. Steel pipe that is galvanized means that it has been coated inside and outside with what substance?

7. In what standard length is galvanized steel pipe manufactured?

8. What weight or strength of galvanized steel pipe is most often used in a plumbing water piping system?

9. Galvanized steel pipe usually has what type of fitting attached to one end?

10. Galvanized steel pipe is especially desirable for what particular use?

11. How does water with a high acid content affect galvanized pipe?

12. What two substances does water contain that causes it to be considered "hard"?

13. What four conditions limit the use of galvanized steel pipe?

14. Where are galvanized malleable iron threaded fittings usually used?

15. What does a tee provide in a water piping system?

16. A tee that has three openings of the same size is called a _____ tee.

17. A tee that has openings of different sizes is known as a _____ tee.

18. Elbows are used to serve what purpose in a rigid pipe line?

19. What makes a street elbow different from a standard elbow?

20. What does a union accomplish when used in an existing line?

21. What is accomplished in a piping installation by using a coupling?

22. What is the name of a coupling that has both male and female threads?

23. How are nipples used?

24. What is the name for a nipple that is threaded almost its entire length?

25. Name two common lengths for nipples used in a plumbing piping system.

26. Plugs are generally used for what purpose?

27. What plug is most widely used?

28. For what two purposes are caps used in a plumbing system?

29. What is the purpose of a bushing when used in a fitting?

30. Where is a faced bushing usually used?

31. Copper is known to resist what action?

32. Rigid copper pipe is usually manufactured in what lengths?

33. Flexible copper tubing is called by what name?

34. Name one main advantage in using flexible copper tubing.

35. Name the three weights or types of copper pipe or tubing used in piping systems.

36. What two types of copper fittings may be used in water piping installations?

37. PVC plastic pipe may be used (where permitted) in what portion of a water supply system?

38. What minimum working pressure is required for plastic pipe used in a water supply system?

39. While all plastic pipes resist rust, rot, scale and corrosion, what additional advantages does polybutylene (PB) pipe have over chlorinated polyvinyl chloride (CPVC) pipe?

16

Installation of Water Systems

Most of the code sections that cover the water supply installation make good sense if you take the time to understand them. Unfortunately, plumbers at the job site often overlook important points in their rush to finish the job on time. As you read through this chapter, ask yourself why the code requires what's being described. In nearly every case, you'll have no trouble guessing the reasons for the rules.

Installing Water Service Supply Pipe

Water service supply pipe can be installed in the same trench as the building sewer. This procedure is common where the soil is hard or rocky or where there is inadequate space to separate the two lines. The following conditions must be met when a single trench is used for sewer and water service pipe:

- The water service supply pipe must be placed above the sanitary line on a solid shelf excavated at one side of the common trench. See Figure 16–1.

- The bottom of the water service supply pipe must be at least 12 inches (some codes permit 10 inches) above the top of the sewer line. See Figure 16–1.

- The number of joints in the water service supply pipe must be kept to a minimum.

The water service supply pipe installed in a separate open trench usually must have a minimum separation of 5 feet from any sewer line. See Figure 16–2. Check with local authorities; they may require some other separation distance.

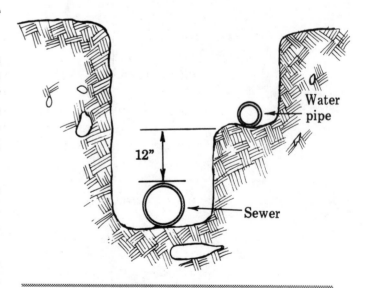

Figure 16–1
Placement of water service supply pipe in sewer trench

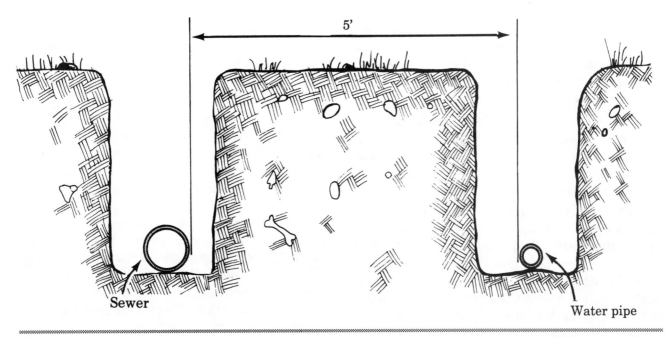

Figure 16–2
Placement of water service pipe in separate trench

The water service supply pipe must have a minimum separation of 5 feet from any septic tank or drainfield.

Water service supply piping should be laid on a firm bed of earth for its entire length. It should be securely supported to prevent sagging, misalignment, and breaking. Installation in open trenches is recommended but not required. Tunneling with care under sidewalks or driveways is permitted. Young trees should not be planted near water service supply piping because growth of the root system may cause the water line to buckle, bend, or break.

Since galvanized steel is strong, durable, and resistant to trench loads, it's considered the best choice for water service supply piping. Because of its resistance to trench loads, the depth of placement isn't critical except as protection against freezing. Avoid backfilling the trench with large boulders, rocks, cinder fill, or other materials which might physically damage or encourage corrosion of the pipe.

The depth for copper pipe or tubing, Type K, L, or M, isn't established in most codes. Copper pipe and tubing is soft and easily damaged. Naturally, the pipe must be placed below the frost line in areas subject to freezing. In warmer climates, depth of placement is critical

where there is likelihood of damage by edgers or other sharp tools. Be careful when selecting trench backfill material. Use fine rather than course materials. Avoid materials such as cinder fill that might promote corrosion.

PVC plastic pipe is more fragile and has to be treated carefully during installation. The bottom quarter of the pipe should be supported continuously and uniformly on the trench bottom by 4 inches of fine, uniform material which should pass through a ¼-inch screen. The fine material should extend 4 inches on each side of the pipe and to a point 6 inches above the top of the pipe. The minimum allowable pipe depth below ground level is 12 inches. The installation method for plastic water service supply piping, when used in open trenches, is the same as for plastic pipe used in building sewers. See Figure 10–14 back in Chapter 10.

Water service supply piping passing under the foundation of a building must have a clearance of 2 inches from the top of the pipe to the bottom of the foundation or footing. (Refer back to Figure 10–27.) Water supply pipe passing through cast-in-place concrete such as a basement wall must be sleeved to provide ½-inch clearance around the entire circumference of the pipe. This avoids damage or breakage due to settling of the building or normal expansion and contraction of the pipe. This clearance also gives protection from the corrosive effects of concrete on metallic pipe.

In some geographic areas, such as near coastal regions, well water may be unsuitable even for irrigation purposes. People in these communities have to depend on the public water supply to meet all their needs.

There's always a danger when lawn sprinkler systems are connected to the potable water supply line. Steps are necessary to prevent a cross-connection. The pipe supplying water to the irrigation system must extend at least 6 inches above the surrounding ground when it's connected to the water service pipe. There must be an approved backflow preventer located on the discharge side of the control valve, located at least 6 inches above the surrounding ground. If the water pressure fails, the breaker will open, bleeding air into the supply line, rather than sucking agricultural water back into the main.

The most commonly-used atmospheric type anti-siphon vacuum breakers (shown in Figure 16–3) incorporate an atmospheric vent in combination with a check valve. Its operation depends on a supply of potable water to seal off the atmospheric vent, admitting the water to downstream equipment. If a negative pressure develops in the supply line, the loss of pressure permits the check valve to drop, sealing the orifice. At the same time, the vent opens, admitting air to the system to break the vac-

uum. See complete installation details in Figure 16–4.

Each building must have a separate water control valve installed in the water service supply pipe. This must be independent of the water meter valve or other interior control valves. The control valve must be an accessible location at or near the foundation line outside the building, either above ground or in a separate approved box with a cover. Look back to Figure 12–1 in Chapter 12.

Some codes require an accessible control valve near the curb and a second control valve with a drip valve near where the water supply pipe enters the building. The drip valve is required in cold climates to drain off water during cold weather. That prevents freezing and bursting of the pipes.

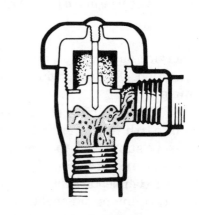

Figure 16–3

Atmospheric type anti-siphon vacuum breaker

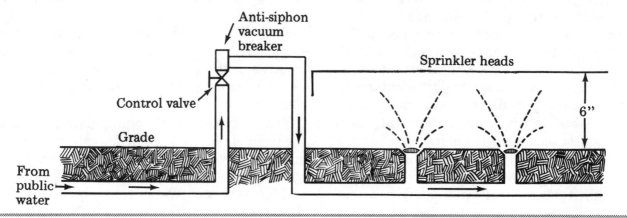

Figure 16–4

Installing an anti-siphon vacuum breaker

Installing Water Distribution Piping

Underground water distribution piping installed under a slab within a building must be firmly supported for its entire length on well-compacted fill. This prevents misalignment, sagging, and the formation of traps or depressions in the pipe that could collect sediment or mineral deposits. Some minerals deposits solidify over a period of time, reducing the flow of water and causing premature failure of the system. Use fine rather than course backfill in the trench so that the pipe isn't damaged when soil is compacted.

Water distribution piping under a building constructed with wood joists should be strapped securely to the joists with pipe straps. Galvanized pipe straps should be used with galvanized pipe. Use copper or copper-plated straps with copper pipe or tubing. Copper straps may also be used to support or secure CPVC plastic pipe (where plastic pipe is permitted), as the outside diameter is approximately the same for the two materials. Other types of supports for small water distribution pipes are illustrated in Figure 16–5.

Water piping inside walls and partitions must be secured by pipe straps (or some other approved method) to keep the pipe from shifting as the pipe expands and contracts. Expansion and contraction can create noises called water pipe creep. Any pipe installed above ground must be securely attached to the building structure.

When pipe is installed in wood partitions, you have to notch these studs to receive the pipe. Most codes don't allow notching *load-bearing* partitions more than 25 percent of the depth. It's important to note also that *nonbearing* partitions should not be notched any deeper than necessary. Cut just deep enough so the outside edge of the pipe is just flush with the edge of the stud. Never under any circumstances notch a stud deeper than 40 percent of its depth.

Copper pipe, tubing, CPVC or PB plastic pipe installed in notched wooden partitions should be protected from later damage by lathers or drywall installers. It's easy to drive a lath nail right through copper pipe. Protect the

pipe by covering the notched opening with a metal stud guard, as in Figure 16–6.

Sometimes you'll have to install piping in a load-bearing partition. To meet code requirements, you may have to drill pipe holes through the centers of the studs. In that case, the pipe must be cut short enough to thread through these holes. This requires extra fittings and joints. A hole can't be larger than 40 percent of the width of the stud.

When you install pipe, remember that the building will settle and the piping will expand and contract with use. Make sure this movement will not damage the pipe. Water supply piping that passes through cast-in-place concrete floor slabs or bearing walls must be sleeved to provide ½-inch clearance around the entire pipe. This sleeve installation gives crawling insects access to the interior of a building unless the space is caulked with an approved flexible sealing compound.

Metal water pipe installed under a concrete slab has to be protected from corrosion. Add corrosion protection even for pipe installed in a building if hydrogen sulfide gas or any other corrosive gas is known to be present. You can protect metal pipe and fittings by coating them with two coats of asphaltum paint. A better choice may be to install water pipe in the attic if you are working in filled ground. Some codes require this procedure.

Don't install metal water pipe where it will touch metal conduit, building steel or reinforcing steel. Keep all metal pipe away from any good conductor of electricity. Where a physical contact is absolutely unavoidable, the pipe must be wrapped with heavy tar paper or other approved material to provide the necessary insulation.

Metal hot and cold water pipes must not contact each other when installed underground or within partitions. Contact (or even close proximity) tends to transfer the heat and cold from one line to the other.

Nearly every plumbing fixture needs an accessible shutoff valve. Refrigerators with automatic ice makers must also be valved. Washing machines don't need separate shutoff valves.

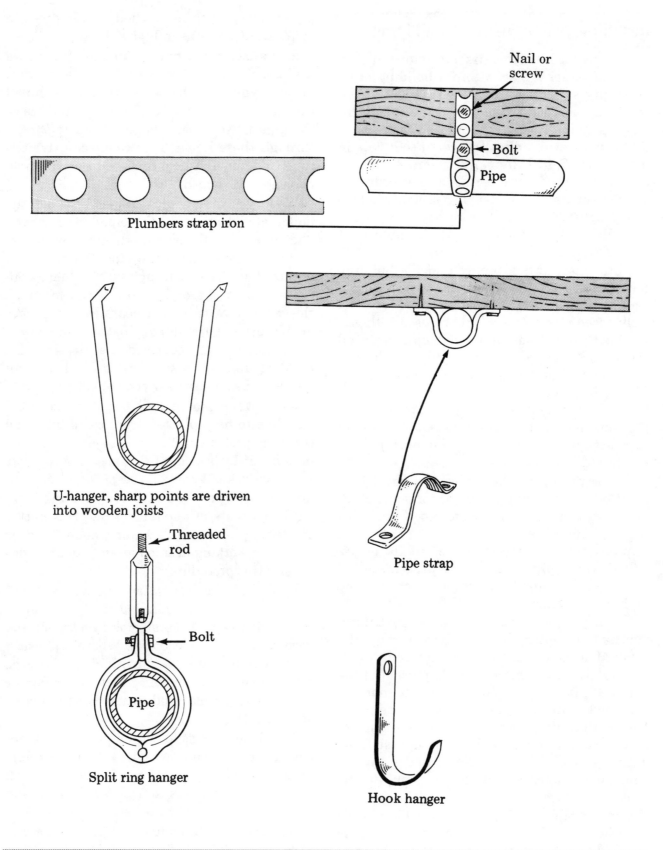

Nail or screw

Bolt

Pipe

Plumbers strap iron

U-hanger, sharp points are driven into wooden joists

Pipe strap

Threaded rod

Bolt

Pipe

Split ring hanger

Hook hanger

Figure 16–5
Pipe hangers

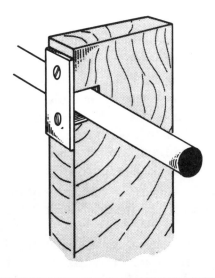

Figure 16–6
Stud guard

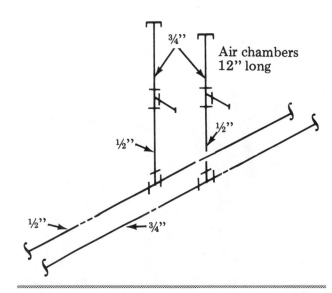

Figure 16–7
Proper location of air chambers

Avoid installing water piping in crawl spaces or other unheated areas in cold climates. If you have to install pipe in crawl spaces, select PB plastic pipe or protect the pipe with adequate insulation.

Most codes now approve the installation of CPVC or PB plastics underground within the walls of a building.

Water pipe installations must be protected from water hammer with properly-located air chambers or other approved devices. Install air chambers on the highest hot and cold water pipes serving a plumbing fixture in each bathroom, preferably the lavatory pipes. See Figure 16–7. Also install air chambers on water pipes serving individual fixtures such as kitchen sinks and washing machines. You can make an air chamber from a piece of capped pipe at least 12 inches long and one pipe size larger than the pipe it serves.

It's good practice to install a valve at the low point on the system so the entire water supply system can drain dry. That makes repairs easier, makes replacing lost air in air chambers a snap, and can prevent freezing of the water pipes in cold weather.

Avoid Cross-Connection

Any arrangement of piping likely to promote connection of the potable water supply system to a contaminated liquid source is a *cross-connection* and is prohibited by code. There are many situations that might create cross-connections. Here are some of the most common:

- A garden hose attached to a hose bibb with the end of the hose lying in contaminated water.

- A garden hose attached to a laundry sink with the end of the hose submerged in a tub of dirty water. See Figure 16–8.

- A private water supply, such as a well, interconnected with the public water supply.

- Failure to keep the correct air gap between fixtures and faucets.

- A leaking water supply pipe in the ground near a leaking sewer line or septic tank.

Cross-connections are much more likely if there's a complete drop in water pressure. That happens occasionally, such as during a fire when pumper trucks place heavy demand

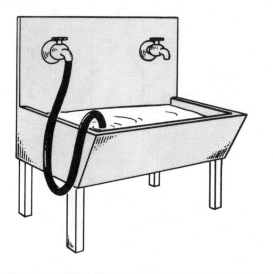

Figure 16–8
A situation that creates a cross-connection

on the system or when the public main is being repaired. When water pressure fails, the main may suck water out of distribution piping in the direction opposite to that normally expected.

The ordinary garden hose is the most common cross-connection offender. It can be connected easily to the potable water supply and used in a variety of potentially dangerous ways. Hoses can be extremely hazardous because they are left submerged in swimming pools, lie in elevated positions (above the sill cock) to water shrubs, are attached to chemical sprayers for weed-killing. They're often left lying on the ground, which may be contaminated with fertilizer, cesspools and garden chemicals. Most codes require that threaded water outlets equipped for hose connections other than for clothes washers have approved backflow preventers.

The purpose of a sill cock is to permit easy attachment of a hose for outside watering purposes. Install a hose bibb vacuum breaker on every sill cock to isolate garden hose applications and protect the potable water supply from contamination (Figure 16–9).

The code requires that the potable water supply outlet terminate at least 1 inch above the overflow rim of each fixture. This provides an air gap which prevents siphoning of the fixture contents back into the water outlet or faucet. See Figure 16–10.

Codes prohibit the use of below-rim water-supplied fixtures, as illustrated in Figure 16–11. A stoppage in the waste line could cause a cross-connection. The liquid in the fixture bowl could rise to the flood-level rim and be siphoned into the water-supply system through the submerged water outlets if a negative pressure develops in the supply piping.

Basic Plumbing Procedures

It's important to know the correct way to measure pipe before you attempt to cut and join it. There are two basic types of pipe measurement — the straight run and the offset — and several ways of measuring individual pieces. But first, fitting pipe exactly is easier if you know the fitting dimensions and the length of thread engagement. See Figure 16–12.

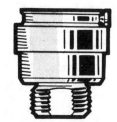

Hose bibb vacuum breaker
for frost-proof hydrants

Hose bibb vacuum breaker

Figure 16–9
Use vacuum breakers to avoid contamination of potable water

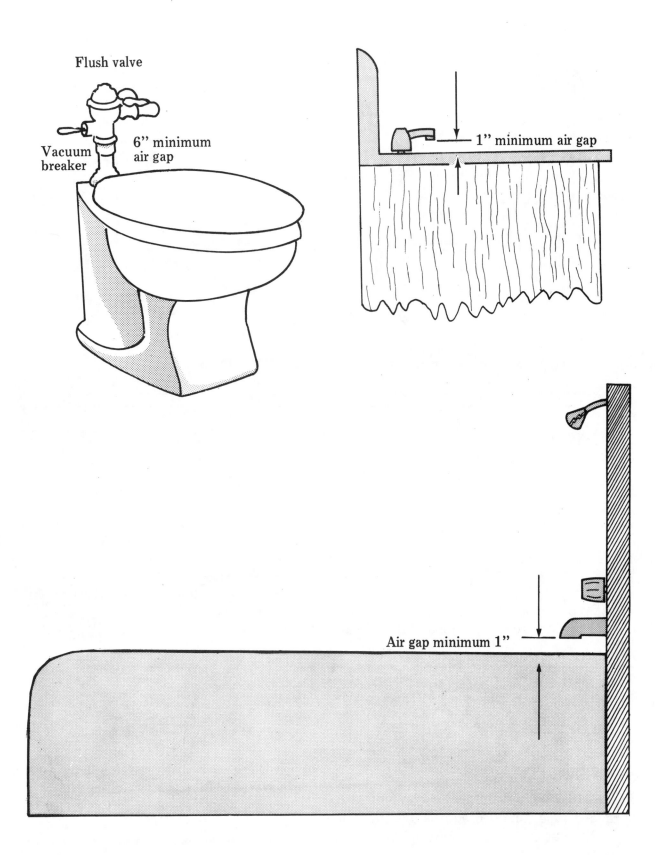

Figure 16–10
Three plumbing fixtures that provide the minimum air gap permitted by code

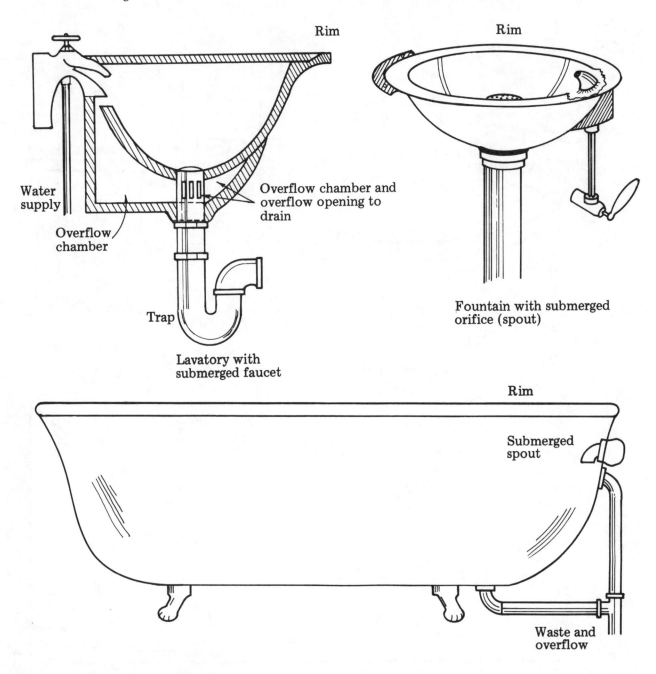

Water supply

Overflow chamber

Overflow chamber and overflow opening to drain

Rim

Trap

Lavatory with submerged faucet

Rim

Fountain with submerged orifice (spout)

Rim

Submerged spout

Waste and overflow

Figure 16–11
Fixtures with supply outlets below the overflow rim are prohibited

▓ Measuring Pipe

Figure 16–13 shows several ways to measure threaded pipe. The procedures are the same regardless of the type of material or joint to be made.

End-to-end measure: Measure the full length of the pipe, including both threads.

End-to-center measure: This is used for pipe which has a fitting screwed on one end only. The pipe length is equal to the measurement *minus* the end-to-center dimension of the fitting *plus* the length of thread engagement.

Face-to-end measure: This is also used for pipe which has a fitting screwed on one end only. The pipe length is equal to the mea-

Pipe Size (inches)	Number of Threads per Inch	Approximate Total Length of Threads (inches)	Approximate Thread Engagement (inches)
$1/4$	18	$5/8$	$3/8$
$3/8$	18	$5/8$	$3/8$
$1/2$	14	$13/16$	$1/2$
$3/4$	14	$13/16$	$1/2$
1	$11^1/2$	1	$9/16$
$1^1/4$	$11^1/2$	1	$5/8$
$1^1/2$	$11^1/2$	1	$5/8$
2	$11^1/2$	$1^1/16$	$11/16$

Figure 16–12
Data for standard pipe threads

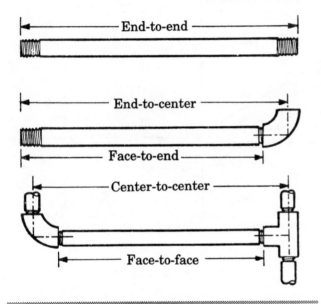

Figure 16–13
Threaded pipe measurements

surement *plus* the length of the thread engagement.

Center-to-center measure: This is used for a length of pipe which has fittings screwed on both ends. The pipe length is equal to the measurement *minus* the sum of the end-to-center dimensions of the fittings *plus* twice the length of the thread engagement.

Face-to-face measure: This is used in the same situation as center-to-center measurement. The pipe length is equal to the measurement *plus* twice the length of the thread engagement.

Measuring Offsets

Offsets are important unless you're making straight runs or 90-degree turns. Use Figure 16–14 to calculate the exact distance between centers of fittings of *offsets*. For example, look at Figure 16–15, which shows a common pipe offset between parallel runs of pipe.

The center-to-center offset (C) is 10 inches. Assume that pipes A and B are to be connected with 45-degree elbows. Use Figure 16–14 to determine the length of pipe required for measurement E in Figure 16–15. The letters in the headings in Figure 16–14 refer to the dimensions in Figure 16–15.

In this case, C is 10 inches. In Figure 16–14 find the row labeled 45 degrees. Look across on that row to the column headed "When C = 1, E = ." Notice that the number in that column is 1.4142. For each inch of dimension C, dimension E will be 1.4142 inches. Since C is 10 inches, E must be 10 times 1.4142 or 14.14 inches.

Use the same procedure to find the length of pipe required for other degrees of offset. For example, C is 10 inches. Assume that pipes A and B are to be connected with a 22 ½-degree offset. Find the number 2.6131 opposite the 22 ½-degree offset in Figure 16–14. Multiply this number by distance C (in this case, 10 inches). The length of pipe required would be 26.13 inches.

The Pipe Vise

Vises are commonly used to hold pipe firmly during cutting, reaming and threading. They

Degree of Offset	When C = 1, D =	When D = 1, C =	When C = 1, E =
60°	0.5773	1.7320	1.1547
45°	1.0000	1.0000	1.4142
30°	1.7320	0.5773	2.0000
22¹/₂°	2.4140	0.4142	2.6131
11¹/₄°	5.0270	0.1989	5.1258
5⁵/₈	10.1680	0.0983	10.2170

Figure 16–14
Finding the length of pipe to connect two parallel runs of piping

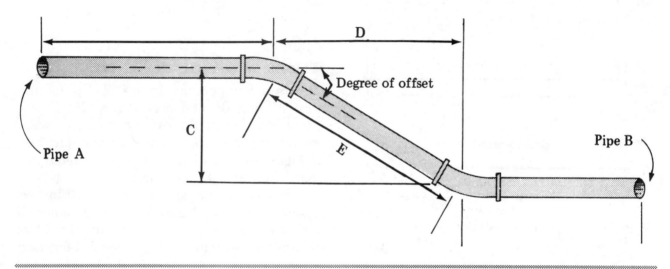

Figure 16–15
Common pipe offset between parallel runs of pipe

can also be used to tighten fittings to one or both ends of a threaded pipe. The two most common types of vises are described here.

The *hinged pipe vise,* known as a *yoke vise,* has a fixed toothed lower jaw and a movable toothed upper jaw. It opens on one side to receive the pipe. Turning the handle clockwise tightens the jaws on the pipe and holds it securely in place. See Figure 16–16.

The *chain vise* has a fixed toothed lower jaw and a chain that's looped over the pipe. The chain can then be hooked into slots on the opposite side of the vise body. Turning the handle clockwise causes the chain to tighten on the pipe and hold it securely in place.

Pipe vises are seldom used or needed to secure fragile or soft metal pipes. But if you must use one, protect the area of the pipe to be secured in the vise jaws. This prevents scratching or marring polished or chrome plated surfaces. Wrap the finished pipe with adhesive tape and tighten the vise with just enough pressure to hold (but not damage) the pipe.

Pipe vises may be mounted on a work bench, but most plumbers use a vise attached to a portable tristand. This three-legged stand can be folded for easy carrying or storing. When open for use, the tristand has a lower shelf for tools and pipe compound.

The procedure for using the pipe vise is the same for both the yoke and the chain vise:

1. Release the pressure on the jaws of the vise (Figure 16–16) by turning the handle counterclockwise.

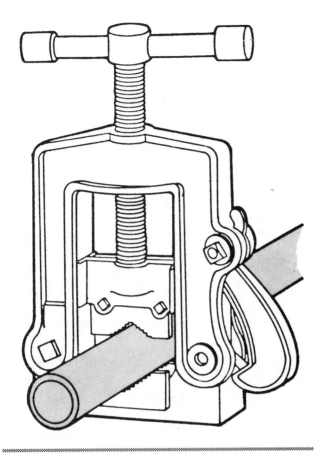

Figure 16–16
Pipe clamped in vise

Cutting, Threading and Joining Steel Pipe

Much pipe is wasted by careless cutting. Improper reaming causes a reduced flow of water through the pipes. Follow the procedures outlined below to avoid cutting waste and reduced water flow.

1. After measuring the pipe to determine the length needed, mark the spot with crayon or lead pencil.

2. Lock the pipe securely in the vise, as described above, with the cut mark up. Provide approximately 8 inches of clearance between the mark and the face of the vise. This distance should provide ample room for working the pipe cutter and the space necessary for operating the die stock.

3. Open the pipe cutter by turning the handle counterclockwise until it clears the diameter of the pipe. Figure 16–17 shows the single-wheel cutter and the three-wheel cutter. The single-wheel cutter is used more often and is the easier of the two models to operate. These are heavy duty pipe cutters and can be used to cut steel pipe.

4. Place the cutter around the pipe from underneath (Figure 16–18) with the cutting wheel exactly on the mark. Note that the illustration shows a single-wheel cutter and a chain vise. The rollers of the single-wheel cutter ensure a square cut. Begin each cut by lightly rotating the cutter completely around the pipe. This gives a "bite" or groove for the cutter wheel to follow.

After each turn of the cutter wheel, tighten the handle slightly. Tightening the handle too rapidly can break the cutter wheel or spring the cutter frame, ruining the cutter. Be sure to use thread-cutting oil occasionally on the cutter wheel and rollers.

If you use a three-wheel cutter, place the two cutter wheels attached to the movable jaw on the mark and make sure that all three wheels lie in line at right angles to the center line of the pipe. Follow the

2. Release the locking lever or chain by pulling outward on it. Then open the vise by tilting the upper body to the side on its pivot.

3. Lay the pipe on the vise on the V-shaped lower jaw. Long lengths of pipe should be supported in some way. This keeps small diameter pipe from bending under its own weight and keeps the tristand from tilting or falling over.

4. After placing the pipe in the vise, tilt the upper body back to the closed position. The lever should lock in place. On the chain vise, the upper body is looped over the pipe and is locked in the slots provided.

5. Turn the handle clockwise until the V-shaped upper jaw or chain presses firmly on the pipe. Tighten the handle only enough to keep it from turning as you thread the pipe.

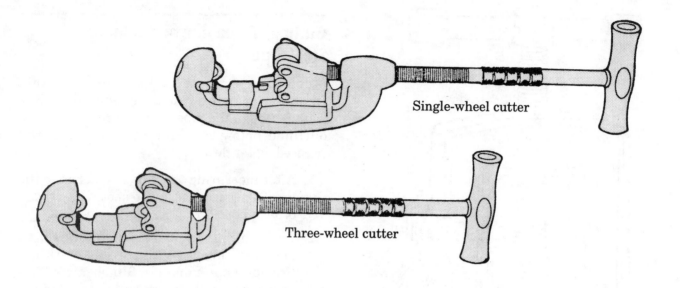

Single-wheel cutter

Three-wheel cutter

Figure 16–17
Pipe cutters

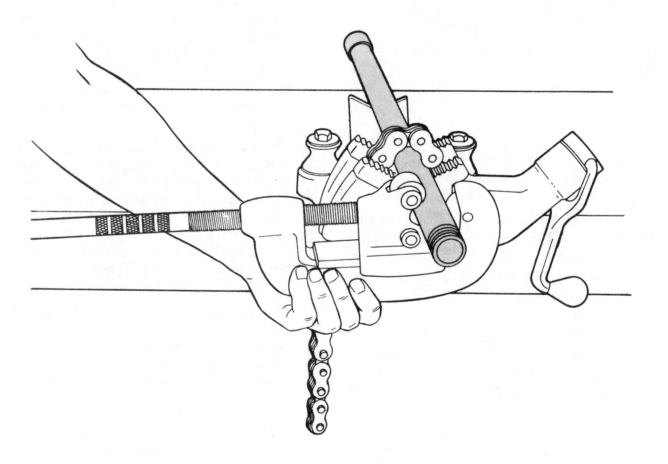

Figure 16–18
Cutting pipe

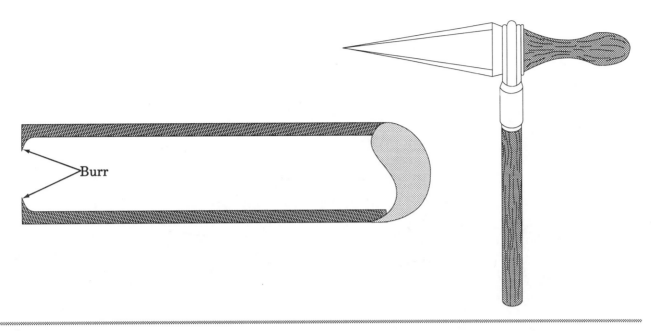

Figure 16–19
Use a reamer to remove the burr left by a pipe cutter

procedure described above for a single-wheel cutter.

5. Rotate the cutter completely around the pipe, making a quarter turn on the handle for each complete revolution until the pipe is cut.

6. The pipe cutter wheel leaves a burr on the inside of the pipe. (See Figure 16–19.) Mineral deposits can collect at the burr and cause premature failure of the line. Use a pipe reamer as shown in Figure 16–20 to remove the burr before threading the pipe.

■ Threading Steel Pipe

The two most common hand operated pipe threading tools are the *three-way threader* and the *ratchet threader.* The ratchet threader is shown in Figures 16–21 and 16–22.

The three-way threader has three pipe-size dies as part of the stock, which has two handles for applying pressures as the pipe is threaded. This type can be used only where there is enough room to make a complete revolution. Typical die sizes are ⅜ inch, ½ inch, and ¾ inch, or ½ inch, ¾ inch, and 1 inch.

The ratchet threader is much easier to use and has dies that are interchangeable within a single stock. The dies can be ⅜ inch up to and including 1 inch. Ratchet threaders for larger pipe sizes are also available. The ratchet threader requires less pressure and can also be used to thread pipe in close places.

Pipe threaders cut standard pipe threads to a taper of ¾ inch per foot. That's the standard taper for pipe and fitting threads.

Here's how to thread the pipe:

1. Fasten the pipe securely in the vise.

2. Slide the stock over the end of the pipe with the guide on the inside (next to the vise). Push the die against the pipe with the heel of one hand. (See Figure 16–21.) Take three or four short, slow clockwise strokes, pressing the die firmly against the pipe. When enough thread has been cut so the die is firmly started on the pipe, apply plenty of thread-cutting oil on the threader dies. This prevents overheating of the dies. Occasionally check to be sure you're cutting clean threads.

NOTE: **Chipped or torn threads indicate that the pipe at this location is too soft, too hard, has**

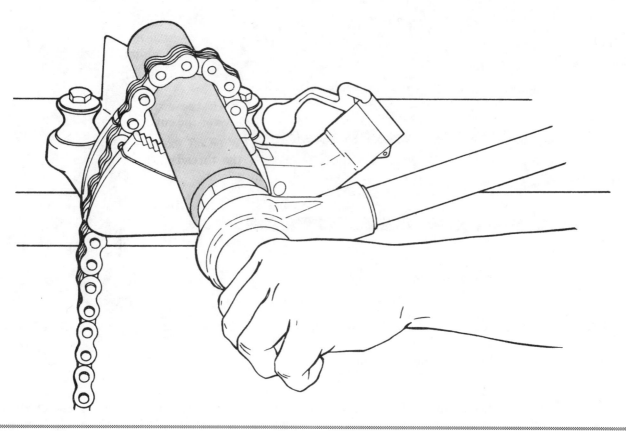

Figure 16–20
Reaming pipe

impurities in it, or that the dies are worn and need replacing. Always cut new threads when this happens. Don't use pipe that has bad threads.

3. Continue to turn the stock with downward strokes on the handle, applying oil often. Back off about a quarter turn forward to clear away pipe chips. Continue until the threaded pipe extends flush to ¼ inch from the die end of the stock. Don't cut threads too long (running threads), as they won't have enough taper at the thread end. The result will be a bad joint. Place a shallow pan covered with a wire screen under the threader. The pan catches dripping oil for use again. The screen catches the pipe chips for safe disposal.

4. To remove the threader after threads are cut, turn the stock counterclockwise until the die is free of the cut threads.

5. Use a rag to wipe away the chips and excess oil that stick to the threads. The pipe is now ready to receive the fitting.

Joining Steel Pipe and Fittings

Fittings are generally screwed to one end of the pipe while it's still in the vise. The assembled pipe and fittings are then screwed into place in the installation. Follow these steps to make leakproof joints:

1. Inspect the female thread on the fitting and, if necessary, clean it with a wire brush. The thread should be free of rust or other foreign substances.

2. Repeat step 1 above for the male thread on the pipe if the thread isn't new.

3. Apply pipe joint compound (*pipe dope*) evenly over the male pipe threads. Don't put pipe joint compound on the female threads of the fitting.

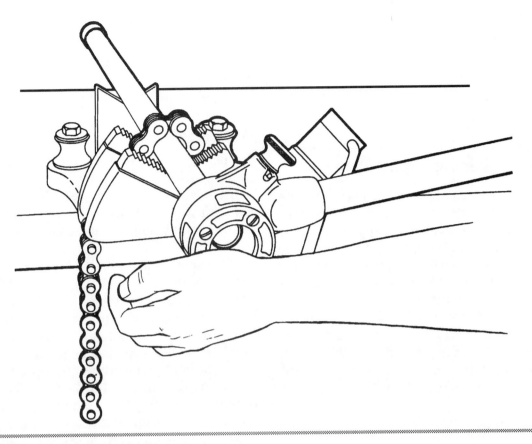

Figure 16–21
Threading pipe with a ratchet threader

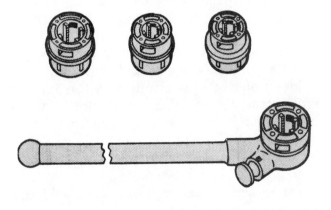

Figure 16–22
Ratchet stock and die set

Pipe Sizes (inch)	Pipe Wrench Size (inch)
$1/8$, $1/4$, $3/8$	6 or 8
$1/2$ and $3/4$	10
1	14
$1\,1/4$ and $1\,1/2$	18

Figure 16–23
Recommended wrench sizes for common pipe sizes

4. Hand screw the fitting onto the pipe. The fitting should turn easily for the length of about three threads.

5. Select the right size pipe wrench when joining fittings to threaded pipe. A wrench too small for the job requires unnecessary effort for your hands, arms, and back. A wrench too large forces the fitting too far onto the threaded pipe. This can result in a bad joint or a cracked fitting. Figure 16–23 shows recommended wrench sizes for common pipe sizes.

When installation is complete, open a hose bibb at the far end of the line. Then open the house valve to flush sand and other foreign matter from the lines. Close the hose bibb, leaving water pressure in the system. Then check for leaks.

Cutting and Joining Copper Pipe

Copper pipe and tubing are measured in the same way as threaded pipe (Figure 16–13) with one exception. Soft-tempered copper tubing is easily bent. Many of the fittings that would be required for offsets with rigid pipe aren't necessary with copper pipe. When using soft tubing, make allowance for fitting dimensions (solder or compression type) and for the distance the pipe or tubing is inserted into the fitting. Offsets for rigid copper pipe, on the other hand, are calculated in the same way as for threaded pipe. See Figures 16–14 and 16–15.

■ Cutting and Reaming

Both rigid and soft copper pipe and tubing may be cut with a tubing cutter (Figure 16–24) or a hacksaw with a fine-tooth blade (24 teeth per inch). Follow the same procedure for cutting iron pipe (Figure 16–18) when using the tubing cutter.

No vise is required to cut most copper tubing. Hold the tubing firmly with one hand. Use your other hand to turn the cutter. It's easier to make a clean, square cut with a tubing cutter. Copper tubing cutters look about like pipe cutters, though they are much smaller. With copper pipe, apply oil to the cutter wheel and rollers sparingly, as excessive oil must be cleaned off completely before soldering. Be

Figure 16–24
Tubing cutter

careful to remove the burr the cutter wheel leaves inside the tubing. Use a reamer attached to the tubing frame for this purpose. When cutting tubing with a hacksaw, remove the rough edge with a round or half-round file.

■ Soldered Joints

Copper tube can be soldered together with a fusible alloy (solder) which has a melting point lower than the copper. The most common joint for copper tubing is the sweat joint. A good sweat joint should last as long as the tubing itself.

The solder recommended for use in copper or brass potable water systems is about 95.5 percent tin, 4 percent copper and 0.5 percent silver. It can be used with any general purpose soldering flux on all materials except aluminum. It melts at 440 degrees F, and gives excellent joint penetration and flow. If the joint is properly cleaned and heated, surface tension spreads the solder to all parts of the joined surfaces. The result is a sound joint.

Soldering is an art. It looks easy. But it takes precision and skill to install a soldered water distribution system in a home. Begin by selecting a torch tip appropriate for the size of copper tubing you're joining. Too large a tip overheats and burns the tubing and fitting. That keeps solder from bonding to the two surfaces. Too small a tip heats the pipe and fitting unevenly. Solder won't be drawn fully into the joint. In either case, the result is a bad joint. If you're lucky, a bad joint will start leaking immediately when tested. If you're not, bad joints will begin leaking a few weeks later when the walls have been installed and finished.

Follow the procedure below when soldering copper pipe and fittings:

1. Inspect the end of the copper tube to make sure it's round and free of burrs. If the end of the tubing is out of round, cut it off.

2. Clean the inside of the copper fitting with a wire brush designed for this purpose, or with emery cloth. The pipe end should be polished bright with emery cloth. All tarnish must be completely removed from the end of the tubing and the inside of the

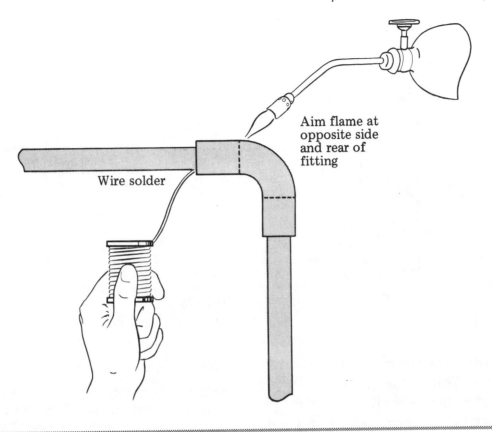

Aim flame at opposite side and rear of fitting

Wire solder

Figure 16–25
Soldering copper fitting

fitting. Doing that promotes even penetration of solder into the joint. When the surfaces are clean and dry, use them as soon as possible. Avoid handling the cleaned ends with dirty or sweaty fingers. The bright surface will oxidize very quickly. Then you've got to clean again.

3. Apply a thin coat of non-corrosive soldering flux to the inside of the cleaned fitting and the outside of the cleaned pipe.

4. Push the tubing into the fitting and turn it a few times to spread the flux evenly. Fitting and tubing are now ready for soldering.

5. Heat the back side of the fitting evenly with a propane-type torch. (Professionals use a special soldering torch fired from a 20-pound tank of gas.) Don't apply the flame directly to the point where the pipe enters the fitting. See Figure 16–25.

6. When the flux begins to bubble, the connection should be hot enough. Remove the torch and touch the wire solder to the point where the pipe enters the fitting. If the fitting and the tubing are hot enough, the solder will melt and be drawn into the joint. When a line of solder shows completely around the joint, the connection is filled with solder.

7. Don't move the tubing or fitting while the solder is cooling. Any movement may result in a faulty joint. Solder will harden in a minute or less if the joint has not been overheated. To speed the cooling process, drape a damp rag over the fitting.

8. When you have to solder near wood or other combustible material, place a metal sheet between the point to be soldered and the combustible material. Always keep a bucket of water handy when soldering copper pipe in wood partitions.

If a leak develops after a system has been pressurized, it's much more difficult to resolder because there's usually water in the pipe. Heat

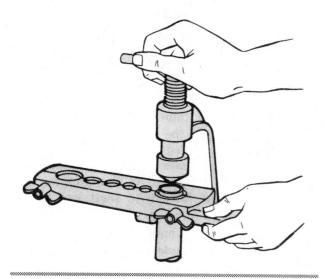

Figure 16–26
Using the flaring tool

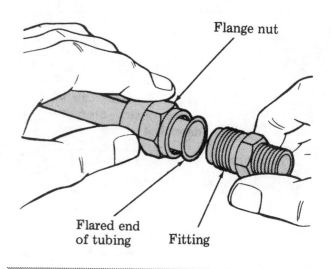

Figure 16–27
Making up a flared fitting

from the torch turns the moisture in the tubing to steam. This makes pinholes in the newly-applied solder before it can harden. Try applying heat to the tubing several inches to each side of the fitting to be soldered. This should dry the pipe long enough to make a good joint, but you must work quickly.

If you can't keep water away from the joint you're soldering, here's a trick I learned. Stuff plain white bread (remove the crust) as far into the pipe as possible in the direction from which the water is flowing. The bread absorbs the water, giving you time to make a good joint. When water pressure is returned to the system, the bread will flush out easily, causing no trouble for the water system.

Use male-threaded or female-threaded brass adapters if you to have to connect copper pipe or tubing to threaded pipe or fittings. Unions in a copper water supply system should be brass. In galvanized water supply systems, unions must have metal-to-metal joints with brass ground seats.

When installation is complete, open a hose bibb at the end of the system. Then open the house valve. Flush sand and other foreign matter from the lines. After flushing, close the hose bibb and check the system for leaks.

Flare Joints

Occasionally a flare joint is required. This type of joint is used on soft flexible copper tubing when connecting certain types of fixtures and appliances. Hard (rigid) copper pipe can't be used to make flare joints, as it's likely to split.

For example, a connector for a gas range or water heater might be joined with a flare fitting. First, cut the tubing to the desired length. Then slip the flare nut on the cut tubing. Use a flaring tool (Figure 16–26) to flare the tubing ends so the fit is perfect. Slide the nuts to each end of the flare tubing and gently bend or shape the tubing by hand on the male thread of each fitting. (See Figure 16–27.) Tighten with a smooth jaw adjustable wrench of proper size until the fitting is snug. Then test the joint for leaks.

Flare fittings are not recommended for general copper installation and should be used sparingly. Never conceal a flare fitting in walls or in any other inaccessible place. Flare fittings are made in ells, tees, adapters, unions, and other shapes. See Figure 15–6 in the previous chapter.

Compression Joints

Compression-type fittings are seldom used in a water supply system. They are most often used to connect fixture supply tubes to an ell or a cutoff valve located under a fixture. Although they're made in ells, tees, couplings, and other shapes, their use is limited. One common application for compression fittings is in connecting the water supply line to an icemaker unit in a refrigerator.

Compression-type fitting are not flanged. They have a threaded nut which forces a compression ring against a ground joint on the fitting. See Figure 16–28. The compression ring bites into the soft copper tubing, making a watertight connection.

First, make any necessary offsets in the fixture supply line tube. Allow the tubing to fit fully into the fitting socket, marking the length with a pencil. Cut the supply line tube square using a small tubing cutter, not a hacksaw. A hacksaw would make a rough cut, leaving the tubing out of round and subject to leaking.

Remove the inside burr with a reamer attached to a tubing cutter frame. Slip the nut on the tubing. Then slip the compression ring onto the tube. To help prevent leaks, apply a small amount of pipe compound over the compression ring.

Insert the tube end into the fitting. Slide the compression ring into the joint, making sure it's squarely aligned. Don't put any undue stress or strain on the tubing.

Slide the compression nut over the fitting thread and screw it down by hand. With a basin wrench or other wrench, tighten the nut at the other end of the tube. Then tighten the compression nut firmly with a smooth-jawed adjustable end wrench. Avoid overtightening. Then turn the water on and check for leaks. If there's any seepage, tighten the nuts just a quarter turn. Then check again for seepage. Continue tightening only one-quarter turn each time after each test until the leaking stops.

Cutting and Joining Rigid Plastic Pipe

Straight plastic pipe is measured the same way as straight threaded pipe. See Figure 16–13. Make an allowance for fitting dimensions and for the length that's inserted into the fitting. Pipe should be cut to fit so that bending isn't necessary. Offsets for plastic pipe are calculated the same way as offsets for threaded pipe. See Figures 16–14 and 16–15.

PVC and CPVC plastic pipe should be cut with a tubing cutter designed for cutting plastic pipe. Don't use a tubing cutter intended for copper tubing. The cutting wheels for plastic pipe are different. A hacksaw with a fine-tooth blade (24 teeth per inch) may also be used. Always make square cuts on the ends. A freehand cut with a hacksaw is an invitation to problems. Use a miter box when you cut plastic pipe with a hacksaw or handsaw. See Figure 10–17 back in Chapter 10.

Be sure to remove the burr left by the cutter wheel inside the pipe. Use a reamer for best results. A hacksaw doesn't leave a burr, but the inside and outside of the pipe will have a rough ridge. Remove this roughness with a knife or small file. See Figure 16–29.

▩ Cemented Joints

Special plastic cement is always used to join plastic pipe and fittings. The joint created is sometimes referred to as a "welded" joint. Use fine sandpaper or a liquid cleaner made for plastic pipe to remove impurities and gloss from the surfaces to be joined. Don't use cement that will not pour from the can or that has discolored. Use only the cement recommended by the manufacturer of the pipe fittings you're installing.

Follow this procedure when preparing and cementing plastic pipe and fittings:

1. Inspect the pipe and fitting for gouges, deep abrasions, or cracks. Pieces with defects should not be used.

2. After the pipe is cut to the proper length and the burrs are removed, check the dry fit before cementing. The pipe must enter the fitting socket smoothly but must not be so loose that the two surfaces don't make good contact. Test to see if the fitting will fall from the pipe. If it does, don't use it.

3. If the fit's correct, adjust the fitting to the position at which it will be cemented. Mark both pieces with a pencil (see Figure 16–30) so they can be quickly repositioned after cement has been applied. Cement sets very quickly once it's applied to plastic surfaces. For this reason, cement only one joint at a time.

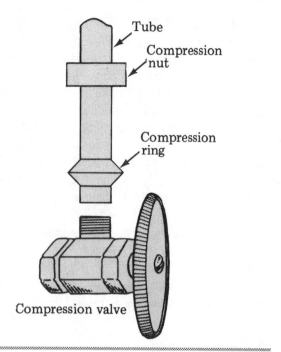

Figure 16–28
Compression-type fitting

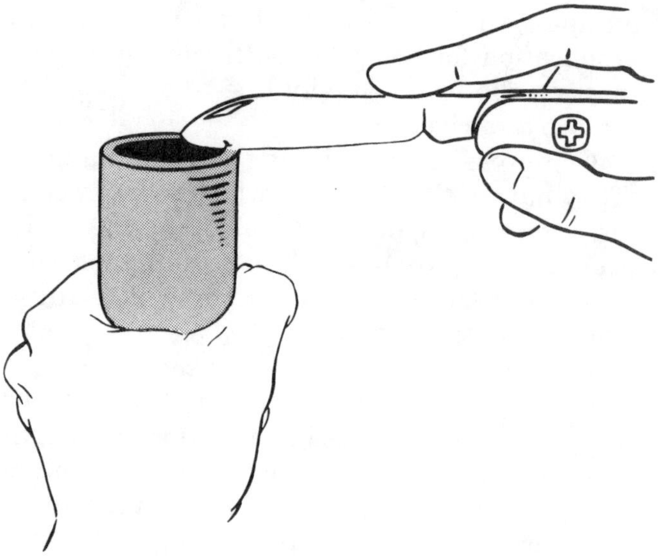

Figure 16–29
Removing burr with knife

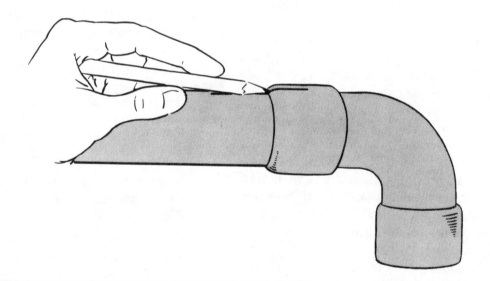

Figure 16–30
Mark both pieces with a pencil

4. Clean the pipe end and the inside of the fitting with a clean rag to remove any grease or moisture.

5. Use liquid cleaner or fine sandpaper to remove surface gloss from the outside of the pipe end and the inside of the fitting socket.

6. Brush a light coat of cement on the outside of the pipe and the inside of the fitting socket. Use the brush supplied with the can. Quickly brush a second coat of cement on the outside of the pipe.

7. Push the pipe fully into the fitting socket and give the pipe a quarter turn. Then set

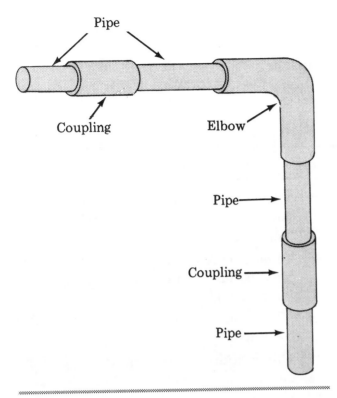

Figure 16–31
Repairing a leak in a joint at the elbow

the pipe and fitting to the pencil marks and check the position. Make any adjustments at once. This is your last chance before the cement sets.

8. Wipe excess cement from the fitting with a rag and close the cement can immediately to keep it from drying up.

9. Hold the pipe and fitting together for approximately 15 seconds until the cement sets.

10. An even bead of cement should appear around the joint if it's properly made. If the bead is incomplete or if there is no bead at all, not enough cement was used. Pull the pipe from the fitting immediately, apply more cement to the outside of the pipe only (not the fitting), and reinsert into the fitting.

Wait at least a half hour after cementing the last joint before testing the system. If time isn't critical, it's best to let the joints harden overnight.

If a leak occurs in a plastic welded system, cut out the bad joint and replace it with a new fitting. Remove enough pipe to allow room for

two couplings and the fitting. Figure 16–31 shows how much pipe should be cut out if a joint at the elbow leaks. In this case three fittings, two short lengths of pipe, and six joints are required. Take your time preparing the pipe and fitting when you cement plastic piping. Plastic fittings can't be reused if they don't work the first time.

When it's necessary to connect plastic pipe to iron pipe valves and fittings, use only approved male-threaded plastic adapters. Use only the thread compound recommended by the manufacturer for these threaded joints.

▓ Installing Polybutylene

Polybutylene, the most up-to-date plumbing pipe, is easy to install. It can be used for every kind of job—new construction, remodeling and repair.

Hot and cold PB piping's versatility makes it desirable for residential and small commercial applications. Because PB piping is flexible at all temperatures, it can be installed in long lengths with a minimum number of fittings.

Polybutylene is lightweight and can be easily handled by one person. You can join PB in four ways:

1. compression fittings

2. flare fittings

3. crimp-type fittings

4. heat fusion

None of these fitting methods requires sweat soldering, open flames (as required for copper joints) or slow solvent curing (as required for PVC or CPVC joints). Adapters can be used for connecting PB piping into copper, galvanized and CPVC systems, making it ideal for replacement or remodeling work.

When the water system is complete, open a hose bibb to flush out sand, pieces of pipe shaving, and other impurities that may have collected inside the pipe during installation. This will save time in cleaning out faucets when fixtures are set, and prevents the damage sand and grit can do to washers.

Regardless of the type of material used, the entire water distribution system must be tested, inspected, and proved to be tight when completed. It must be tested at a water pressure not less than the maximum working pressure under which it will be used.

Remember to install air chambers and secure all water supply pipes. A system with water shock or hammer will not pass final inspection.

Questions

1. What two procedures are required when installing a water service pipe in the same trench with a building sewer?

2. What is the minimum separation distance between a water service pipe and a sewer line installed in separate trenches?

3. What is the minimum separation between a water service pipe and a septic tank or drainfield?

4. Why is it important to support water service supply piping for its entire length?

5. Why should young trees not be planted near a water service pipe?

6. What characteristics make galvanized steel pipe superior for use in outside trenches?

7. What should be avoided when backfilling a trench containing water service piping?

8. How deep should a water service piping trench be?

9. Why must PVC plastic pipe be treated carefully during installation?

10. What precautions should be taken when backfilling over PVC plastic pipe in water service supply piping trenches?

11. What clearance is required between the top of a water service supply pipe and the bottom of a building footing?

12. What clearance is required around the circumference of a water supply pipe which passes through cast-in-place concrete?

13. When a lawn sprinkler system is connected to a potable water service supply pipe, what device is used to prevent a cross-connection?

14. This device must be located a minimum of how many inches above the highest sprinkler head?

15. To meet code requirements, what device must be provided on each building water service supply pipe?

16. Where must a water supply control valve be located?

17. Why is a drip valve required in cold climates?

18. What type of pipe straps should be used to secure copper water lines to wooden partitions?

19. What problem is likely to occur if water line pipes are not properly strapped?

20. What is the maximum depth a wooden partition may be notched to conceal water piping?

21. What is the standard acceptable method of providing openings to receive water piping in a load-bearing partition?

22. What should be provided to protect soft piping installed in notched wooden partitions from damage by lath nails?

23. What must be done to protect piping installed in filled ground where hydrogen sulfide gas is known to be present?

24. What must be done to protect metal water piping from other metals or ferrous pipes within a building?

25. What happens when hot and cold water piping within a building make contact with each other?

26. In what common areas of a building should water piping not be installed when subject to freezing?

27. What protects water pipe installation from water hammer?

28. Why must water supply systems be installed to drain dry?

29. What is a cross-connection?

30. Name two of the most common causes of cross-connection.

31. What triggers back-siphonage (cross-connection) in a water supply system?

32. Except for clothes washing machines, what must each threaded water outlet have to prevent cross-connection?

33. What is the minimum *air gap* required from water supply outlets to the overflow rim of each fixture?

34. Why are below-rim water supplied fixtures prohibited?

35. Name the two basic types of pipe measuring for plumbing installation.

36. What is meant by end-to-center measure?

37. What mathematical figure must be known in order to find the length of pipe between two 45-degree ells?

38. Name the two common types of pipe vises used to hold pipe.

39. In what direction do you turn a vise handle to tighten the jaws on the pipe?

40. When a vise is used to secure a chrome pipe, how should the finish be protected?

41. What is left inside a pipe when a pipe cutter is used?

42. To cut steel pipe square with a pipe cutter, how should you begin the operation?

43. What can happen when a pipe cutter handle is tightened too rapidly?

44. What adverse effect may occur in a water pipe if the burr isn't removed?

45. What is the name of the tool used to remove a burr from the inside of a water pipe?

46. What are the two most common types of pipe threaders in use?

47. How do they differ?

48. Describe the type of pipe threader that can be used only where there is enough room to make a complete circumference.

49. What is the taper per foot of standard pipe threads?

50. What is the first thing you must do before threading a pipe?

51. Describe briefly the other steps in threading pipe.

52. How can used cutting oil be reclaimed?

53. Describe how joint pipe compound should be used when joining an elbow to a threaded pipe.

54. Why is it important to select the right size pipe wrench when joining fittings to threaded pipe?

55. What should be done after the installation of a piping system is complete?

56. What procedure is used in measuring copper pipe?

57. What makes soft-tempered copper tubing desirable for some uses?

58. What allowances must be considered when measuring copper pipe for cutting?

59. A fine-toothed saw blade is considered to have how many cutting teeth per inch?

60. What is the name of the most common joint used in a copper water system?

61. Name two other types of joints frequently used in a copper system.

62. Why is it important to solder copper pipe and fittings immediately after they are cleaned?

63. Why is it important to select the proper size tip when soldering copper pipe and fittings?

64. Describe the soldering process.

65. What precautions should be taken when soldering is to be done near combustible material?

66. Why is it more difficult to resolder a leaking joint than it was to solder it the first time?

67. When it's impossible to keep water away from a joint you are trying to solder, what "trick of the trade" should be tried?

68. What type of seat should a galvanized union have in a water supply system?

69. What should you check for, once a water system is complete?

70. With what type of copper should flare fittings be used?

71. What type of wrench should be used to tighten flare nuts?

72. Flare fittings, when used in copper installations, should never be _____.

73. Compression-type fittings are seldom used in a water supply system. When they are used, how are they used?

74. Describe how a compression-type fitting works.

75. Why should a tubing cutter be used to cut supply line tubes to fixtures?

76. Why should a freehand cut not be used when cutting plastic pipe?

77. What tool may be used to ream plastic pipe?

78. What piece of equipment should be used with a hacksaw to ensure a square cut for plastic pipe?

79. Cement used to create plastic pipe and fitting joints may be referred to as a _____ joint.

80. Liquid cleaner or sandpaper removes what from the surface of plastic pipe and fittings?

81. After completion of a plastic system, what is the minimum waiting time recommended before water testing?

82. In a plastic welded system, what procedure must be followed in repairing a leaking joint?

83. What is the name of the newest plastic pipe now approved for most water supply systems?

84. What fitting must be used to connect plastic pipe to iron pipe valves and fittings?

85. What is the required test pressure for a newly-completed water system?

86. What must *not* be present in a water system if it's to pass inspection?

Private Water Wells and Sprinkler Systems

A few years ago, the Water Quality Association estimated that nearly one home in five is not connected to a public water supply. This means that about 47 million people depend on water from a private system. Many suburban homes and most rural homes use water from private wells. In some rural areas, lakes and streams make wells unnecessary.

The rapid spread of urban communities around large metropolitan areas has often outstripped the installation of public water distribution systems. Many smaller towns and cities have population densities low enough to make the cost of public water systems prohibitive.

Either your local health department or the Department of Environmental Resource Management (DERM) enforces rules on approval and inspection of potable water wells. Irrigation wells may or may not be covered by the same rules.

DERM or your health department has set guidelines on depth and separation distance of potable water supply wells (and in some cases irrigation wells) from sources of contamination. A permit is required for any drilled or driven well, regardless of whether the water is intended for domestic or irrigation purposes.

Each well is inspected to ensure that the owner has complied with applicable regulations.

Well water is generally classified as hard water because of its high mineral content. Many people have to acquire a taste for untreated well water. It has a distinctly different taste than "city water." It has none of the chlorine or other chemicals that city dwellers are accustomed to. It also requires more soap to make a lather.

The minerals in untreated well water (especially the iron) often leave a dark reddish brown stain on the surface of plumbing fixtures, exterior building walls, and sidewalks. These stains are virtually impossible to remove. An easy way to avoid stains from irrigation water is to place sprinkler heads where they won't sprinkle walls and walks.

Dissolved minerals can also cause a buildup of scale within water heaters and water distribution pipes. The result of this buildup can be a premature failure of the system. You can reduce this buildup of scale and rust and improve the taste by installing a water softener on the building service line. See Figure 17–1. On existing installations, you must install a full bypass. A full bypass is not required on new installations. The pressure drop through the softener will

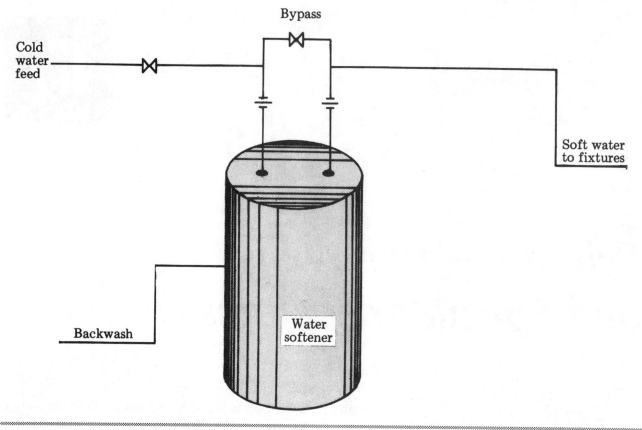

Figure 17–1
Water softener installed in building water service line

reduce the quantity of water available to operate all plumbing fixtures, so be sure to consider this pressure drop when sizing the water pipe system. Backwash disposal methods should be approved by the local authority.

Well water is ground water that has filtered down through the soil to the water table level. It's known as *meteoric* water and makes up most of the estimated two million cubic miles of ground water in the upper crust of the earth. Rainwater soaks into the ground and moves slowly down to this underground water reservoir, which may be a few feet or hundreds of feet below the surface. Since it has been filtered through sand and rock, it's usually cool, low in harmful bacteria, and high in dissolved minerals.

The Well

When a well is driven or drilled, the bottom of the well casing must extend into the dry weather water table. Otherwise, during prolonged droughts, the water table may fall below the level of the well. The result is a dry well. See Figure 17–2.

Most codes require that potable water supply wells, suction lines, pumps, and water pressure tanks be installed by certified and licensed professional well drillers or plumbing contractors. Irrigation wells may be installed by homeowners or other nonprofessionals.

The depth of a potable water supply well is determined by the local authority and by the depth of the water table. Even though the underground water table may be within a few feet of the ground surface, the local authority can require a 30-feet minimum depth. The authority can also require a separation of the potable water supply well from a source of contamination: 50 feet from a septic tank, 100 feet from a drainfield. Irrigation wells don't have the same depth and separation requirements as potable water supply wells. But it's prudent to follow the stricter guidelines for irrigation wells too.

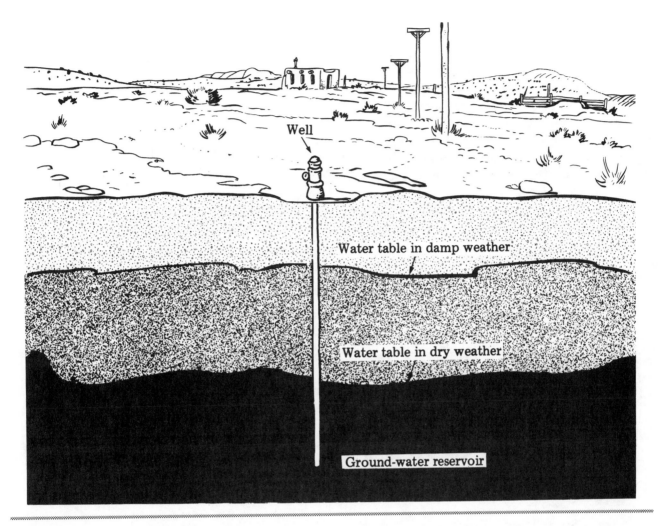

Figure 17–2
Well extended to dry weather water table

■ Deep and Shallow Wells

Wells which must be drilled are generally classified as *deep wells*. They may penetrate hundreds of feet into the earth. Water from deep wells is more desirable because there's less chance of contamination. The water level in these wells is little affected by seasonal rainfall or dry years.

Wells which are dug or driven are classified as *shallow wells*. Dug or driven wells are used where the water table is within 22 feet of the ground surface. The well must penetrate deep enough below the existing water table to assure a dependable supply of water even in very dry seasons.

There are two types of driven wells: those with an open end casing and those with a casing equipped with a well point.

An *open end casing* can supply both potable water and irrigation water in areas where the water table is close to the ground surface and is in good rock formations. In some types of soil, rock or corrosion would clog the protective screening of a well point in a short time. When this happens, the well won't draw water and becomes useless. The open end well casing is preferable under these conditions.

If an open end casing is used, the pipe is driven to the desired depth. Then loose soil and rock that have collected within the pipe are flushed from the driven section. Flushing for these shallow wells may be done by inserting a ½- or ¾-inch pipe into the larger pipe casing (usually 1½ or 2 inches in diameter). The smaller pipe has one end reamed and cut on an angle (see Figure 17–3) and is used to chip a pocket into the rock base. Attach a garden hose to the other end of the small pipe. Turn the water on and work the smaller pipe up and down inside the larger casing. Water pressure will flush loose debris out of the casing.

When most of the loose debris is flushed out and a good well has been installed, connect a 3 horsepower gasoline-driven centrifugal pump to the top of the well casing. Pump water out of the well until the water is free of rocks and sand. Check the clarity of the water by catching samples in a glass jar. When no sand or rock fragments show up in the water samples, the well is considered good. Disconnect the temporary pump and install the suction line to the permanent pump.

A *well point* is used in areas where the groundwater reservoir ends in loose shale or sand. The well point is screwed to the threaded well casing and is driven to the desired depth. Two types of well points are available in lengths that fit the use and water demand. One type has fine perforations and a fine mesh screen and is used in sand. The other type has larger perforations and a coarse mesh screen and is used in gravel or loose rock. No flushing is required if a well point is used. See Figure 17–4.

▨ Code Requirements for Wells

Potable water supply and irrigation wells must meet the following requirements to be acceptable by most codes:

- Unless specifically approved by the local authority, a well must not be located within any building or under the roof or projection of any building or structure.

- The well casing must be continuous new pipe and must terminate in a suitable aquifer. Well casing pipe for a residential potable or irrigation water supply must be galvanized steel.

- Pour a concrete pad around the well casing. The pad must be a minimum of 4

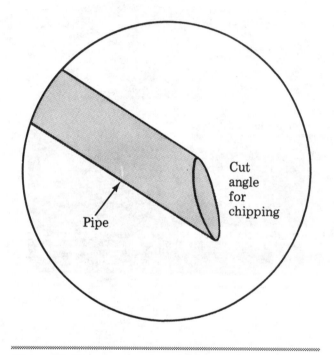

Figure 17–3
Pipe designed to chip into rock base

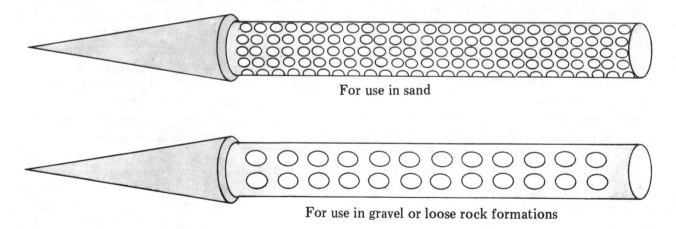

For use in sand

For use in gravel or loose rock formations

Figure 17–4
Well points

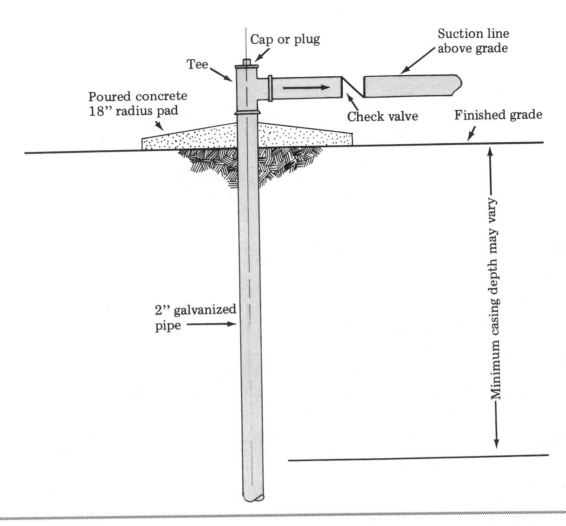

Figure 17–5
Supply well detail (irrigation water)

inches thick and measure at least 18 inches from the center of the well to the edge (36 inches in diameter). The concrete pad must be placed immediately below the tee if the suction line is above grade. See Figure 17–5. This type of installation is usually used for irrigation water supply wells, or if the well location is very near the pump.

If the suction line is below grade, the concrete pad must be placed immediately above the tee and an extension piece with a plug must be extended to grade. This type of installation is usually used for potable water wells or when the well is located some distance from the pump. See Figure 17–6. The concrete pad must slope away from the casing to prevent surface water from carrying pollutants down the well casing to the water reservoir.

- Install a tee with a plug the same size as the well at the top of the well casing. This gives access to the well for inspections, adding disinfecting agents, measuring well depth, and testing the static water level.

Suction Lines

Suction lines for potable water or irrigation must have a soft seat check valve rated at 200 pounds which must be installed as close as practical to the well. The check valve may be either the spring-loaded or flapper type. The suction line must have a union or an approved slip coupling installed just ahead of the pump.

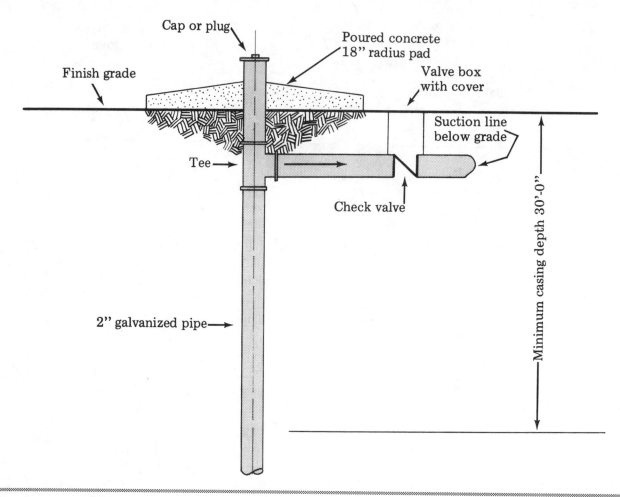

Figure 17–6
Supply well detail (potable water)

This allows disconnecting the pump for repairs or replacement without disturbing the suction line. Also install a hose bibb above grade ahead of the pump. Use this bibb to add water if the pump needs to be primed. Look ahead to Figure 17–9.

▉ Suction Line for Potable Water Supply System

The suction line must be large enough to provide the water volume and pressure required to operate the plumbing fixtures in the building. The suction line or water service pipe from the well to the pump must not be smaller than 1 inch. It must be installed with a slight pitch toward the well. If the well requires a suction line longer than 40 feet, you'll have to increase the suction pipe to the next pipe size shown in

Figure 17–7. Suction lines may be galvanized unless the code permits Schedule 40 PVC pressure-rated pipe.

As an example, consider a residence which has 30 fixture units and a suction line 65 feet long. The next larger suction pipe size in Figure 17–7 would be 1½ inches. Therefore, a 1½-inch suction line pipe should be used.

▉ Suction Line for Irrigation Supply Well

The suction line must be as large as the pump suction inlet, never smaller. For example, a 2-horsepower pump would require a 2-inch suction line; a 1½-horsepower pump would require a 2-inch suction line. It must be installed with a slight pitch toward the well. Separation distance between pump and well is not a factor in sizing. On irrigation systems,

Fixture Units	Supply Required (gph)	Diameter of Suction Pipe	Diameter Pressure Pipe	Diameter Service Pipe	Size of Tank	HP	Well Size
23	720	1	¾	¾	42	½	1½
30	900	1¼	1	1	82	¾	2
40	1200	1½	1	1	120	¾	2

Figure 17–7
Tank and pump size requirements (predominantly for flush tanks)

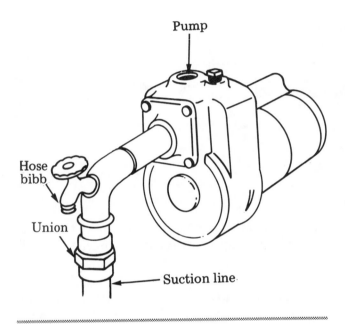

Figure 17–8
Irrigation pump with suction line detail

the hose bibb can be used to prime the pump and supply pesticides or liquid fertilizer through the sprinkler system. See Figure 17–8.

Pressure Tank Water Systems

The water piping from the pump to the pressure tank should not be smaller than the discharge outlet of the pump. Install a gate valve with the handle removed between the pump and the tank if the tank has capacity of more than 42 gallons. A minimum ¾-inch gate valve must be installed on the discharge side of the tank. This serves as the house valve to control water in the building. See Figure 17–9. Two types of pressure tanks are commonly used today.

Hydropneumatic Tank

The *hydropneumatic tank,* with its pressure switch, has been the dominant well water supply system since the 1920s. While the tank itself has given good service, unattended systems aren't reliable. Air pressure and volume controls need regular service. Hydropneumatic tanks use compressed air to control water pressure. Compressed air acts like a huge spring to drive water out of the tank and to where it's needed. When air pressure drops in the system, the pump starts forcing more water into the tank, compressing the air again.

Unfortunately, air in the tank is gradually absorbed by water passing through the tank. If this lost air isn't replaced, eventually there won't be enough air available in the tank. Water itself is nearly incompressible. If there isn't enough air, the pump cycles on and off too often.

Air in hydropneumatic tanks should be maintained automatically with air chargers and air volume controls. The tank can be manually drained of water. Doing that refills the tank with air at atmospheric pressure. As the pump fills the tank with water, it will again have the normal supply of trapped air to compress and expand for properly operating the water supply system. Figure 17–9 shows a typical pressure tank.

Hydropneumatic tanks should be large enough to prevent excessive cycling of the pump. Consider three capacities when sizing the tank:

1. The draw-off capacity of the tank between the "cut-out" and "cut-in" limits. (The tank size should provide a draw-off of 6 gallons of water while maintaining an operating range of 20 to 40 psi water pressure.)

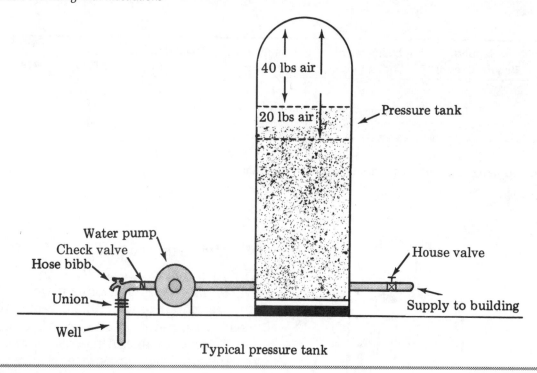

Figure 17–9
Typical pressure tank

2. The pump capacity should be great enough to quickly refill the tank.

3. The minimum size hydropneumatic tank for a single-family residence is 42 gallons. Tanks are sized by the fixture units of a building. That's the best indicator of the rate of water use. For other sizes of tanks and specific requirements, see Figure 17–7.

▨ Breather Tank

The second type of pressure tank, the *breather tank,* has been around for about 30 years. See Figure 17–10. It has been approved for use by most codes. This tank has a built-in elastic diaphragm membrane which stores and releases water in response to pressure variations in the system. This type of tank with its breather-type air-charging device is suitable for most small systems. It gives consistent service and requires no air to be injected into the tank. Too-frequent pump starts (excessive cycling) is kept to a minimum.

The tank volume for a home may be as small as 14 gallons or as large as 36 gallons, depending on the water need. The sizing procedure is the same as for a hydropneumatic tank and

should consider water-use habits of the occupants. Allowing a little extra capacity usually adds very little to tank cost.

Install the pump and pressurizing system so that equipment is reasonably accessible for repair or replacement. Interior water piping, materials, and installation methods are the same as described in Chapters 15 and 16 for water distribution pipes.

▨ Installation Problems

Inexperienced plumbers occasionally make mistakes when installing pumps in well systems. Some of these mistakes may not be obvious at first glance. In the majority of the cases where pumps were installed incorrectly, the plumber didn't follow the instructions that came with the pump. Any of the following circumstances could cause problems:

- Connecting the suction line from the well to the discharge side of the pump

- Trying to use 115 volts on a pump wired for 230 volts. (Naturally the pump would not perform properly and will burn out if used too long.)

Heavy-Duty Butyl Diaphragm

- *Flexes, never stretches or creases—*
  Seamless construction. No weak stress areas.

- *Conforms exactly to shell configuration—*
  No bubbles or corners to trap water or sediment.

- *High abrasion resistance—*
  Tough Butyl Compound withstands all normal grit in water.

- *Meets FDA minimum requirements—*
  For potable water supply.

- *Thirteen years of field use—*
  in over one and one half million installations.

Polypropylene Liner and Acceptance

- *Separate, rigid water reservoir—*
  100% non-corrosive. Not bonded to shell walls. Will never crack, pit, chip, craze or flake.

- *Mechanical "O" ring seals—*
  Exclusive mechanical clamping ring permanently bonds diaphragm and liner to shell groove.

  Swedged "O" ring seal on water side of acceptance fitting.

  Water never touches steel.

- *Copper-lined acceptance fitting—*
  Silver brazed for watertight seal.

- *Listed with NSF—*
  Polypropylene liner tested and accepted by the National Sanitation Foundation.

Figure 17–10
Parts of a breather tank

• Trying to use a pump too small to supply a particular need

• Installing the well improperly

At the end of this chapter you'll find a troubleshooting guide for jet and submersible well pumps. This guide points out the mistakes commonly made by inexperienced installers, lists what to look for and the procedures to follow when checking an installation. It also explains how to make corrections once a problem is identified. This guide can also be used to check similar symptoms for the centrifugal (irrigation) pumps.

Lawn Sprinkler Systems

When planning and designing a lawn sprinkler system, consider first the type of sprinkler heads to be used. After selecting the heads, consider these factors:

• The amount of water required

• The horsepower of the pump (if well water will be used)

• The number of circuits necessary

• The pipe size

There are two general types of sprinkler heads in use: the flush spray and the pulsating head.

Flush Spray Heads

This is the most popular head for the average lawn system. It's manufactured in many patterns so you should have no trouble finding heads for odd corners or hard-to-reach areas. This type of head provides a soaking spray. Therefore, the system does not have to operate for long periods of time. Some heads are provided with adjustment screws to control the distance of the spray (see Figure 17–11).

All spray heads provide about the same amount of water and coverage under the same installation conditions, whether made from metal or plastic. (See Figure 17–12 for important data on flush spray heads.)

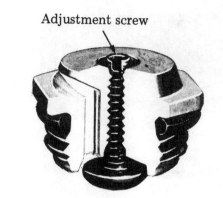

Figure 17–11
Adjustable spray head insert fitting

Pulsating Heads

The pulsating head sprinkler is commonly referred to as a *rain bird* and is a good choice for watering large open areas. This type of head may pop up from the ground (as in play areas) or it may be secured to a pipe riser. There are two kinds of pulsating heads commonly used—full circle and the adjustable part-circle.

These heads need more pressure to operate than spray heads, but each one covers about three times as much ground area as a spray head. Pulsating heads must be in operation for considerably longer periods of time to supply the same amount of water as the spray heads provide. (See Figure 17–13 for important data on pulsating heads.)

Figures 17–12 and 17–13 show the approximate lawn area covered with several head configurations based on 20 to 25 psi at the head, and the gallons delivered per minute (gpm). Naturally, the greater the pressure, the greater the coverage. More water per minute makes for faster soaking, thus less time is required for pump operation. Figure 17–14 illustrates the various head patterns and degrees for flush spray heads.

Sample Plan for a Sprinkler System

Step by step, here's how to plan a sprinkler system. Let's assume that for this job you plan to use flush spray heads.

Type of Pattern	psi	gpm	Diameter (feet)	Radius (feet)
Full head	20	3.4	24	12
¾ head	20	2.2		12
½ head	20	1.7		12
¼ head	20	.8		12
End strip head	20	.6	3 × 12 strip	
Full strip head	20	1.1	3 × 24 strip	

Figure 17–12
Data for flush spray heads—standard brass nozzle

Type of Pattern	psi	gpm	Diameter (feet)	Radius (feet)
Full circle	25	2.9	44	22
Part circle	25	2.9	44	22

Figure 17–13
Data for pulsating heads–½", ⁹⁄₆₄" nozzle

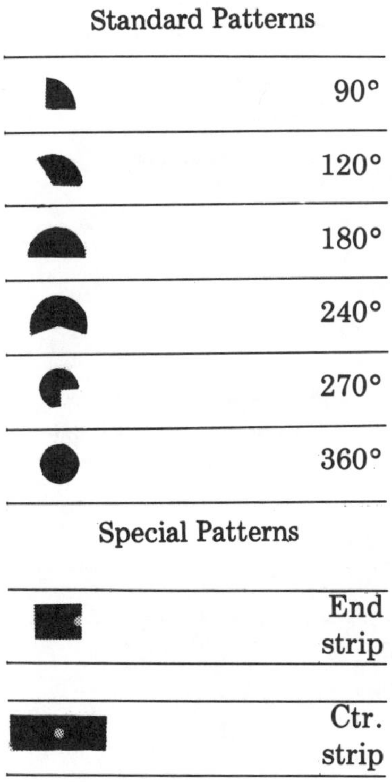

Standard Patterns

	90°
	120°
	180°
	240°
	270°
	360°

Special Patterns

| | End strip |
| | Ctr. strip |

Figure 17–14
Flush spray head patterns

Step 1

Determine the patterns required and the number and types of heads needed for complete coverage. Figure 17–12 shows that the coverage diameter for full heads is 24 feet. Notice that the coverage radius of all special pattern heads is 12 feet. Subtract 4 feet from the coverage of each type of head listed in Figure 17–12. This allows for sufficient overlap of the spray and avoids dry spots. This means full heads can be placed 20 feet apart and special pattern heads (such as ¾, ½, ¼) 8 feet apart. In some cases, the shape of the areas to be covered or the design of the head might require even closer spacing.

Step 2

Begin your layout of sprinkler heads from the wall line of the building or other permanent structure. Use small stakes driven into the ground to mark each head location.

Along a building wall, use half heads. Locate heads to spray away from building walls, fences, driveways and walkways to avoid rust stains. After laying out the special pattern heads around the building walls, driveways, walkways and the like, lay out the full heads. Figure 17–15 shows a sample sprinkler layout.

The full heads should not be placed in a straight line with one-half heads or other full heads. Instead, stagger full heads off-center with other heads. See Figure 17–15. This helps assure good overlaps. You don't need excessive overlap, but be sure all irrigated area is within 8 feet of a partial head or 20 feet of a full head.

When the plan is complete, count the number of full heads and partial heads shown on the plan. Total the number of heads for each type. The next step is to figure the amount of water needed so we can calculate the pump size. The figures below show our water use calculations for the plan in Figure 17–15.

15 full heads	x	3.4 gpm	=	51.0 gallons
2 ¾ heads	x	2.2 gpm	=	4.4 gallons
27 ½ heads	x	1.7 gpm	=	45.9 gallons
7 ¼ heads	x	.8 gpm	=	5.6 gallons
		Total gallons needed	=	106.9 gallons

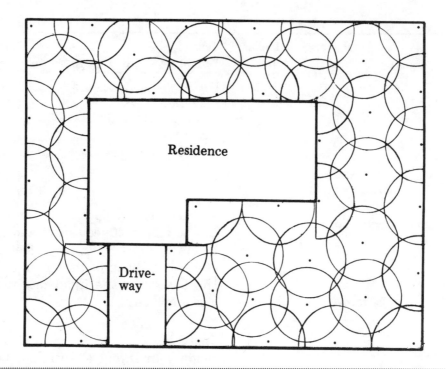

Figure 17–15
Flush spray head installation

You can see that this plan will require 107 gallons per minute.

Step 3

If well water is to be used for irrigation, choose a pump to supply this volume of water. Consider the depth of the well when sizing the pump. The deeper the well, the larger the pump capacity must be to deliver the quantity of water needed. The well in this example is 60 feet. That's a common depth for sprinkler wells in most parts of the southern U.S.

Before buying the pump, compare capacities of several centrifugal pumps. Generally, the higher the horsepower rating, the greater the pumping capacity.

Consider also the water pressure (in pounds per square inch) required for the heads you've selected. No head works as designed if water pressure is too low. Water pressure that's too high causes excessive misting. Water mist will eventually come back to earth, but may be blown quite a ways before reaching the ground. For best results, water pressure should be within the limits recommended by the manufacturer.

Pump Size (horsepower)	Approximate Capacity 60' well (gal)	Reserved for 20 psi (gallons)	Volume Available for Use (gal)
1	20	4	16
1½	60	12	48
2	94	19	75

Figure 17–16
Data on pumps

Flush spray heads need 20 psi, as in Figure 17–12. This means that approximately 20 percent of the pump's capacity must remain uncommitted so that it can generate the necessary 20 psi. Review the capacity of each pump and select the one appropriate for your job requirements.

Figure 17–16 shows that a 1 horsepower pump won't deliver enough water for our sample layout. We need 107 gallons of water per minute. The 1½ horsepower pump could be used if we divide the layout into three separate systems, each controlled by a valve. With a 1½ horsepower pump, one third of the lawn could be watered at a time. With automatic sprinklers,

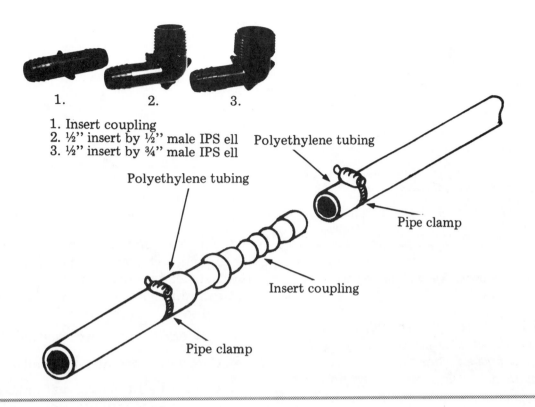

1. Insert coupling
2. ½" insert by ½" male IPS ell
3. ½" insert by ¾" male IPS ell

Figure 17–17
Assembling polyethylene plastic tubing

the owner should have no objection to three separate circuits.

The ideal pump for this layout would probably be the 2 horsepower pump. The sprinkler system would operate efficiently on two separate circuits. Using the 2 horsepower pump cuts the watering time by one third. The higher cost of the larger pump would be offset by a considerable long-term savings from reduced pump wear and lower electricity use.

■ Materials and Pipe Sizes

Materials used in a potable water supply system can also be used for piping and fittings in a sprinkler system. Note, however, that there are two types of materials used exclusively for sprinkler systems:

1. PVC Schedule 160 rigid pipe and fittings. This is a thin wall, lightweight plastic pipe which can be used only on open sprinkler systems. It is not designed to work under constant pressure. This type

of pipe is cut and assembled the same way as PVC Schedule 40 pipe. See Chapter 16.

2. Flexible polyethylene plastic tubing. This type of tubing comes in 100-foot coils and can be cut to length with a hacksaw or knife. It's assembled with insert fittings held securely in place with special pipe clamps. See Figure 17–17.

Other types of plastic fittings are now available for use with polyethylene plastic tubing. They come in different sizes and patterns such as ells, tees, couplings and the like. See Figure 17–18.

These fittings have a special taper designed on each fitting opening. A tight joint is formed when the correct size plastic tubing is inserted into the fitting.

They do, however, have one disadvantage. Once the tubing is inserted into the fitting, it can't be withdrawn. If you make a mistake, the fitting can never be used again. As with PVC cemented joints, the adjoining tubing has to be cut close to the fitting and discarded.

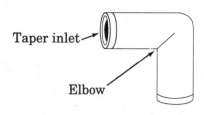

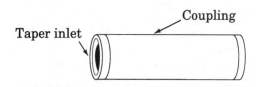

Figure 17–18
Plastic fittings for polyethylene plastic tubing

Flow (gpm)	Pipe Size
0–5	½"
5–10	¾"
11–15	1"
16–25	¼"
26–40	½"
41–60	2"

Figure 17–19
Sizing guide for PVC and polyethylene pipe

Rigid PVC plastic pipe should be sized from Figure 17–19. *Flexible polyethylene pipe should be sized one pipe size larger due to the restriction placed on the water flow by the insert fittings.*

Place plastic sprinkler piping in the trench deep enough to protect it from sharp tools. At this depth a nipple 2 inches long screwed into the threaded fitting on the pipe should reach approximately 1 inch below the grass surface. On this nipple, screw a flush head. This should leave the head flush with the grass surface. Backfill the trench with earth and replace the sod. For extra protection, place a concrete collar around each sprinkler head. Do not install the last head at the end of each run of pipe until the lines have been thoroughly flushed clean of sand and other foreign matter.

Sprinkler circuits may be controlled by:

- Gate valves operated manually

- Special ball-type valves that automatically change circuits when the pump is manually shut off momentarily, then turned on again.

- Special circuit controls equipped with electric timers so that the lawn may be watered automatically. (Most people prefer this type.)

Questions

1. Approximately how many people in the United States depend on private water to supply their needs?

2. Name the two agencies responsible for approving potable water supply wells.

3. With what guidelines are these two agencies mainly concerned?

4. Why is well water generally classified as hard water?

5. What is the main reason that well water has a different taste than public water?

6. What does well water require more of than public water when used for bathing purposes?

7. What causes well water to stain other surfaces a dark reddish brown?

8. What device may be used to reduce scale buildup and iron in a plumbing piping system?

9. To what depth must the bottom of a well casing extend so that it can provide a continuous supply of water?

10. Most codes require that what professional person install a potable water supply well and its equipment?

11. Most authorities require what separation distance between a potable water supply well and a source of contamination?

12. How are dug or driven wells classified?

13. How are drilled wells classified?

14. Why is water from deep wells more desirable than water from shallow wells?

15. Compare the two types of driven wells.

16. Why is it recommended for some areas that wells with well points not be installed?

17. Explain how to flush rock and sand from a shallow well.

18. How is a well checked to determine whether the water is free of sand?

19. What type of well point is recommended when the water reservoir ends in sand?

20. What type of well point is recommended when the water reservoir ends in gravel?

21. What material is required for well casing pipe used for residences?

22. What must be installed around the top of a well casing?

23. What purpose is served by having a tee screwed to the top of a well casing?

24. What is the minimum size suction line permitted for supplying water to plumbing fixtures?

25. In which direction must a suction line pitch?

26. For supplying water to a sprinkler system, what size must the suction line be?

27. What device must be installed on all suction lines as near the well as possible?

28. What device must be installed on all screw pipe suction lines as near the pump as possible?

29. What are some of the advantages of installing a hose bibb on the suction line above grade on an irrigation well pump?

30. What is the minimum size of the discharge pipe from the pump to the pressure tank?

31. What is the minimum size hydropneumatic tank permitted for a single-family residence?

32. What determines the size of a hydropneumatic tank?

33. Name two advantages that the newer breather pressure tank has over the older hydropneumatic tank.

34. When initially installing the pump and equipment for a single-family pressurized system, what important consideration must be made?

35. What are possible outcomes when a pump wired for 230 volts is connected to wires supplying 115 volts?

36. What are the two types of sprinkler heads that may be used for a sprinkler system?

37. What is the first factor to be considered in planning and designing a lawn sprinkler system?

38. What two factors determine the size pump to be used on an irrigation water well?

39. Where are pulsating sprinkler heads generally used?

40. What is the main disadvantage in using pulsating sprinkler heads in an average-size residential yard?

41. What is accomplished when pressures exceed those listed in Figures 17–12 and 17–13?

42. What is the average radius coverage for a ½ flush spray head at 20 psi?

43. What spray pattern should flush heads have when installed along a building wall?

44. Why is it better not to install flush spray heads in a straight line with other heads?

45. Why is the depth of the well an important factor in sizing the well pump?

46. What type of well pump is usually installed for irrigation purposes?

47. What two types of materials are used exclusively for sprinkler systems?

48. What should be used to protect sprinkler heads?

49. What should be done to a newly-installed sprinkler system before putting it in use?

50. What source other than a public water supply do many suburban and rural homes use for their water supply?

TROUBLE SHOOTING GUIDE — JET PUMPS

PUMP WON'T START OR RUN

CAUSE OF TROUBLE	HOW TO CHECK	HOW TO CORRECT
1. Blown fuse.	Check to see if fuse is OK.	If blown, replace with fuse of proper size.
2. Low line voltage.	Use voltmeter to check pressure switch or terminals nearest pump.	If voltage under recommended minimum, check size of wiring from main switch on property. If OK, contact power company.
3. Loose, broken, or incorrect wiring.	Check wiring circuit against diagram. See that all connections are tight and that no short circuits exist because of worn insulation, crossed wire, etc.	Rewire any incorrect circuits. Tighten connections, replace defective wires.
4. Defective motor.	Check to see that switch is closed.	Repair or take to motor service station.
5. Defective pressure switch.	Check switch setting. Examine switch contacts for dirt or excessive wear.	Adjust switch settings. Clean contacts with emory cloth if dirty.
6. Tubing to pressure switch plugged.	Remove tubing and blow through it.	Clean or replace if plugged.
7. Impeller or seal.	Turn off power, then use screwdriver to try to turn impeller or motor.	If impeller won't turn, remove housing and locate source of binding.
8. Defective start capacitor.	Use an ohmmeter to check resistance across capacitor. Needle should jump when contact is made. No movement means an open capacitor; no resistance means capacitor is shorted.	Replace capacitor or take motor to service station.
9. Motor shorted out.	If fuse blows when pump is started (and external wiring is OK) motor is shorted.	Replace motor.

MOTOR OVERHEATS AND OVERLOAD TRIPS OUT

CAUSE OF TROUBLE	HOW TO CHECK	HOW TO CORRECT
1. Incorrect line voltage.	Use voltmeter to check at pressure switch or terminals nearest pump.	If voltage under recommended minimum, check size of wiring from main switch on property. If OK, contact power company.
2. Motor wired incorrectly.	Check motor wiring diagram.	Reconnect for proper voltage as per wiring diagram.
3. Inadequate ventilation.	Check air temperature where pump is located. If over 100°F., overload may be tripping on external heat.	Provide adequate ventilation or move pump.
4. Prolonged low pressure delivery.	Continuous operation at very low pressure places heavy overload on pump. This can cause overload protection to trip.	Install globe valve on discharge line and throttle to increase pressure.

TROUBLESHOOTING GUIDE — JET PUMPS

PUMP STARTS AND STOPS TOO OFTEN		
CAUSE OF TROUBLE	**HOW TO CHECK**	**HOW TO CORRECT**
1. Leak in pressure tank.	Apply soapy water to entire surface above water line. If bubbles appear, air is leaking from tank.	Repair leaks or replace tank.
2. Defective air volume control.	This will lead to a waterlogged tank. Make sure control is operating properly. If not, remove and examine for plugging.	Clean or replace defective control.
3. Faulty pressure switch.	Check switch setting. Examine switch contacts for dirt or excessive wear.	Adjust switch settings. Clean contacts with emory cloth if dirty.
4. Leak on discharge side of system.	Make sure all fixtures in plumbing system are shut off. Then check all units (especially ballcocks) for leaks. Listen for noise of water running.	Repair leaks as necessary.
5. Leak on suction side of system.	On shallow well units, install pressure gauge on suction side. On deep well systems, attach a pressure gauge to the pump. Close the discharge line valve. Then, using a bicycle pump or air compressor, apply about 30 psi pressure to the system. If the system will not hold this pressure when the compressor is shut off, there is a leak on the suction side.	Make sure above ground connections are tight. Then repeat test. If necessary, pull piping and repair leak.
6. Leak in foot valve.	Pull piping and examine foot valve.	Repair or replace defective valve.

PUMP WON'T SHUT OFF		
CAUSE OF TROUBLE	**HOW TO CHECK**	**HOW TO CORRECT**
1. Wrong pressure switch setting or setting "drift".	Lower switch setting. If pump shuts off, this was the trouble.	Adjust switch to proper setting.
2. Defective pressure switch.	Arcing may have caused switch contacts to "weld" together in closed position. Examine points and other parts of switch for defects.	Replace switch if defective.
3. Tubing to pressure switch plugged.	Remove tubing and blow through it.	Clean or replace if plugged.
4. Loss of prime.	When no water is delivered, check prime of pump and well piping.	Reprime if necessary.
5. Low well level.	Check well depth against pump performance table to make sure pump and ejector are properly sized.	If undersized, replace pump or ejector.
6. Plugged ejector.	Remove ejector and inspect.	Clean and reinstall if dirty.

TROUBLESHOOTING GUIDE — JET PUMPS

PUMP OPERATES BUT DELIVERS LITTLE OR NO WATER		
CAUSE OF TROUBLE	**HOW TO CHECK**	**HOW TO CORRECT**
1. Low line voltage.	Use voltmeter to check at pressure switch or terminals nearest pump.	If voltage under recommended minimum, check size of wiring from main switch on property. If OK, contact power company.
2. System incompletely primed.	When no water is delivered, check prime of pump and well piping.	Reprime if necessary.
3. Air lock in suction line.	Check horizontal piping between well and pump. If it does not pitch upward from well to pump, an air lock may form.	Rearrange piping to eliminate air lock.
4. Undersized piping.	If system delivery is low, the discharge piping and/or plumbing lines may be undersized. Refigure friction loss.	Replace undersized piping or install pump with higher capacity.
5. Leak in air volume control or tubing.	Disconnect air volume control tubing at pump and plug hole. If capacity increases, a leak exists in the tubing of control.	Tighten all fittings and replace control if necessary.
6. Pressure regulating valve stuck or incorrectly set. (Deep well only.)	Check valve setting. Inspect valve for defects.	Reset, clean, or replace valve as needed.
7. Leak on suction side of system.	On shallow well units, install pressure gauge on suction side. On deep well systems, attach a pressure gauge to the pump. Close the discharge line valve. Then, using a bicycle pump or air compressor, apply about 30 psi pressure to the system. If the system will not hold this pressure when the compressor is shut off, there is a leak on the suction side.	Make sure above ground connections are tight. Then repeat test. If necessary, pull piping and repair leak.
8. Low well level.	Check well depth against pump performance table to make sure pump and ejector are properly sized.	If undersized, replace pump or ejector.
9. Wrong pump-ejector combination.	Check pump and ejector models against manufacturer's performance tables.	Replace ejector if wrong model is being used.
10. Low well capacity.	Shut off pump and allow well to recover. Restart pump and note whether delivery drops after continuous operation.	If well is "weak," lower ejector (deep well pumps), use a tail pipe (deep well pumps), or switch from shallow well to deep well equipment.
11. Plugged ejector.	Remove ejector and inspect.	Clean and reinstall if dirty.
12. Defective or plugged foot valve and/or strainer.	Pull foot valve and inspect. Partial clogging will reduce delivery. Complete clogging will result in no water flow. A defective foot valve may cause pump to lose prime, resulting in no delivery.	Clean, repair, or replace as needed.
13. Worn or defective pump parts or plugged impeller.	Low delivery may result from wear on impeller or other pump parts. Disassemble and inspect.	Replace worn parts or entire pump. Clean parts if required.

TROUBLESHOOTING GUIDE — SUBMERSIBLE PUMPS

FUSES BLOW OR CIRCUIT BREAKER TRIPS WHEN MOTOR IS STARTED		
CAUSE OF TROUBLE	**HOW TO CHECK**	**HOW TO CORRECT**
1. Incorrect line voltage	Check the line voltage terminals in the control box (or connection box in the case of the 2-wire models) with a voltmeter. Make sure that the voltage is within the minimum-maximum range prescribed by the manufacturer.	If the voltage is incorrect, contact the power company to have it corrected.
2. Defective control box: a. Defective wiring.	Check out all motor and powerline wiring in the control box, following the wiring diagram found inside the box. See that all connections are tight and that no short circuits exist because of worn insulation, crossed wires, etc.	Rewire any incorrect circuits. Tighten loose connections. Replace worn wires.
b. Incorrect components	Check all control box components to see that they are the type and size specified for the pump in the manufacturers' literature. In previous service work, the wrong components may have been installed.	Replace any incorrect component with the size and type recommended by the manufacturer.
c. Defective starting capacitor (skip for 2-wire models).	Using an ohmmeter, determine the resistance across the starting capacitor. When contact is made, the ohmmeter needle should jump at once, then move up more slowly. No movement indicates an open capacitor (or defective relay points); no resistance means that the capacitor is shorted.	Replace defective starting capacitor.
d. Defective relay (skip for 2-wire models).	Using an ohmmeter, check the relay coil. Its resistance should be as shown in the manufacturer's literature. Recheck ohmmeter reading across starting capacitor. With a good capacitor, no movement of the needle indicates defective relay points.	If coil resistance is incorrect or points defective, replace relay.
3. Defective pressure switch.	Check the voltage across the pressure switch points. If less than the line voltage determined in "1" above, the switch points are causing low voltage by making imperfect contact.	Clean points with a mild abrasive cloth or replace pressure switch.
4. Pump in crooked well.	If wedged into a crooked well, the motor and pump may become misaligned, resulting in a locked rotor.	If the pump does not rotate freely, it must be pulled and the well straightened.
5. Defective motor winding or cable: a. Shorted or open motor winding.	Check the resistance of the motor winding by using an ohmmeter on the proper terminals in the control box (see manufacturer's wiring diagram). The resistance should match the ohms specified in the manufacturer's data sheet. If too low, the motor winding may be shorted; if the ohmmeter needle doesn't move, indicating high or infinite resistance, there is an open circuit in the motor winding.	If the motor winding is defective — shorted or open — the pump must be pulled and the motor repaired.
b. Grounded cable or winding.	Ground one lead of the ohmmeter onto the drop pipe or shell casing, then touch the other lead to each motor wire terminal. If the ohmmeter needle moves appreciably when this is done, there is a ground in either the cable or the motor winding.	Pull the pump and inspect the cable for damage. Replace damaged cable. If cable checks OK, the motor winding is grounded.
6. Pump sand locked.	Make pump run backwards by interchanging main and start winding (black and red) motor leads at control box.	Pull pump, disassemble and clean. Before replacing, make sure that sand has settled in well. If well is chronically sandy, a submersible should not be used.

TROUBLESHOOTING GUIDE — SUBMERSIBLE PUMPS

PUMP OPERATES BUT DELIVERS LITTLE OR NO WATER		
CAUSE OF TROUBLE	**HOW TO CHECK**	**HOW TO CORRECT**
1. Pump may be air locked.	Stop and start pump several times, waiting about one minute between cycles. If pump then resumes normal delivery, air lock was the trouble.	If this test fails to correct the trouble, proceed as below.
2. Water level in well too low.	Well production may be too low for pump capacity. Restrict flow of pump output, wait for well to recover, and start pump.	If partial restriction corrects trouble, leave valve or cock at restricted setting. Otherwise, lower pump in well if depth is sufficient. Do not lower if sand clogging might occur.
3. Discharge line check valve installed backward.	Examine check valve on discharge line to make sure that arrow indicating direction of flow points in right direction.	Reverse valve is necessary.
4. Leak in drop pipe.	Raise pipe and examine for leaks.	Replace damaged section of drop pipe.
5. Pump check valve jammed by drop pipe.	When pump is pulled after completing "4" above, examine connection of drop pipe to pump outlet. If threaded section of drop pipe has been screwed in too far, it may be jamming the pump's check valve in the closed position.	Unscrew drop pipe and cut off portion of threads.
6. Pump intake screen blocked.	The intake screen on the pump may be blocked by sand or mud. Examine.	Clean screen, and when reinstalling pump, make sure that it is located several feet above the well bottom—preferably 10 feet or more.
7. Pump parts worn.	The presence of abrasives in the water may result in excessive wear on the impeller, casing, and other close-clearance parts. Before pulling pump, reduce setting on pressure switch to see if pump shuts off. If it does, worn parts are probably at fault.	Pull pump and replace worn components.
8. Motor shaft loose.	Coupling between motor and pump shaft may have worked loose. Inspect for this after pulling pump and looking for worn components, as in "7" above.	Tighten all connections, setscrews, etc.

TROUBLESHOOTING GUIDE — SUBMERSIBLE PUMPS

PUM STARTS TOO FREQUENTLY		
CAUSE OF TROUBLE	**HOW TO CHECK**	**HOW TO CORRECT**
1. Pressure switch defective or out of adjustment.	Check setting on pressure switch and examine for defects.	Reduce pressure setting or replace switch.
2. Leak in pressure tank above water level.	Apply soap solution to entire surface of tank and look for bubbles indicating air escaping.	Repair or replace tank.
3. Leak in plumbing system.	Examine service line to house and distribution branches for leaks.	Repair leaks.
4. Discharge line check valve leaking.	Remove and examine.	Replace if defective.
5. Air volume control plugged.	Remove and inspect air volume control.	Clean or replace.
6. Snifter valve plugged.	Remove and inspect snifter valve.	Clean or replace.

MOTOR DOES NOT START, BUT FUSES DON'T BLOW		
CAUSE OF TROUBLE	**HOW TO CHECK**	**HOW TO CORRECT**
1. Overload protection out.	Check fuses or circuit breaker to see that they are operable.	If fuses are blown, replace. If breaker is tripped, reset.
2. No power.	Check power supply to control box (or overload protection box) by placing a voltmeter across incoming power lines. Voltage should approximate nominal line voltage.	If no power is reaching box, contact power company for service.
3. Defective control box.	Examine wiring in control box to make sure all contacts are tight. With a voltmeter, check voltage at motor wire terminals. If no voltage is shown at terminals, wiring is defective in control box.	Correct faulty wiring or tighten loose contacts.
4. Defective pressure switch.	With a voltmeter, check voltage across pressure switch while the switch is closed. If the voltage drop is equal to the line voltage, the switch is not making contact.	Clean points or replace switch.

TROUBLESHOOTING GUIDE — SUBMERSIBLE PUMPS

FUSES BLOW WHEN MOTOR IS RUNNING		
CAUSE OF TROUBLE	**HOW TO CHECK**	**HOW TO CORRECT**
1. Incorrect voltage.	Check line voltage terminals in the control box (or connection box in the case of 2-wire models) with a voltmeter. Make sure that the voltage is within the minimum-maximum range prescribed by the manufacturer.	If voltage is incorrect, contact power company for service.
2. Overheated overload protection box.	If sunlight or other source of heat has made box too hot, circuit breakers may trip or fuses blow. If box is hot to the touch, this may be the problem.	Ventilate or shade box, or remove from source of heat.
3. Defective control box components. (skip this for 2-wire models).	Using an ohmmeter, determine the resistance across the running capacitor. When contact is made, the ohmmeter needle should jump at once, then move up more slowly. No movement indicates an open capacitor (or defective relay points); no resistance means that the capacitor is shorted. Using an ohmmeter, check the relay coil. Its resistance should be as shown in the manufacturer's literature. Recheck ohmmeter reading across running capacitor. With a good capacitor, no movement of the needle indicates relay points.	Replace defective components.
4. Defective motor winding or cable.	Check the resistance of the motor winding by using an ohmmeter on the proper terminals in the control box (see manufacturer's wiring diagram). The resistance should match the ohms specified in the manufacturer's data sheet. If too low, the motor winding may be shorted; if the ohmmeter needle doesn't move, indicating high or infinite resistance, there is an open circuit in the motor winding. Ground one lead of the ohmmeter onto the drop pipe or shell casing, then touch the other lead to each motor wire terminal. If the ohmmeter needle moves appreciably when this is done, there is a ground in either the cable or the motor winding.	If neither cable or winding is defective—shorted, grounded, or open—pump must be pulled and serviced.
5. Pump becomes sand-locked.	If the fuses blow while the pump is operating, sand or grit may have become wedged in the impeller, causing the rotor to lock. To check this, pull the pump.	Pull pump, disassemble, and clean. Before replacing, make sure that sand has settled in well. If well is chronically sandy, a submersible should not be used.

TROUBLESHOOTING GUIDE — SUBMERSIBLE PUMPS

PUMP WON'T SHUT OFF

CAUSE OF TROUBLE	HOW TO CHECK	HOW TO CORRECT
1. Defective pressure switch.	Arcing may have caused pressure switch points to "weld" in closed position. Examine points and other parts of switch for defects.	Clean points or replace switch.
2. Water level in well too low.	Well production may be too low for pump capacity. Restrict flow of pump output, wait for well to recover, and start pump.	If partial restriction corrects trouble, leave valve or cock at restricted setting. Otherwise, lower pump in well if depth is sufficient. Do not lower if sand clogging might occur.
3. Leak in drop line.	Raise pipe and examine for leaks.	Replace damaged section of drop pipe.
4. Pump parts worn.	The presence of abrasives in the water may result in excessive wear on the impeller, casing, and other close-clearance parts. Before pulling pump, reduce setting on pressure switch to see if pump shuts off. If it does, worn parts are probably at fault.	Pull pump and replace worn components.

18

Roughing-in

A typical single-family residence has about 300 feet of concealed piping under the floor and in the walls. This piping is known as the rough plumbing and includes the drainage and waste piping, the vents and the hot and cold water lines.

The roughing-in brings waste and supply piping to the point where connections can be made to plumbing fixtures. Roughing-in for the fixtures is the most critical part of a plumber's work. It's here that a plumber displays either good professional skill or a lack of it. Plumbing knowledge and good workmanship are required to properly rough-in waste and supply lines. The plumbing fixtures will be used regularly during the entire life of the building. Do the rough-in correctly and those lines should be nearly trouble-free for decades.

Plumbing Fixture Clearance

One of the most critical parts of roughing-in is laying out and spacing the plumbing fixtures you plan to install. Every fixture must be spaced and installed so that it can be used for its intended purpose and is accessible for cleaning and repairs.

Observe the minimum clearances recommended in this chapter so the plumbing system you install will pass final inspection. Typical clearances are shown in Figure 18–1.

Water closets must be set a minimum of 15 inches from the center of the bowl to any finished wall or partition. Where a bidet is installed next to a water closet, there must be a minimum spacing of 30 inches center to center. A water closet installed next to a bathtub must have a minimum of 12 inches from the center of the bowl to the outside edge of the tub apron. A water closet must also have a minimum clearance of 21 inches from the front of the bowl to any finished wall, door, or other plumbing fixture.

Bidets must have the same minimum spacing as water closets.

Lavatories are manufactured in various designs and widths, so center-to-center measurements do not apply. The minimum clearance is measured from the edge of the lavatory to the nearest obstruction. A lavatory must have a minimum clearance of 4 inches from its edge to any finished wall. It must have a minimum clearance of 2 inches from its edge to the edge of a tub. A minimum clearance of 4 inches is required from the edge of the lavatory to the

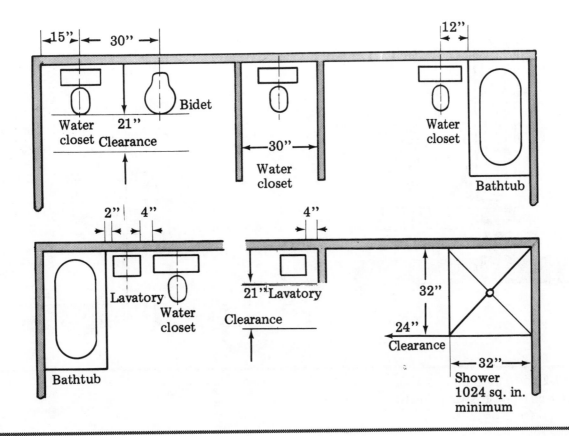

Figure 18–1
Plumbing fixture clearances

edge of a water closet tank. Allow 21 inches from the front of the lavatory to any finished wall, door, or other plumbing fixture.

Shower openings must allow easy entrance and exit. The compartment or stall must have a minimum clearance of 30 inches from any finished wall, door, or other plumbing fixture. The minimum floor area for a shower stall is 1024 square inches.

Kitchen sinks and laundry trays don't have any spacing requirements. But be sure to provide enough space for easy access for cleaning and repairs as well as for the intended use.

■ Roughing-in Measurements

All professional plumbers should know the typical rough-in measurements for various types of plumbing fixtures. Know the correct heights, distances, and locations for waste and water outlets for both wall-hung and floor-mounted fixtures. Memorize the measurements for the most com-

mon fixtures. If you have trouble remembering, jot them down in a notebook that you can keep in your tool box. Complete rough-in information is usually available from fixture manufacturers and distributors.

At the end of this chapter you'll find typical rough-in dimensions and installation data for the more common plumbing fixtures. Included are measurements for bathtubs, water closets, lavatories, bidets, and kitchen sinks. The rough-in measurements are for American Standard fixtures but will be similar or identical to the measurements for like plumbing fixtures from other manufacturers.

Note that the standard roughing-in measurement for a floor-mounted water closet is 12 inches from a finished wall. If you make an error and the rough-in is either too close or too far from the finished wall, the code does not permit the use of an offset closet flange to correct the mistake. This would restrict the discharge flow from the water closet and could cause an over-

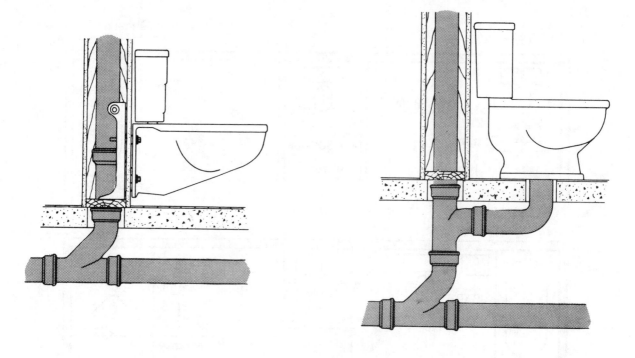

Figure 18–2
Two tank-type residential water closets

flow or a stoppage. Special water closets are available that can be set at 10 or 14 inches from the wall.

Plumbing Fixture Carriers

Bathrooms are the most susceptible of all rooms to unsanitary conditions. The bases of floor-mounted fixtures are natural areas for accumulated filth that's nearly impossible to remove. Bathroom floors made of wood tend to deteriorate under toilet fixtures. Wall-mounted water closets can help solve this problem. Maybe that's why more and more modern homes use wall-mounted water closets. A water closet carrier supports the water closet at the desired height above the finished floor.

Figure 18–2 shows two tank-type water closets, one for mounting on the floor and the other for mounting on the wall. Both of these are commonly used in residential buildings. The drawings clearly show the difference between the two installations—and some of the advantages of off-the-floor installation. With on-the-floor water

closets, slabs or floors must be penetrated at each fixture to accommodate waste piping. But with off-the-floor fixtures, the waste piping doesn't penetrate the floor. A clear, unobstructed floor is available for cleaning, and deterioration isn't a problem. The contractor gets a faster and easier installation and the homeowner gets years of easy cleaning and minimum maintenance.

Residential carriers have fewer parts and are therefore easier to assemble than commercial carriers. They are designed to receive the waste from a single water closet or at most two water closets (when back-to-back installations are used, as shown in Figure 18–3.) Battery-type water closet installations are used in larger commercial buildings.

Residential carriers are designed to be compatible with newer piping materials (like PVC plastic and no-hub cast iron pipe and fittings). Figure 18–4 shows an off-the-floor residential closet carrier with the parts identified. Figure 18–5 shows the roughing-in dimensions for residential plumbing fixtures from American Standard.

Some apprentices may never have an opportunity to work with off-the-floor plumbing fixtures

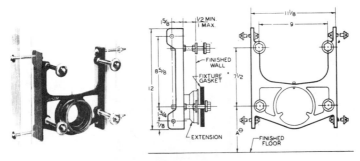

3" single, copper adapter type

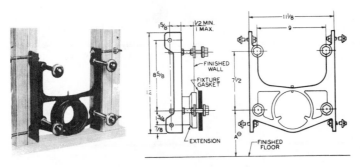

3" single, PVC plastic, adaptor type

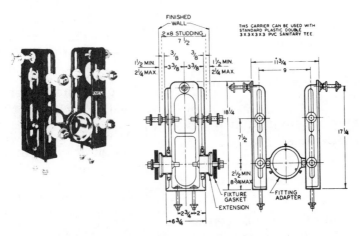

3" back-to-back, PVC plastic, adaptor type

Figure 18–3
Residential carriers

because of the limited type of work their company performs. If you're not familiar with this type of fixture, illustrations on the following pages may be of interest. Carriers for commercial use are shown in the more advanced *Plumbers Handbook Revised* by this author. There's an order form for it, and other construction references, bound into the back of this book.

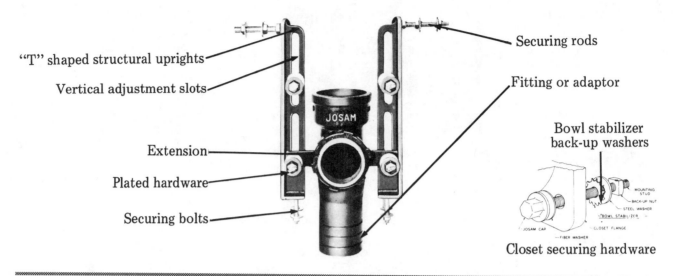

"T" shaped structural uprights

Vertical adjustment slots

Securing rods

Fitting or adaptor

Extension

Plated hardware

Securing bolts

Bowl stabilizer back-up washers

Closet securing hardware

Figure 18–4
Josam water closet carrier

Questions

1. The concealed piping of a single-family residence represents approximately how many feet?

2. How does the term "roughing-in" differ from the term "rough plumbing"?

3. Why is it so crucial for a plumber to know installation dimensions when roughing-in the plumbing for fixtures?

4. What is the major consideration when installing plumbing fixtures?

5. How many inches should be allowed from the center of a water closet bowl to any finished wall?

6. What is the minimum center-to-center spacing for a bidet installed next to a water closet?

7. What is the minimum clearance between the center of a water closet bowl and the edge of a bathtub?

8. What is the minimum clearance from the front of most fixtures to any finished wall?

9. Why are minimum center-to-center spacing requirements not applicable to lavatories?

10. What is the standard height from finished floor to the overflow rim of a lavatory?

11. What is the distance from the center of most water closet bowls to the center of the water supply outlet?

12. What is the minimum distance from the edge of lavatory to the nearest obstruction?

13. What is the standard height from finished floor to the center of a waste outlet for a wall-hung lavatory?

14. What is the standard measurement, center-to-center, of hot and cold water outlets for a lavatory?

15. What is the minimum clearance from the nearest obstruction for the entry/exit point of a shower?

16. What is the minimum floor area acceptable for a shower?

17. What is the standard roughing-in measurement for a water closet from the finished wall to the center of the waste outlet?

18. What is the standard height from the finished floor to the top of a water closet bowl?

19. What is the standard height from the finished floor to the overflow rim of a kitchen sink?

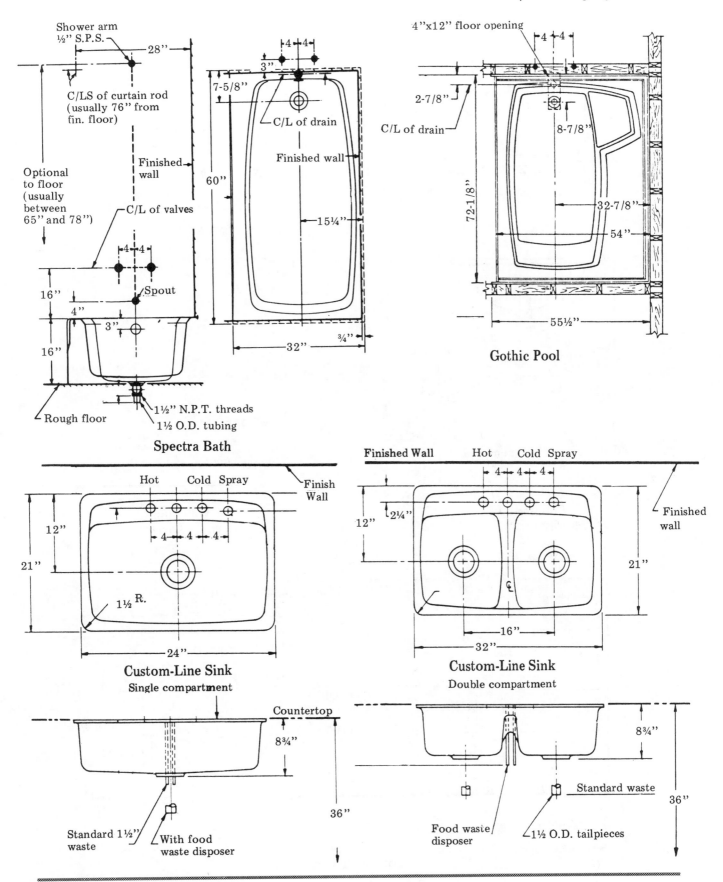

Figure 18–5
Roughing-in dimensions

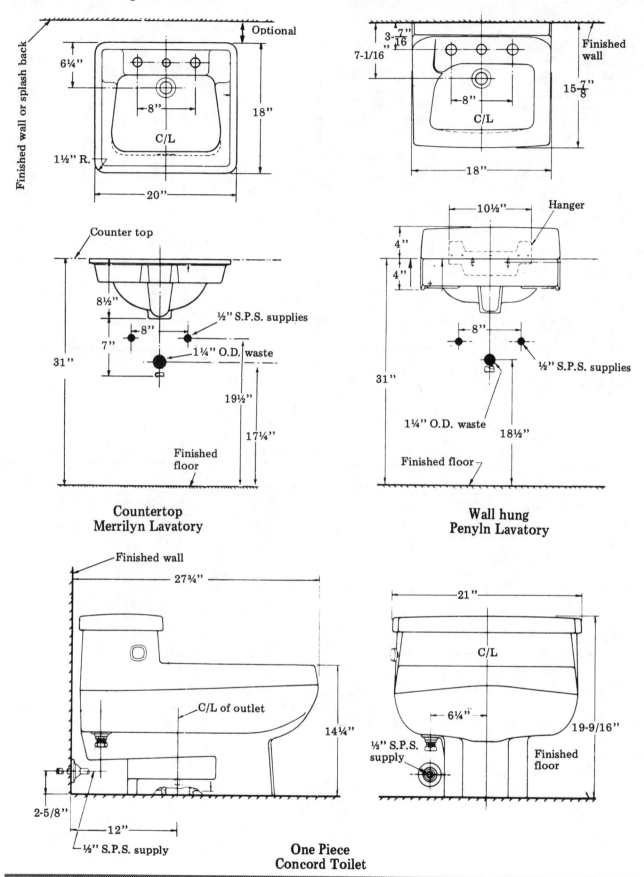

Countertop
Merrilyn Lavatory

Wall hung
Penyln Lavatory

One Piece
Concord Toilet

Figure 18–5 (continued)
Roughing-in dimensions

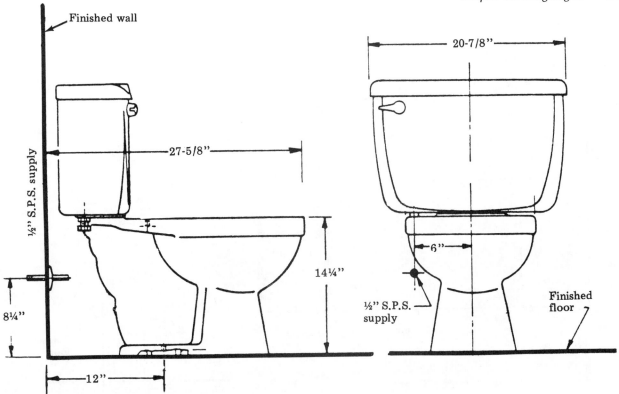

Closed coupled combination round front
Plebe Toilet

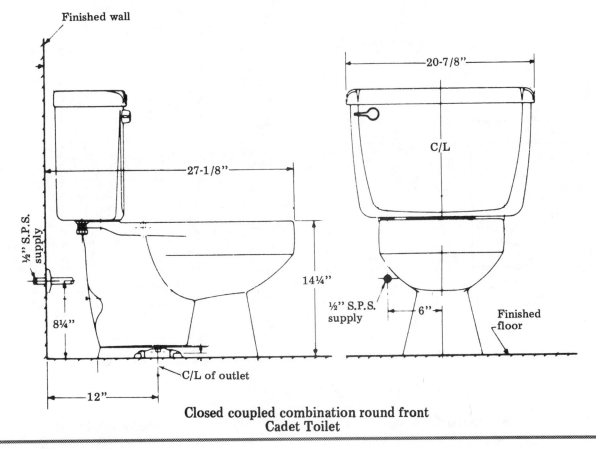

Closed coupled combination round front
Cadet Toilet

Figure 18–5 (continued)
Roughing-in dimensions

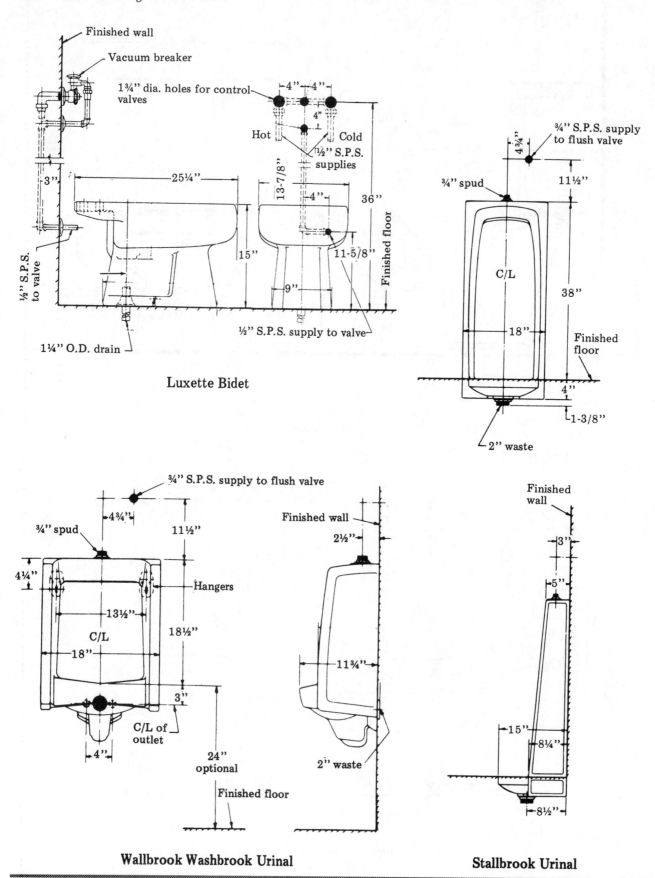

Luxette Bidet

Wallbrook Washbrook Urinal

Stallbrook Urinal

Figure 18–5 (continued)
Roughing-in dimensions

20. When the waste outlet for a water closet is roughed-in too close to a finished wall, what is the only permitted method of correcting the error?

21. To what conditions are bathrooms more susceptible than probably any other room?

22. What must be used to support off-the-floor water closets?

23. Name two newer piping materials that are compatible with specially-designed residential carriers.

24. With what size drain is a bidet generally equipped?

Plumbing Fixtures

Over the last 100 years, standards have been developed for plumbing fixtures. These standards are written into the code to control both quality and design of all plumbing fixtures in common use today.

Plumbing fixtures are the end of the potable (drinking) water supply system and the beginning of the sewage system. Liquid waste flows into the fixture before being released into the drainage system. The most common residential fixtures are water closets, bidets, bathtubs, shower baths, kitchen sinks and laundry trays.

By code, plumbing fixtures must be made of high quality materials. They must be free of defects and concealed fouling surfaces. Fixtures must have surfaces that are smooth and nonabsorbent and must be designed so that all surfaces are available for cleaning.

Fixtures designed to meet U.S. standards are commonly made of enameled cast iron, enameled pressed steel, vitreous china, or stainless steel. Fixtures constructed of pervious materials (such as Roman baths or tile showers) must not have waste outlets that can retain water.

Plumbing fixtures should be located in adequately lighted and ventilated rooms. If natural ventilation from a window is not available, a fan and duct are required. The lack of adequate lighting or ventilation promotes unsanitary conditions. Most codes prohibit locating fixtures where there isn't enough light and ventilation.

Minimum Fixture Requirements

The type and number of fixtures required by most codes for residences depends on the type of occupancy and the number of people expected to use the toilet facilities. Codes vary considerably (especially for commercial use) in the number of fixtures needed

■ Residential Fixtures

For residential buildings, the following fixture requirements should be adequate. Refer to your local code for exact requirements.

Single-family residences: The minimums are one kitchen sink, one water closet, one lavatory, and a bathtub or shower unit. Provision must also be made for a clothes washing machine. Hot water is optional in some codes and is required by others.

Duplex residential units: The minimum requirements are: one kitchen sink, one water closet, one lavatory, and one bathtub or shower

Males				Females		
Users	Water Closets	Urinals	Lavatories	Users	Water Closets	Lavatories
1–15	1	1	1	1–15	1	1
16–35	2	1	1	16–35	3	1
36–55	3	2	2	36–60	4	2

Figure 19–1

Fixtures for employees of office or public buildings (Uniform Plumbing Code)

unit for each dwelling unit. Provision must also be made for a clothes washing machine for each unit. One washing machine is enough for two units if it's available to all residents. Hot water is optional in some codes and mandatory in others.

■ Light Commercial Fixtures

The number of toilet facilities for light commercial buildings is based on the assumed number of employees. Separate facilities may be required for males and females. The code sets guidelines for the ratio of fixtures in male toilets to fixtures in female toilets. See Figure 19–1. The ratio and types of fixtures required for males and females may be changed by the plumbing plans examiner. The examiner in your area will consider altering the requirements in Figure 19–1 (or your code) if you can provide data showing that some other fixture ratio is more appropriate.

Consider an example: Assume that toilet facilities are needed for an office building employing 30 persons. Some building codes require a ratio of 50 percent male and 50 percent female facilities. Obviously, if these ratios were used rigidly, an imbalance of toilet fixtures would result in many cases. In the example, the plumbing plans examiner could request a notarized letter from the owner giving the probable maximum number of male and female employees in the office building. If the letter stated that 25 females and 5 males would be employed there, then the correct number and type of plumbing fixtures could be determined from Figure 19–1 (or a similar

table in your code). More complete tables may be found in your local code.

Here are other essential requirements for light commercial buildings:

A *drinking fountain* must be provided for up to 150 persons. The fountain must be accessibly located within 50 feet of all operational processes. Drinking fountains must not be located in any restroom or vestibule to a restroom.

Wash-up sinks may be substituted for lavatories where the type of employment warrants their use.

Manufacturing plants that may subject their employees to excessive heat, infection, or irritating materials must provide a *lavatory* for every 5 persons.

Small office buildings or similar establishments that employ 100 persons or more must provide a *service sink*.

Where more than one person can use the toilet facilities at a time, water closets must be separated from the rest of the room and from each other by *stalls* made of some impervious material.

Toilet rooms connected to public rooms or passageways must have a *vestibule* or must otherwise be screened or arranged to ensure decency and privacy. This vestibule must not be common to the toilet rooms of both sexes.

Toilet bowls must be of the elongated type having seats with open fronts.

Establishments employing 9 persons or fewer that do not cater to the public (such as storage warehouses and light manufacturing buildings) have less rigid requirements. Here, some codes require only one water closet and one lavatory

Males				Females		
Users	Water Closets	Urinals	Lavatories	Users	Water Closets	Lavatories
1-50	1	1	1	1-50	1	1
51–150	2	1	1	51–150	2	1
151–300	3	2	3	151–300	4	3

Figure 19–2
Fixtures for employees of office or public buildings (Uniform Plumbing Code)

for both sexes. But consider the following code conditions for such establishments:

- If the minority sex exceeds 3 persons, separate toilet facilities are required. For example, where 4 males and 5 females (or vice versa) are employed, separate toilet facilities must be provided.

- If the number of males employed exceeds 5, a urinal must be provided

Fixtures for Retail Stores

Establishments frequented by the public must provide toilet facilities for the number of employees and the public reasonably anticipated. Retail stores of 1,500 square feet or less (allowing for displays and storage) are usually required to have one water closet and one lavatory to adequately serve both employees and the public. See your local code.

Fast Food and Small Restaurants

Fast food restaurants that offer seating for the public must have toilet facilities available for use by the public. The number of fixtures depends on the maximum number of people that can be served at one time. The ratio of fixtures observed by most codes is 50 percent male and 50 percent female.

Use Figure 19–2 or a similar table in your code to determine the minimum toilet facilities required where food, drink and/or alcoholic beverages are served and consumed on the premises.

More complete tables are in *Plumbers Handbook Revised*. Your code will also have more complete information.

Here are other essential fixture requirements for fast food and small restaurants:

- Establishments serving drive-in customers must provide toilet facilities at the ratio of one person for each parking space. (Refer to Figure 19–2 or the similar table in your code.) For example, a drive-in restaurant with 100 parking spaces would need to provide toilet facilities for 50 males and 50 females, plus employees.

- Public food service establishments that offer only take-out service need not provide toilet facilities for customers. Toilet facilities are required here only for employees.

- The floors and walls of *public* toilet rooms in fast food and small restaurants must be covered with tile or other impervious materials to a height of 5 feet. These toilet rooms must have convenient access for both patrons and employees. The restrooms must be located within a 50 foot line of travel from the nearest exit to the dining room or food service area. Toilet rooms must be located on the same floor as the area they serve.

- A *dishwashing machine* or suitable *three-compartment sink* must be installed in food or drink establishments where dishes, glasses, or cutlery are to be reused.

- In establishments where food or drink are prepared or served, a *hand sink* must be installed for employees' use. Lavatories in adjoining toilet rooms may not serve this purpose.

- Water closets must be separated from the rest of the room and from each other by *stalls* made of an impervious material. A

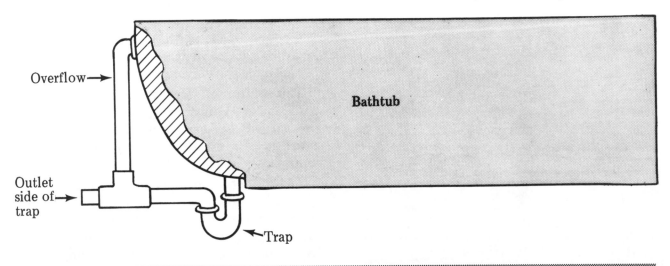

Figure 19–3
Prohibited overflow connection

privacy lock is not permitted on the entrance door to a public toilet room.

- Toilet rooms connected to public rooms or passageways must have a *vestibule* or must otherwise be screened or arranged to ensure decency and privacy. This vestibule must not be common to the toilet rooms of both sexes.

- *Toilet bowls* must be of the elongated type and must have seats with open fronts.

- All public toilet rooms must have adequate built-in provisions for the *handicapped.*

Plumbing Fixture Overflows

Bathtubs and lavatories are two of the most common fixtures provided with overflows. The code doesn't require overflows on lavatories, however, so many modern lavatories don't have any. Integral overflow passageways offer secondary protection against self-siphonage of the fixture trap seal. They also keep water from overflowing onto the floor if the fixture gets too full of water.

Here are a few points to remember when installing fixtures with an overflow. The waste pipe must be designed to prevent water from rising into the overflow when the stopper is closed. The waste pipe must also prevent water

from remaining in the overflow when the drain is open for emptying.

The overflow pipe or passageway from a fixture must be connected on the inlet side of the fixture trap. This prevents sewer gases and odors from entering the room through the overflow. In fact, the code prohibits connecting a fixture overflow to any other part of the drainage system. See Figure 19–3.

Fixtures must have durable strainers or stoppers. (An exception is made for fixtures with integral traps.) The strainer or stopper must not prevent rapid drainage of the fixture. The strainer should not be smaller than the fixture waste outlet it serves and (except for fixed strainers) should be easy to remove for cleaning.

▓ Bathtubs

Bathtubs come in many styles, designs, and colors. The enamel coating is generally acid-resistant and has an easy-to-clean glasslike surface. Bathtubs are manufactured in enameled pressed steel, enameled cast iron, or in gel-coated fiberglass.

Recessed tubs are built into the floor and walls at two ends and the back. Bathtubs that recess into tile or other finished wall materials must have waterproof joints. Such tubs are either right-hand or left-hand, depending on which way they fit into the recess.

Corner tubs are built into the floor and walls at one end and at the back.

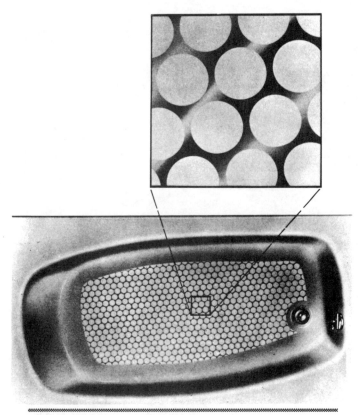

Figure 19–4
Slip-resistant surface for bathtub

Modern bathtubs have slip-resistant surfaces for safe tub bathing and a secure standing area for showering. See Figure 19–4.

The minimum size waste and overflow for bathtubs is 1½ inches. The code has approved several ways of arranging the tub waste and overflow. Figure 19–5 shows the three most common types.

Tip-toe waste and overflow (A in Figure 19–5). This is the most trouble-free of the three wastes shown. There are no internal moving parts to wear or break. Accumulated hair or other debris won't cause a stoppage.

Drain plugs for this waste are manufactured in two types. One is the *tip-toe* drain plug. To close the drain, press the stopper down with a toe. To release the water, press the stopper down a second time and the drain opens.

The second type of stopper (drain plug) operates like the basket strainer in a kitchen sink. You lift with your fingers and turn to open. Turn and let it drop to close. If these parts ever need replacing, unscrew the stopper with your fingers (the only tool necessary) and replace it with a new stopper of the same type.

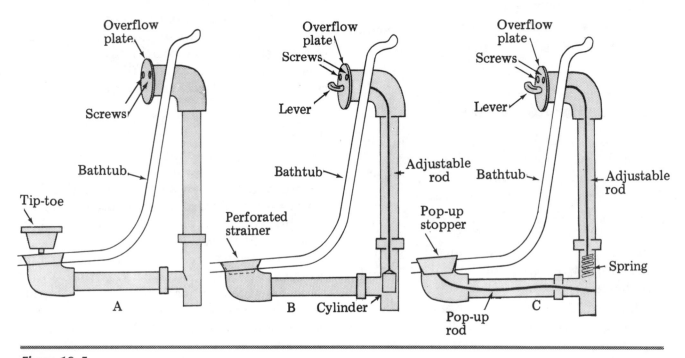

Figure 19–5
Bathtub waste and overflow fixture

Trip waste and overflow (B in Figure 19–5): This type has several disadvantages and some codes will no longer permit its use. The water stop is controlled by a heavy brass cylinder attached to a small adjustable rod. This rod connects to a lever on the overflow plate. The overflow plate is the large chrome disk near the top of the tub. The waste opening at the bottom of the tub is covered by a perforated strainer. When the lever is pressed to the "down" position, the rod lifts the cylinder and the waste water drains away. To close the waste opening and retain the water, pull the lever up, lowering the cylinder to a ground seat.

One disadvantage of the trip waste is that hair, lint, and other debris tend to accumulate on the cylinder surfaces. After a while, the cylinder won't seat properly to close the drain and won't lift completely, allowing free drainage. To clean the cylinder, remove the two screws in the overflow plate and pull it up and out. Clean the cylinder of foreign matter, cover it with petroleum jelly and replace as before.

Another disadvantage of the trip waste is that the rod must have threads and a small lock nut for adjustment. If the cylinder is adjusted downward too much, the waste water won't drain away freely. If the cylinder is adjusted upward too much, the tub won't retain water long enough for a leisurely bath. Very accurate adjustments are required for proper operation.

The adjusting rod is only ⅛ inch in diameter and tends to deteriorate over the years. If it breaks, the cylinder will drop to the bottom and partially block the opening. If there is no access panel, there are only two ways to retrieve this cylinder. The first is to break out the bathtub waste pipe. The second is to remove the overflow plate and fashion a hook at one end of a length of strong wire, such as a clothes hanger. Then "fish" for the cylinder. With patience, the cylinder can usually be removed.

Pop-up waste and overflow (C in Figure 19–5): This is the most popular waste and overflow now in use. The pop-up stopper and the trip waste operate the same way. Each has a lever attached to an adjustable rod, which is in turn attached to the overflow plate. Turn the lever to the left to close and to the right to open the drain. The difference in the pop-up

system is the coiled spring attached to the end of the adjusting rod. Pressing this spring down raises the stopper and opens the drain. Raising the spring closes the drain opening.

An advantage of the pop-up is that the adjustment on the rod does not have to be as accurate to give good performance. A disadvantage is its tendency to collect hair, lint, and other foreign matter. To clean the pop-up, raise the tub stopper to an open position and pull it out with your fingers. Foreign substances can then be removed and the stopper can be worked back to its original position. Then remove the two screws located on the overflow plate, pull the mechanism up and out, and clean the spring. Replace as before.

The spring and the raised back portion of the overflow plate will eventually wear out with use. This wear causes a loss of the tension necessary to keep the stopper or cylinder in an open or closed position. Your dealer may offer replaceable overflow plates that can be installed without disturbing the rest of the waste and overflow mechanism. As you read these directions for a pop-up waste and overflow installation, as shown in C in Figure 19–5, follow along with the illustrations in Figure 19–6:

1. Place slip nut and gasket on drain tube.

2. Screw the tailpiece up into the bottom of the drain tee. Use pipe dope on threads. Tighten securely.

3. Place the drain tube into the drain tee. Hold the drain against the bottom of the tub at the opening and tighten the slip nut hand tight.

4. This positions the waste assembly so you can connect it to the waste pipe opening, by slip nut, by direct connection to a 1½-inch waste pipe thread, or other method of your choice.

5. Place the slip nut and gasket on the overflow tube. The bevel on the gasket should face down. Slip the upper tube into the lower tube.

6. Center the upper overflow ell with the tub opening. Place the tapered gasket between the flange on the ell and the tub opening, narrow side down.

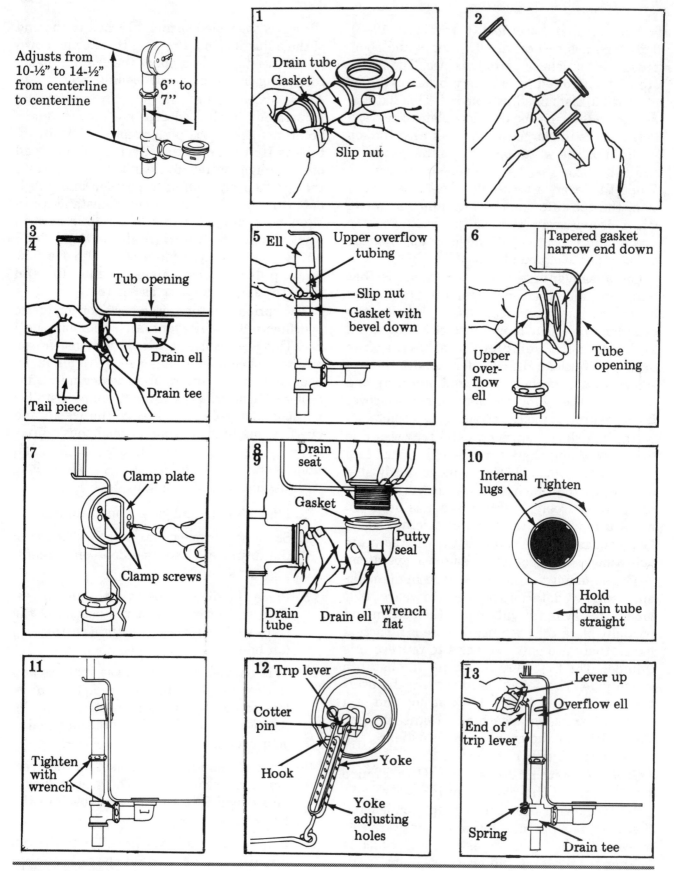

Figure 19–6

Pop-up waste overflow installation

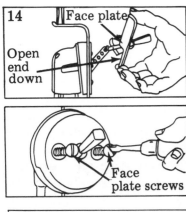

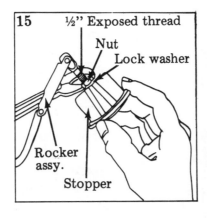

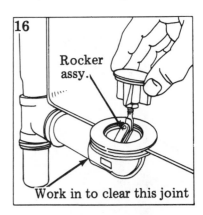

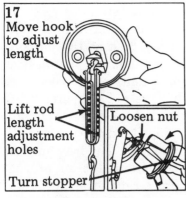

Figure 19-6 continued
Pop-up waste overflow installation

7. Center the clamp plate over the tub opening and thread the brass clamp screws through it into the overflow ell. Use the 4 o'clock and 10 o'clock positions. Tighten the screws, making sure all parts are aligned.

8. Next, install the waste seat. Thoroughly clean the tub around the drain opening. Apply a liberal amount of plumber's putty around the bottom of the flange on the drain seat.

9. Position the drain ell and tube under the tub, with the gasket on the top of the flange. Use the thinner gasket if the tub is cast iron. Use the thicker gasket if the tub is steel.

10. Thread the drain seat into the drain ell. Hold the ell on the wrench flats. Tighten the seat using a block of wood or a spud against the internal lugs provided for this purpose. Make sure the drain tube is straight.

11. Now tighten both slip nuts, the one on the overflow tubes and the one on the drain tube. Test all joints with water.

12. Remove the cotter pin in the top of the yoke in the lift assembly. Assemble the lift lever to the trip lever using a cotter pin.

13. Put the trip lever up and the lift assembly down. Then hold the lift assembly against the overflow tubes. Hold the end of the trip lever in the center of the overflow ell. The end of the spring should be in the center of the drain tube and tee. If not, adjust the lift length by moving the hook in the yoke adjusting holes.

14. With the open side of the face plate down, push the lift assembly down into the overflow tube. Thread the chrome-plated face plate screws into the overflow ell at 9 o'clock and 3 o'clock.

15. Remove the brass nut and star lock washer from the end of the rocker assembly. Replace these fittings, putting the nut on first and lock washer last. Leave about 14 inches of thread exposed on the end. Thread on the stopper and tighten the nut against it.

16. Thread the rocker assembly into the drain tube. You'll have to work it in to clear the tubing joint.

17. Adjust the lift rod length until the lever opens and closes freely. Then adjust the stopper screw. If the stopper raises too far and won't drop down, loosen the nut and screw the stopper down a couple of turns at a time until it drops freely.

18. Fill the tub. Then drain the tub and check all joints.

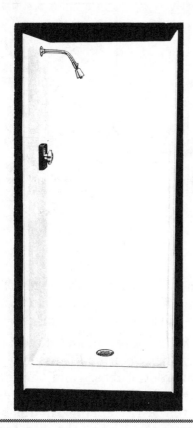

Figure 19–7
Fiberglass shower stall

Chain and stopper waste and overflow: This old-fashioned type of waste and overflow isn't shown in Figure 19–5. The stopper must be manually placed in the waste opening to retain water, and manually removed to release water. This type is used today only as a replacement in the older tubs.

Showers

Showers may be installed in the bathtub or in a separate stall.

Shower Baths

Shower baths are usually installed against a wall at one end of the bathtub. A diverter valve or diverter spout supplies water to the tub and can transfer the flow to the shower head. This type of shower may use two separate valves and a diverter spout, but the mixing valve with a diverter spout is more popular. You adjust the water temperature with the valve and then pull up the knob located on the tub spout to divert water to the shower head.

Shower Stalls

Shower stalls have many advantages. They take up less space and use less water than bathtubs. Manufactured stall showers are made from either fiberglass or porcelain-coated steel.

The porcelain-steel shower stall is a prefabricated unit with three sides and a base. The sides and base are assembled at the job site. These are usually very small units, just meeting minimum space requirements of the code. The sides are made of thin sheets of steel grooved so that they fit together and make a watertight joint. But the metal sides tend to rust and it's hard to keep the joints clean. The base may be made from heavier porcelain-steel or precast concrete or other approved material. A shower curtain or glass door closes off the open side.

Fiberglass shower stalls are usually made in one piece, including the base. These come in various sizes and colors and have replaced tile showers in many homes. See Figure 19–7. The fiberglass stall has three sides. The entrance may have a glass or plastic door or shower curtain. Fiberglass stalls have a smooth, watertight surface that's easy to keep clean.

Stall shower bases usually come with the shower drain attached to an opening in the base. Accuracy is critical when roughing in stall showers. They are usually installed after the floor is poured and the rough partitions are in place.

Tiled Showers

Tiled showers have tile or marble walls on three sides and a floor of similar material. These showers are found in most homes and can be designed and sized to fit any situation. Tiled showers can be constructed level with the existing bathroom floor or recessed to the most usable depth. Recessed tiled showers are called *sunken* showers. Tile showers may have shower curtains or, more likely, doors of plastic or safety glass. The waste opening may be installed in any location within the shower as long as it's low enough so that the shower floor slopes toward the drain from all angles. Roughing in the waste opening for a tiled shower is not as critical as for a stall shower.

Good waterproofing is essential in tile shower stalls. The walls of a tile shower must be waterproof, smooth, noncorrosive, and non-absorbent to 6 feet above the floor. There must be an impervious waterproof base under the tile on the shower floor. Any water on the tile surface should empty into the drain. Water left in puddles on the tile floor can seep through and may drain into the floor framing.

■ Other Shower Requirements

Waste outlets for shower compartments must be a minimum of 2 inches in diameter. The free area of the shower strainer must be a minimum of 3½ square inches. The strainer must be removable to allow easy cleaning of the shower trap. Shower traps must not be smaller than the waste outlet pipe used in the shower compartment.

Shower compartments need a minimum floor area of 1,024 square inches. This requires a minimum 32-inch span between walls, a space considered adequate for adult use. Floors of shower compartments must be smooth and sound.

Shower pans of lead or copper are required by some codes. Lead pans should not weigh less than 4 pounds per square foot. Copper pans must not weigh less than 12 ounces per square foot. Shower pans of lead or copper must be painted with asphaltum paint inside and outside to protect the pan from corrosion where it joins concrete or mortar.

The carpenter should install the wall framing and curb before you install the shower pan. If the building has wood floors, the carpenter must provide a solid base of subflooring for the shower pan. In buildings with concrete floors, a layer of 15-pound saturated asphalt felt or a ½-inch layer of sand is required under the pan. This protects the pan against rough surfaces that could puncture the membrane.

Some shower pan material is soft and flexible. Support it with backing secured to the partition studs. This should keep the pan sides from sagging until the interior of the shower compartment is in place to hold the pan rigid. If you have to cut the shower pan material, the penetration can't be lower than 1 inch from the top of the pan's turnup. Figure 19–8 shows an installed pan.

Cut the shower pan large enough so it laps at least 2 inches over on all sides when folded to fit the compartment. The lap should be at least 3 inches above the finished curb or 4 inches above the rough curb. Cut a hole where the drain is located and then put the shower pan in place. Securely fasten the shower pan to the shower strainer base at the invert of the weepholes.

At the drain opening, paint the top of the shower base with pipe joint compound. Paint the bottom side of the clamping ring with more pipe joint compound. Place the clamping ring on top of the shower base and screw them together. Clamp the pan material between the two flanges, making a watertight joint between the shower waste outlet stub and the pan. Screw the strainer portion of the shower drain down into the threaded flange body to the desired height of tiling (generally 1 inch above bottom of pan at this point).

See the shower pan installation detail in Figure 19–9. The pipe between the trap and shower drain can be threaded as illustrated, or it can be made of other code-approved drainage materials.

Each shower pan must be tested for inspection. Remove the strainer plate and plug the waste outlet. Fill the pan with water. The pan must be full and ready for inspection during the tub and water pipe inspection. Otherwise, the contractor you work for may have to pay for a reinspection. While shower pans are not required for prefabricated shower stalls, each stall requires approval by the plumbing inspector for watertightness.

You can omit shower pans in shower compartments built on a concrete slab on the ground floor, provided the bottom, sides, and curbs of the shower compartment are poured at the same time the floor slab is poured. A curb 1 inch higher than the existing slab must be poured around three sides of the shower compartment. This usually keeps the water level below the height of any surrounding

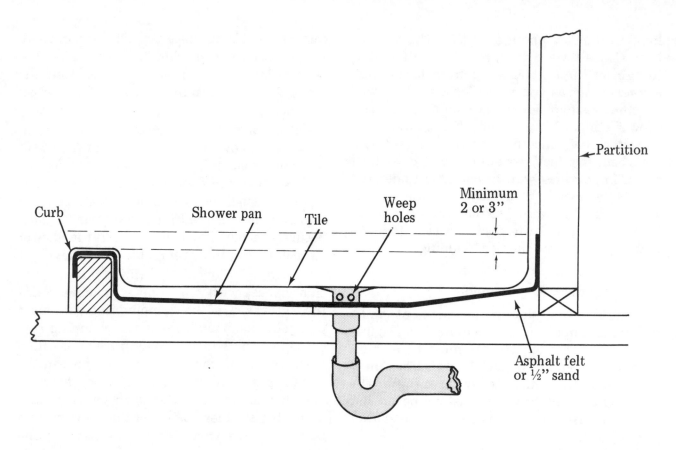

Figure 19–8
Installed shower pan

wood plates or studs and helps keep the compartment watertight. See Figure 19–10.

Shower Rods

Shower rods are generally installed 6½ feet above the finished floor. The rod should be centered over the bathtub rim or shower entrance curb so that the shower curtain drains splashed water back into the fixture. Cut the shower rod with a hacksaw to fit the opening. The end flanges which support the rod and shower curtain must be held securely in place by the screws supplied with the flanges. The screws must pass through the finished wall and penetrate the backing material installed during the rough partition work.

Shower Enclosures

In many parts of the country, shower enclosures are now installed by plumbers. They aren't difficult to install. The only tools needed are a hacksaw, a screwdriver, a level, and a masonry hand drill or a ¼ horsepower electric drill with a masonry bit.

The enclosure comes in a kit which generally includes a can or tube of sealing compound with instructions on how to use it. The kit should contain screws and shields for fastening the uprights. Enclosures are manufactured with either plastic or safety glass doors

Enclosures are designed for installation either on the rim of a bathtub or on the curb of a shower stall. Since most bathtubs are 5 feet long, no measurement of dimensions between walls is necessary. For shower enclosures, measure from finished wall to finished wall along the top of the curb.

Manufacturers are usually generous with the length of the bottom and top track or rail they supply. This rail must be cut with a hacksaw at least ¼ inch shorter than the overall

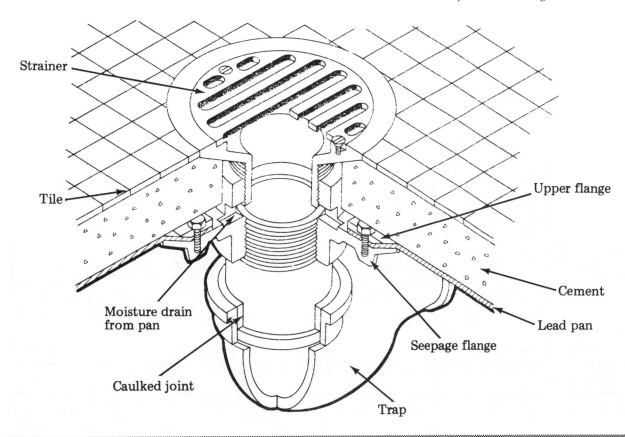

Strainer

Tile

Moisture drain
from pan

Caulked joint

Upper flange

Cement

Lead pan

Seepage flange

Trap

Figure 19–9
Shower pan installation

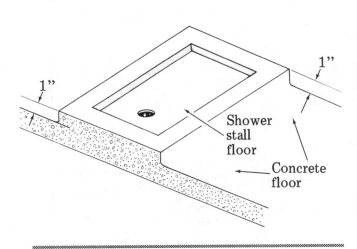

1"

1"

Shower
stall
floor

Concrete
floor

Figure 19–10
A monolithic poured shower stall

length between the end walls. This is necessary so that the uprights (which are a standard height) fit snugly between the bottom track and the vertical tile wall.

Install the bottom track first with the drain holes facing into the shower compartment. Apply a generous amount of sealing compound on the center line of the bathtub rim or shower stall curb the full length of the bottom track. Then press the bottom track firmly down and against the rim or curb (Figure 19–11 A). Place the uprights against the end walls and plumb them straight with the level, as shown in Figure 19–11 B. Use a pencil to mark the tile through the preformed screw holes in the uprights.

Remove the uprights and prepare to drill the holes in the tile wall. The masonry bit should be the same size as the enclosed shields. Before drilling, punch a small hole in the glazed tile finish with a nail punch or large nail. This gives the bit a bite and keeps it from slipping.

Drill the holes (there are usually three in each upright) and insert the shields (Figure 19–11 C). Spread sealing compound on the wall side of the uprights and set the uprights in place. Secure them to the wall with screws. Drop the top rail into place as shown in Figure 19–11 D.

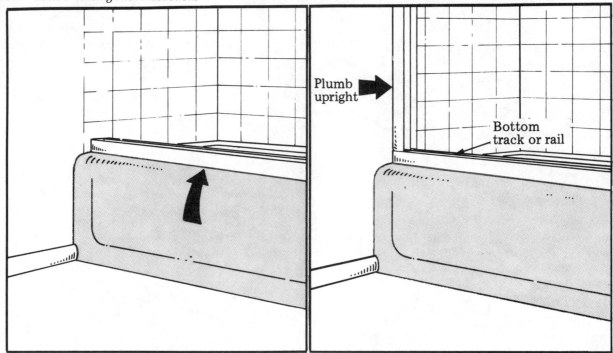

A Attach bottom track to tub with sealing compound

B Plumb uprights with level and mark holes for drilling

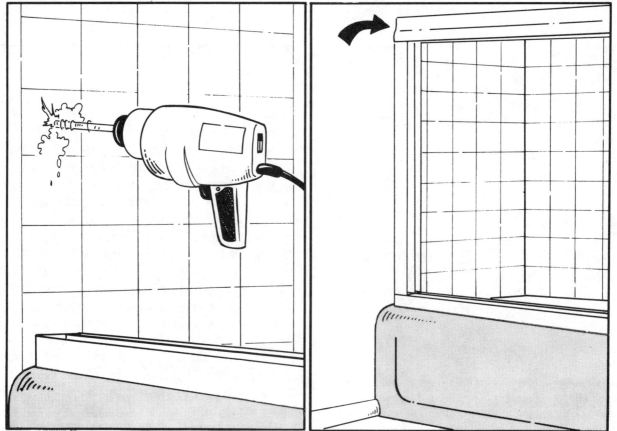

C Use masonry drill to make holes in the wall. Insert shields in holes and attach uprights to wall with screws. Seal uprights to tile.

D Drop top rail into place on uprights and hang doors

Figure 19–11

Installing a tub shower enclosure

When attaching the door rollers to the top rail, install the *smooth* side of the glass or plastic doors toward the *inside* of the bathtub or shower stall to permit easy cleaning.

Lavatories

Lavatories are commonly made of enameled cast iron, enameled pressed steel, vitreous china, stainless steel, or acrylics. Lavatories are manufactured in various designs, widths, and sizes. The bowl may be round, square, or oval and may hold from 1 to 2 gallons of water. Most lavatories are supplied with fittings for both hot and cold water. Wall-hung, pedestal type, and countertop lavatories are available. See Figure 19–12.

The countertop lavatory is the most popular of the three for new construction. It is designed for both above-counter and under-counter installations. Acrylic lavatories are available as either a basin alone or as a one-piece unit with counter.

The pedestal lavatory is supported by both a wall bracket and a pedestal base. Once commonly used in residences, this type has been almost entirely replaced by the countertop lavatory.

The wall-hung lavatory hangs from a steel or cast iron wall hanger secured to the wall with screws. It is used more often in older homes and in commercial buildings.

▓ Installing the Countertop Lavatory

Many types of clamp assemblies are available to secure lavatories to countertops. The type of fixture and construction of the cabinet will determine installation procedure. Regardless of the type you install, follow instructions provided by the manufacturer. Figure 19–13 shows two types of clamp assemblies. In the first (Figure 19–13 A), apply sealing compound around the edge of the hole. Drop the lavatory into the hold, install clips as shown, and tighten firmly. Then remove the excess sealing compound. For the second type (Figure 19–13 B), apply sealing compound to the underside of the lavatory rim. Place the lavatory over the cutout, make the final fitting connection and align the lavatory. Install the clamp assemblies and tighten the wing nuts. Add sealing compound around the base of the lavatory. In a 20-inch countertop, also add sealant on top of the back edge of the lavatory, forming a fillet with your wet finger or a sponge. Finally, carefully wipe excess compound from the countertop and lavatory rim with a damp cloth or sponge.

▓ Installing the Wall-Hung Lavatory

Wall-hung lavatories for residential and commercial use are generally supported by a hanger screwed securely to wooden backing material fastened to the bathroom partition studs. The mounting board, usually a 1 x 6, must be installed flush with the face of the studs during the rough carpentry work. The board should be installed above the center of the waste outlet for the lavatory and 31 inches from the finished floor to the center of the board. The standard height from the finished floor to the overflow rim of the lavatory is 31 inches. The extra width from the center of the mounting board to its edge will generally allow some leeway in case the lavatory back varies in height.

It's best to set the lavatory at standard height. When the lavatory is ready to be hung, make the necessary allowance for back height so the lavatory overflow rim will be 31 inches above the finished floor.

Place the hanger directly over the center of the waste pipe and use a pocket level to level the hanger. With a pencil, mark the holes on the backing through those provided in the hanger. (Use two at each end and one in the center.) If the wall is tile or other hard surface, use an electric drill with a masonry bit to drill through the finished wall to the mounting board. Use wood screws long enough (usually 2 to 2½ inches) to penetrate the mounting board at least 1 inch. See Figure 19–14.

Wall-hung lavatories need enough wall support so that no strain is transmitted to the fixture pipe connection or to the finished wall. There is little possibility that a lavatory so secured can pull away from the wall.

The weight of the countertop lavatory is transferred to the cabinet top and places no strain on the fixture piping.

Self-rimming lavatory

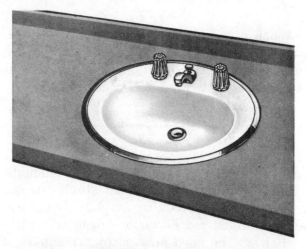

Stainless rim lavatory

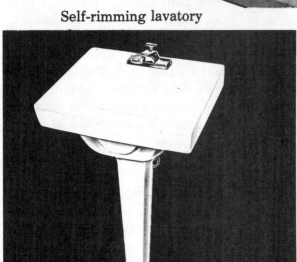

Pedestal lavatory

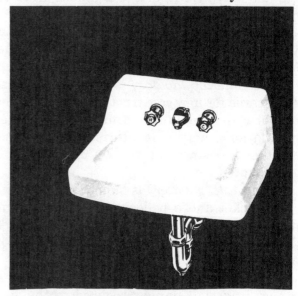

Wall hung lavatory

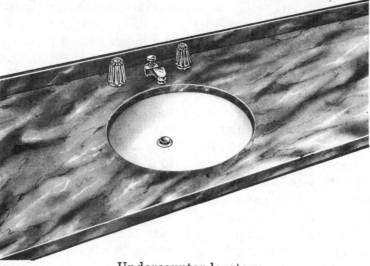

Undercounter lavatory

Figure 19–12
Kinds of lavatories

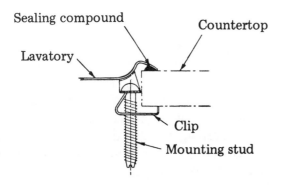

A Clamp assembly for thin material
(enameled pressed or stainless steel)

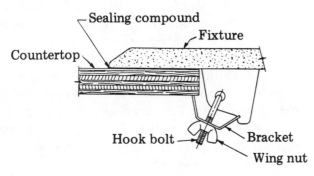

B Clamp assembly for thick material
(enameled cast iron or vitreous china)

Figure 19–13
Two types of clamp assemblies

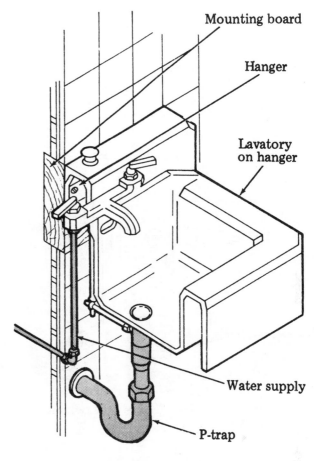

Figure 19–14
Wall-hung lavatory installation

▇ Lavatory Faucets

The type of faucet installed depends on the design of the lavatory. Many wall-hung lavatories in residences have factory-installed faucets, as shown in Figure 19–12. Wall-hung lavatories for commercial use are usually designed for separate hot and cold water faucets.

Lavatories in some commercial buildings have only a single cold water faucet. The opening for the hot water faucet is concealed by a faucet hole cover. A lavatory with separate hot and cold water is less desirable because the water can't be blended before it enters the basin.

Some lavatories in commercial buildings have self-closing faucets to save water. The faucet has a spring which snaps the handle closed when released, shutting off the supply of water. A disadvantage to this type of faucet is that one hand must remain on the handle to hold it open as long as flow is desired. And repeated quick closing of the valve can set up a water hammer in supply pipes.

The combination faucet with either a mixing valve or separate hot and cold water valves is the best choice for use in residences. The combination lavatory faucet usually allows either a 4-inch or an 8-inch distance between the centers of the handles. This has partially standardized the holes in lavatories. The combination faucet has a chamber for mixing hot and cold water before it comes out of the central spout. That makes it easy to adjust the water temperature.

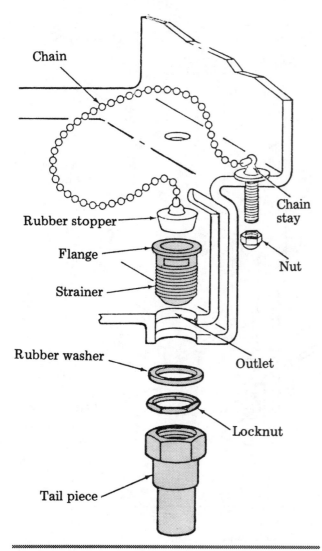

Figure 19–15
Detail of pull-out

■ Lavatory Drain Assemblies

There are two types of drain assemblies for lavatories.

1. The chain and rubber stopper pull-out (P.O.) plug type is generally used in commercial buildings. It consists of an open drain with a flange on the upper end and a threaded tube projecting from the lower end. The rubber stopper is inserted in the drain opening. Remove the stopper to release water down the drain. This type of stopper is inexpensive, usually trouble-free and can be used on lavatories that have single hot and cold water faucets. See Figure 19–15.

2. The pop-up drain stopper has a combination faucet and a lift rod connected to a pop-up rod in the waste. When the lift rod is pressed down, the pop-up stopper is pushed up, releasing the lavatory's contents to drain away. When the lift rod is pulled up, the pop-up stopper closes to seal the opening in the drain. See Figure 19–16. The pop-up drain assembly is more sanitary than the pull-out plug drain.

There are several styles of lavatory pop-up drain stoppers.

The *pop-up drain stopper* (Figure 19–17 A) can be lifted up and out with the fingers when it's open. The stopper isn't attached to any part of the drain assembly. Its weight rests on the pop-up rod and either rises with the rod or drops into place when the rod is lowered. This is the least likely pop-up stopper to cause problems because the concealed parts don't collect debris as easily. The rod that extends into the drain assembly on which the stopper rests may catch long hairs or strings. These can easily be removed with tweezers.

The *pop-up drain stopper* (Figure 19–17 B) is more difficult to remove as it has a slot in the side that locks directly to the pop-up rod. To remove it, push the lift rod down all the way. This raises the stopper. Twist to unlock it and lift the stopper up and out. This style of pop-up stopper is a round, hollow tube. Hair tends to wrap around the outside of the tube. This makes stopper operation difficult and slows the flow of waste water. When this happens, remove all hair and other foreign matter from the stopper and rod. Then replace the stopper.

The pop-out drain stopper (Figure 19–17 C) is the most difficult of the three to remove for replacing or cleaning. This style of pop-up stopper is designed with a hole on the opposite end through which the pop-up rod extends. Figure 19–16 shows more detail. Loosen the compression nut with pliers so the pop-up rod can be pulled toward the rear of the lavatory bowl, freeing it from the stopper. The stopper can now be removed for cleaning. Reassemble as before, tighten the compression nut, and run water through the lavatory waste. Check the compression nut for leaks.

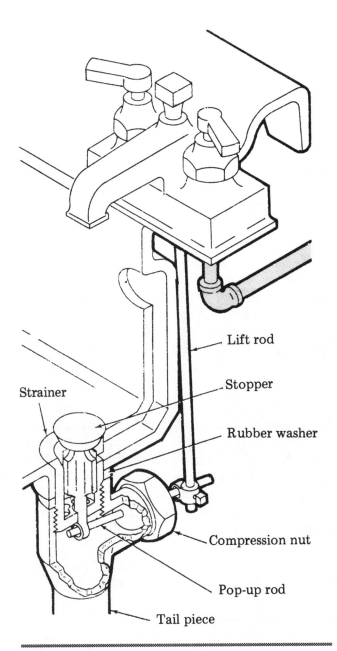

Figure 19–16
Detail of pop-up drain

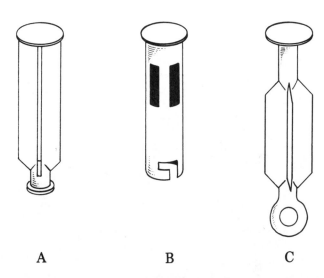

Figure 19–17
Lavatory pop-ups (drain stoppers)

▦ Installing Lavatory Trim

It's easier to fit faucets and waste assemblies on a lavatory before the lavatory is installed. Protect the lavatory from scratches by resting it on rags or cardboard during faucet installation. Place a ring of putty on the undersides of single or combination faucets before securing them in place. This keeps water from seeping under the faucet body through the preformed holes in the lavatory and leaking onto the cabinet below.

Also, put a ring of putty under the drain assembly flange before it's secured in place.

A mack washer seals the underside of the drain assembly to the bottom of the lavatory. The mack washer is held in place by a metal washer and lock nut. Screw the rest of the drain assembly to the threaded tube projecting below the lavatory body. The rubber stopper and chain may be secured to a chain stay fastened to the lavatory. See Figure 19–15. Secure the lift rod to the pop-up rod and adjust it so the pop-up stopper opens and closes completely. See Figure 19–16.

Then place the wall-hung lavatory on the hanger and push it down into place. Countertop lavatories should be fastened securely to the countertop with special rim clips. Make the joint watertight with a caulking compound or the recommended adhesive. Seal the joint between a wall-hung lavatory and the finished wall with white cement or other suitable material.

Once mounted, it's time to connect the lavatory with the water supply piping. The lavatory supply tube probably has a compression ell or stop to connect the water supply rough-in to the faucet. Lavatory supply tube is soft and flexible and can be bent to any offset by hand and then cut to the desired length. The top of the supply tube is secured to the faucet by a coupling nut supplied with the faucet. The bottom of the tube

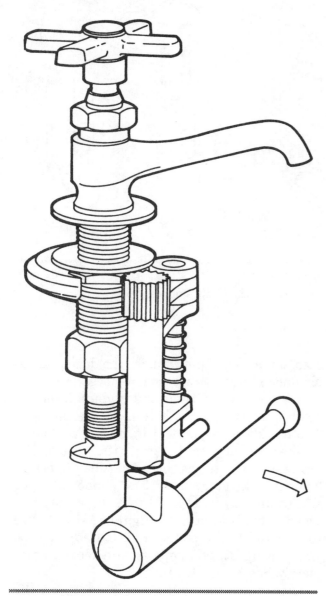

Figure 19–18
Basin wrench

is usually secured to the ell or stop by a compression nut and ring (shown in Chapter 16, Figure 16–28). The coupling nut is tightened by a basin wrench, which is used to reach less accessible places. See Figure 19–18. The compression nut may be tightened into place with a 6- or 8-inch adjustable smooth-jawed wrench.

Waste outlets for lavatories must be a minimum of 1¼ inches outside diameter. A 1¼-inch P-trap installed on the fixture drain extending through the finished wall completes the installation. The P-trap and the waste line must be of a similar material.

Water Closets

Credit for inventing the first water closet goes to an Englishman, Joseph Bramah. The valve closet he developed in 1788 worked with water pressure supplied by a pump. About 1830 the pan closet was introduced, followed by the long hopper water closet around 1850. The more popular plunger type water closet came into use about 1870.

By 1890 the washdown water closet, which originated in the United States, was introduced. Its design included the siphon action, greater water coverage of interior bowl surfaces, and complete scouring of all interior bowl surfaces with each flushing. The washdown water closet made obsolete the other types of water closets.

The modern water closet first came into use during the early part of the twentieth century. Wall-hung water closets first appeared about 1905. Around 1915 the water tank came down from the wall with the introduction of the low-down flush tank combination water closet. By the early 1920s, the reverse-trap and siphon jet water closets were introduced. These offered quieter flushing action. In the 1930s, the one-piece water closet appeared. Improvements in water closet design will continue, making them quieter, more sanitary, more reliable and easier to repair.

▓ Types of Water Closets

Water closets carry body waste to the drainage system. They're made from vitreous china and are available in various designs and colors. Common designs include the common washdown bowl, the reverse trap bowl, the siphon jet bowl, and the siphon action bowl. Figure 19–19 illustrates each of these. A separate trap is not needed, since all of them have a built-in trap to provide a seal. Although all of these water closets are installed in the same way, they differ in flushing action.

Washdown bowl: This is the least expensive and simplest type of water closet. The trap is at the front of the bowl and the bowl is flushed by small streams of water running down from

Standard

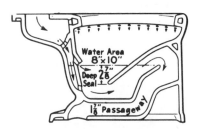

A *Washdown* action closets are noisy.

Quiet

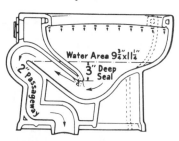

B *Reverse Trap* closets are quieter than the wash-down models.

Quieter

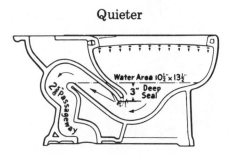

C *Siphon Jet* closets are the next in quietness.

Quietest

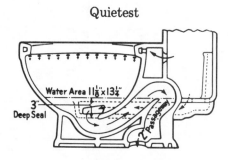

D *Siphon Action* closets are the quietest toilets.

Figure 19–19
Types of water closets

the rim. Some codes now prohibit use of the washdown bowl water closets.

Reverse trap bowl: This type is similar to the washdown bowl except that the trap is at the rear of the bowl, making the bowl longer. This bowl holds more water than the washdown bowl and operates more quietly.

Siphon jet bowl: This looks like the reverse trap bowl but flushes differently. The unit has a small hole in the bottom that delivers a direct jet of water into the trap. This, combined with more water from the flushing rim, starts a siphoning action with each flush.

Siphon action closet: This is the most efficient, the quietest, and the most expensive of the water closets. It looks like the siphon jet bowl but holds almost a full bowl of water. More of the inside surface of the bowl is submerged when the water is at rest in the bowl

The first three water closet bowls in Figure 19–19 are *two-piece*. That means the tank is separate from the bowl. These are usually referred to as "closet combinations." The fourth one is of one-piece construction.

■ Bowl Installation

The type of piping material used in a drainage system determines the type of closet floor flange used to hook up the water closet. Where lead stubs are used to secure the fixture to the drainage system, a brass or hard lead flange must be soldered securely to the stub. In a copper drainage system where copper stubs are used to secure the floor flange, a brass flange should be soldered securely to the copper stub. In a cast iron drainage system, a cast iron flange with a lead and oakum joint secures the floor flange to the cast iron stub. In a plastic

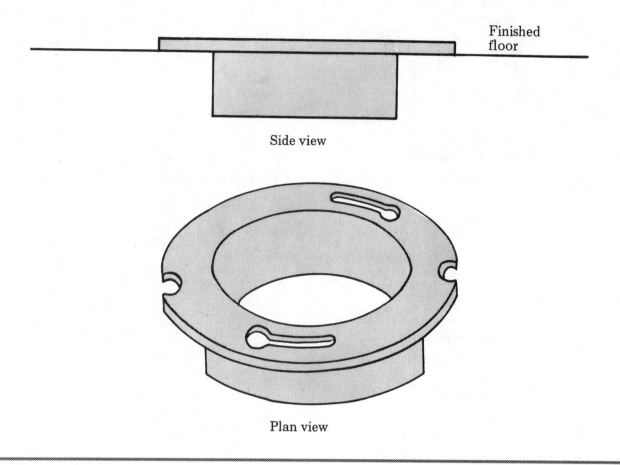

Finished
floor

Side view

Plan view

Figure 19–20
Water closet floor flange

drainage system, a plastic flange with a cement-welded connection secures the floor flange to the plastic stub.

Slip the water closet flange (Figure 19–20) over the stub and slide it down until the flange portion rests level with the finished floor.

Make up the joint just like any other joint of like material. Cast iron stubs that extend above the top of the closet flange must be broken off with a hammer and chisel. Copper, lead, and plastic stubs that extend above the top of the closet flange may be cut flush with the top of the flange with a hacksaw.

Put two brass closet bolts in the closet flange slots provided, threaded ends up. If the water closet bowl needs four bolts, as some do, place it properly on the flange and mark the spots on the floor with a pencil for the two additional (front) bolts. For wood floors, screw the closet screw into the floor, leaving the

machine threads to secure the front of the bowl to the floor with closet nuts and washers. For tile or concrete floors, drill holes into the floor where marked and insert the proper size and length lead shield into the hole. Tighten the closet screw on the lead shield as described above for wood floors.

Turn the closet bowl upside down, carefully setting it on wood strips or cardboard to avoid breaking or scratching it. Place a preformed wax setting seal over the discharge opening of the horn.

Turn the water closet bowl right side up and set it on the flange with the horn projecting down into the closet stub. Guide the two closet bolts up through the bolt holes on either side of the base of the water closet. Place one hand on each side of the closet rim, rock the bowl gently, and press it firmly down into the wax at the same time.

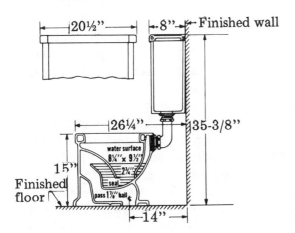

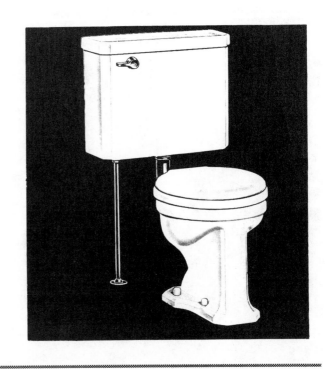

Figure 19–21
Wall-hung closet tank

Place the closet washers and nuts on the closet bolts and tighten them. Lay a 2-foot level alternately across the width and then the length of the bowl. The bowl must be level in each direction. Alternate this tightening procedure from one closet nut to the other until the closet bowl is firmly set on the water closet flange. The wax setting seal makes the joint gasproof and waterproof. Don't overtighten the bolts as this could crack the base of the closet bowl.

When you're satisfied that the bowl is securely in place, grout the bowl with white cement or another suitable material at the floor line. This prevents accumulation of odor-causing materials and keeps insects away from these ideal breeding areas.

Water Closet Tanks

Most residential water closets have closet tanks. A minimal amount of water flushes the water closet quietly and effectively.

In commercial and other public buildings, where the toilets have to serve more people, water closets with flush valves are preferred over water closets with tanks. This is because

a flush valve is ready for use again immediately. There is no wait for a tank to refill. That's an important consideration in high-occupancy buildings.

Closet tanks and bowls are manufactured of vitreous china. Specify the closet tank you need by the height at which it is to be installed above the water closet bowl.

There are two types of low closet tanks; the *wall-hung type* (Figure 19–21) and the *close coupled type* (Figure 19–22). The wall-hung type is usually found in older homes. The close coupled type is used almost exclusively in new construction. Close coupled closets are easy to install and aren't subject to shifting or sagging on the wall. The tank is supported entirely by the water closet bowl, to which it is attached by two tank bolts.

Water closets with tanks designed to use ballcocks should refill after each flushing and then close tight when the tank is full. The ballcock must have a refill tube reaching and turning down into the overflow tube. Water from this tube automatically restores the closet bowl water seal. An anti-siphon valve must be built into the unit to prevent contamination of

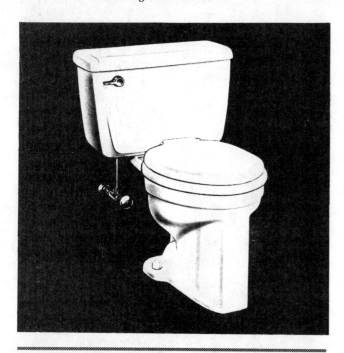

Figure 19–22
Close coupled tank closet

the water supply system. The flush valve is operated manually but the flushing operation must be automatic after manual activation. Each tank must have an overflow tube adequate to prevent tank overflow and remove excess water at the rate it enters the tank. (Figures 21–26 and 21–27 show more detail.)

Consider what would happen if the flush ball is securely in place on the flush valve seat and the ballcock should become locked in an open position. The flush valve seat must be a minimum of 1 inch above the rim of the bowl.

The flushing device and the connection between the tank and the bowl should have enough flow capacity to allow the water to flush all surfaces of the bowl.

Flushometer Valves

The diaphragm-type flushing valve (Figure 19–23), also called a flushometer, is a compact and efficient device for delivering water under pressure directly into the water closet bowl. The flushing action is quick and automatic and the amount of water delivered can be adjusted. This valve has an automatic device which shuts off the water after approximately 10 seconds.

Flushing valves are the best choice for toilet rooms serving large numbers of people.

The flushometer valve must have a vacuum breaker a minimum of 6 inches above the rim of the bowl. This prevents back-siphonage of the bowl contents into the water system if the water pressure should drop while there is a stoppage in the water closet bowl. The diaphragm flush valve has two chambers separated by a relief valve mounted on a rubber diaphragm. The upper chamber is directly connected to the main water supply by a small bypass. The lower (or flushing) chamber is connected to the 1-inch water supply line. When pressure is applied to the handle, water is released on the inlet side of the relief valve. The unequal pressure which results lifts the diaphragm and lets water flow into the water closet bowl. Within about 10 seconds the water forces itself around the bypass and equalizes pressure on the two sides, forcing the diaphragm down on its seat and shutting off the flow of water. Only one water closet can be served by a single flushometer. Flushometer-type valves are also used to supply water to urinals.

Tank Installation

Closet tanks are installed after the closet bowl has been set in place. Methods of installing the two types of low tanks differ greatly. Both tanks arrive from the manufacturer with all interior parts assembled.

Close Coupled Tanks

1. Place the tank gasket over the portion of the flush valve body that extends through to the underside of the tank. Coat the outside of the gasket with pipe compound.

2. Place the tank carefully on the water closet bowl. The tank gasket will fit the opening of the inlet to the closet bowl.

3. Place rubber sealing washers on the tank bolts. Insert the tank bolts from inside the tank through the preformed holes in the closet bowl. Install the washers and nuts and tighten them until the closet tank is drawn down firmly on the gasket against the closet bowl. Use a level to be sure the nuts have tightened the tank into a horizontal position. Don't overtighten the

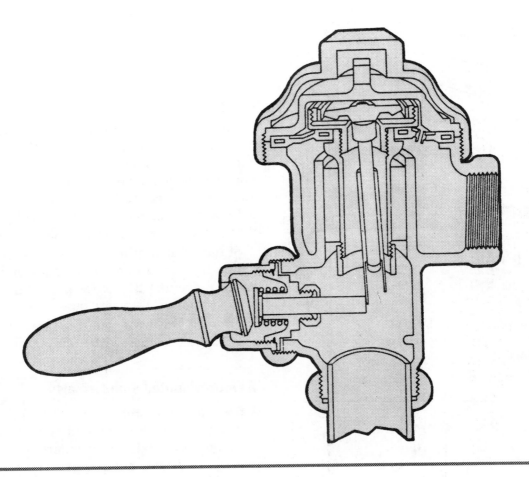

Figure 19–23
Diaphragm-type flushing valve

nuts, as this can crack either the closet tank or the closet bowl.

4. Connect the water supply tube to the inlet side of the ballcock valve that extends through the underside of the tank. The water supply tube is assembled and connected as described previously for lavatories.

5. Fill the tank with water and adjust the float rod for the correct water level. Flush the water closet and check all connections for leaks.

Wall-Hung Closet Tanks

1. A mounting board for this type of tank should have been installed during rough carpentry work. The mounting board is usually cut from a 2 x 4 stud and should finish flush with the face of the partition studs. When the closet tank is ready to be installed, screw two wood screws approximately 1½ inches long through preformed openings at the rear of the tank. The screws anchor the tank firmly to the mounting board. This places the weight of the tank and its contents on the mounting board and not on the closet elbow (flush ell), which could cause these joints to leak.

2. Place one end of the closet elbow (flush ell) into the inlet opening on the back of the water closet bowl and the other end into the opening in the closet tank. Coat both slip joint washers with pipe compound and tighten both slip joint nuts until they form watertight joints.

3. Connect the water supply tube to the inlet side of the ballcock valve, as described above for the close coupled tank.

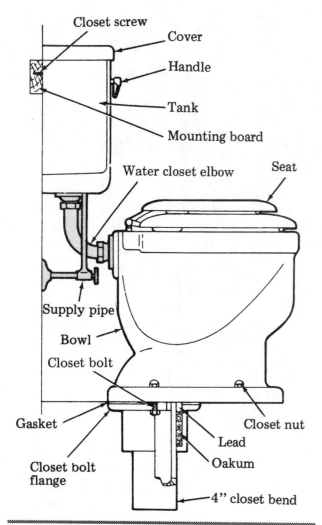

Figure 19–24
Detail of water closet and wall-hung tank in place

4. Fill the tank with water and adjust the float rod for the correct water level. Flush the water closet and check for leaks.

See Figure 19–24 for a detailed view of a wall-hung closet tank and bowl installed in place.

Seats for water closets must be constructed of smooth, nonabsorbent materials and must fit the water closet bowl. Don't install a round front seat on an elongated bowl.

Urinals

The two general designs for urinals that meet most code standards today are the wall-hung type and the floor-mounted stall type. They're constructed of vitreous china and operate with flushometer valves.

▨ Wall-Hung Urinals

A wall-hung urinal must be rigidly supported by a concealed metal carrier or other approved backing so that no strain is transmitted to the pipe connection. The joint between the urinal and the finished wall surface must be grouted with white cement or other suitable material that can provide a watertight seal.

Standard mounting requires that the waste piping be a minimum of 2 inches in diameter and the opening 21 inches from finished floor to the center of the waste pipe.

As with the water closet, most wall-hung urinals are designed with an integral trap.

▨ Floor-Mounted Stall Urinals

A floor-mounted stall urinal must be recessed slightly below the finished floor to provide drainage. Its weight is transmitted to the subflooring and does not require any other support. The back is recessed slightly into the finished wall and therefore must be grouted to provide a watertight joint.

The waste piping must be no less than 2 inches in diameter, and the opening should be 8½ inches from the finished wall to the center of the waste pipe.

A 2-inch trap must be installed in the waste line directly under the bottom of the urinal. The waste opening should be provided with a beehive-type strainer.

Flushometer-type valves are used with both types of urinals. These valves generally require a minimum ¾-inch water line to supply their water needs. (Some types may have only a ½-inch water supply opening, but this size isn't accepted by some codes.) The flushometer must complete the normal flushing cycle automatically after manual activation. It should deliver water at a rate that will flush all surfaces of the urinal. The valve must open fully and close tightly at normal water pressure. The flushometer must also have some means of regulating water flow. Only one urinal can be served by a single flushometer.

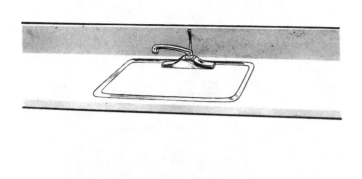

Figure 19–25
Laundry sink supported by countertop

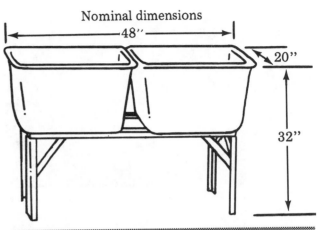

Nominal dimensions

Figure 19–26
Double laundry sink supported by cast iron stand

See Chapter 18 for the roughing-in dimensions of wall-hung and floor-mounted urinals.

Laundry Sinks

Laundry sinks or wash tubs are usually placed in the utility area or in a room near the clothes washing machine. This is both convenient and economical because clothes washers and laundry sinks can use the same supply and waste piping.

Laundry sinks may be made of concrete, enameled pressed steel, enameled cast iron, synthetic composition material or vitreous china. They may be mounted on a cabinet or supported by a metal frame. Both single and double-compartment styles are available. Figure 19–25 shows a countertop installation. Figure 19–26 illustrates two laundry sinks supported by a metal stand.

The waste pipe for laundry sinks or tubs must be a minimum of 1½ inches in diameter. Traps, tailpieces, and continuous waste pipe must be a minimum of 1½ inches outside diameter. Each compartment in a laundry sink should have a waste outlet with a suitable stopper for retaining water.

A double-compartment sink with two adjacent laundry sinks can use waste outlet piping connected by a continuous waste (Figure 19–26). This provides a single outlet and trap to connect to the drainage system. The faucet

is usually of the swing spout type and may or may not have threads for attaching a hose.

Kitchen Sinks

Kitchen sinks are usually made of enameled cast iron, enameled pressed steel, or stainless steel. Sinks are manufactured in various designs and lengths. There may be one, two, or three bowls. The most popular kitchen sink for new construction is the countertop sink, either double or triple bowl type. A sink may require a rim to secure it to the countertop, as shown in Figure 19–27. Another popular design is the self-rimming type, as shown in Figure 19–28.

The waste pipe for any sink must be a minimum of 1½ inches in diameter. Traps, tail pieces, and continuous waste pipe must be a minimum of 1½ inches outside diameter. A double-compartment sink (as in Figure 19–28) may have a continuous waste connecting the two bowls together. The continuous waste may be of the end outlet or the center outlet type with a single trap. See Figure 7–6 back in Chapter 7. The faucet may or may not have a spray for rinsing purposes.

Install kitchen sinks in a countertop the same way as described previously for installing countertop lavatories.

Domestic sinks need a waste opening at least 3½ inches in diameter. This is necessary to fit the standard size sink basket strainer as

Figure 19–27
Countertop sink secured by rim

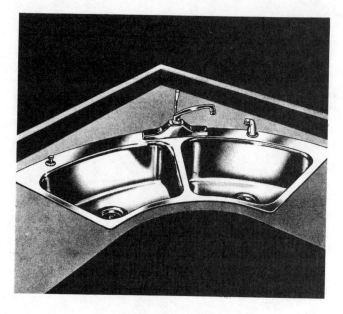

Nominal dimensions

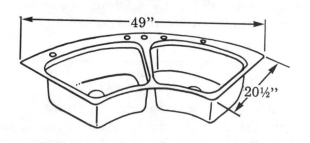

well as to provide an opening large enough for a garbage disposer.

Food Waste (Garbage) Disposers

Food waste disposers are installed in nearly all residences except where prohibited by the local authority. Contrary to what some may believe, food waste disposers won't clog up the drainage system if you follow the manufacturers' instructions. In fact, food waste disposers probably help keep drains flowing freely because they discharge waste under pressure, forcing waste down the line.

Some codes have special requirements for waste disposers installed in two-compartment sinks. The code may require that the disposer discharge waste through a separate trap and waste line. In that case, you'll have to install two traps and two waste lines, as shown in Figure 19–29. Some codes permit the disposer and second sink compartments to use one trap and one waste line. Where two traps and waste lines are required, use a hi-lo fitting 2 inches in diameter with two 1½-inch double vertical tappings not more than 6 inches apart. See Figure 19–30.

Food waste disposers may be installed on single compartment sinks using one trap and waste line. There are no special requirements.

Figure 19–28
Corner stainless steel double bowl self-rimming countertop sink

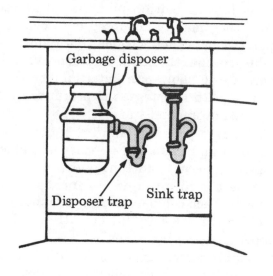

Figure 19–29
Food waste disposer installation

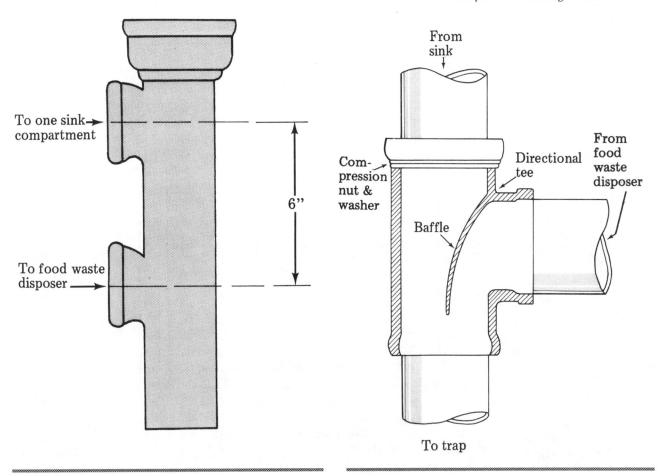

Figure 19–30
Hi-lo fitting for garbage disposer

Figure 19–31
Food waste disposer directional tee

A food waste disposer can be installed on a double compartment sink in an existing home if a second waste opening is not available. But a special directional tee must be used to channel the flushed garbage away from the other sink compartment. See the directional tee in Figure 19–31. A 1½-inch trap is required for a food waste disposer.

Under-Counter Dishwasher

Under-counter dishwashers can be installed with either side against the wall at the end section of a kitchen cabinet (right or left), but never directly under a sink. In new construction, the cabinet opening for a dishwasher is provided when kitchen cabinets are installed. For existing cabinets, the opening may be installed by following the instructions in Figure 19–32. The cabinet opening must be square, level and the proper size. The dishwasher must not be installed or secured with the frame distorted or twisted.

▓ Hot Water Supply Line

The hot water supply line to the inlet valve of the dishwasher may be of any approved water piping material. However, hard or soft copper with brass or copper fittings are generally preferred because they're easier to work with. Do not use rubber hose or connectors. Constant pressure and high temperature will eventually rupture this material. The drain hose supplied with some units must not be used as a water supply or coupler.

Follow the guidelines below when installing the hot water supply line:

- An accessible hand shut-off valve must be installed in the adjacent sink cabinet.

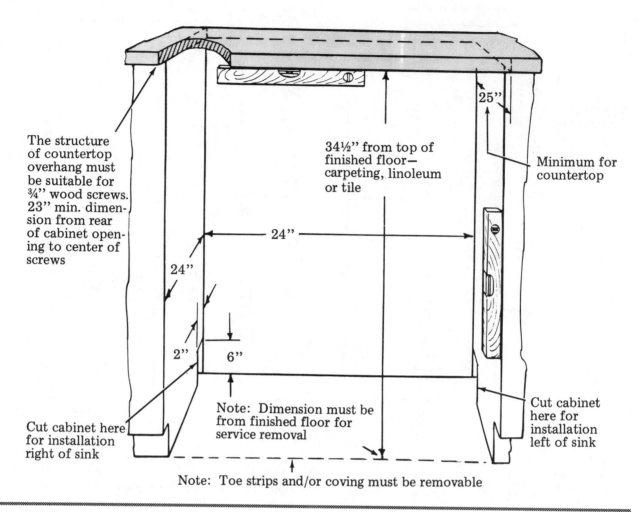

The structure of countertop overhang must be suitable for ¾" wood screws. 23" min. dimension from rear of cabinet opening to center of screws

34½" from top of finished floor—carpeting, linoleum or tile

25"

Minimum for countertop

24"

24"

2"   6"

Cut cabinet here for installation right of sink

Note: Dimension must be from finished floor for service removal

Cut cabinet here for installation left of sink

Note: Toe strips and/or coving must be removable

Figure 19–32
Cabinet opening

- The water supply line must not be smaller than ⅜ inch inside diameter.

- A compression fitting or union must be used to make the final connection of the inlet valve to the dishwasher. Never use a solder fitting; the heat may damage the inlet valve.

- For best results, the water heater should supply the dishwasher with water heated to approximately 150 degrees. (See Figure 19–33 for the hot water supply connection.)

▓ Dishwasher Installation

The drain hose from a dishwasher with a pump discharge must rise to a height equal to that of the underside of the dishwasher top. It is absolutely essential that you install the high loop fitting provided with each dishwasher drain hose. This prevents backups into the dishwasher if the sink should become clogged. Install the drain hose so that it's not kinked. Securely clamp the hose to the underside of the countertop.

Some plumbing codes require that an air gap fitting be installed in the drain line. When this is the case, mount the air gap assembly on the sink or countertop as shown in Figure 19–34. The air gap assembly, when required by code, must be mounted above the sink overflow rim. Drill one 1¼-inch diameter hole in the sink or countertop. Secure the drain gap assembly in this hole with the collar and nut at the proper height so that the cover can snap into position over the assembly.

Rough-In Fill Line

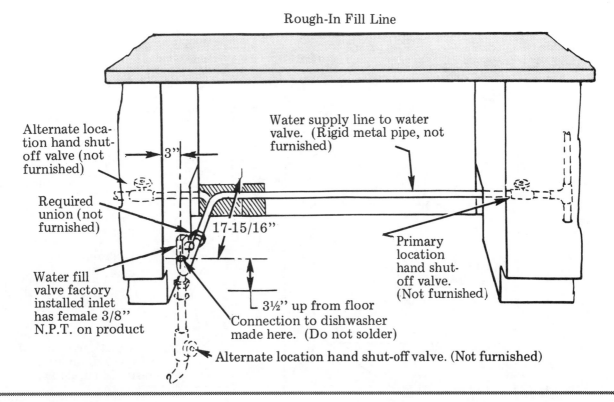

Alternate location hand shut-off valve (not furnished)

Water supply line to water valve. (Rigid metal pipe, not furnished)

3"

Required union (not furnished)

17-15/16"

Water fill valve factory installed inlet has female 3/8" N.P.T. on product

Primary location hand shut-off valve. (Not furnished)

3½" up from floor

Connection to dishwasher made here. (Do not solder)

Alternate location hand shut-off valve. (Not furnished)

Figure 19–33
Hot water supply connection

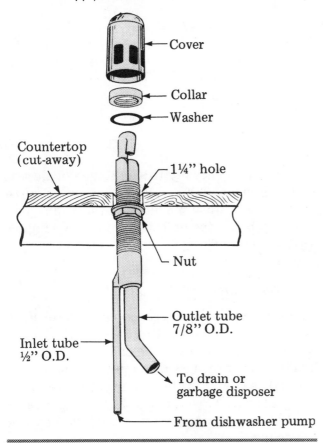

Cover

Collar

Washer

Countertop (cut-away)

1¼" hole

Nut

Outlet tube 7/8" O.D.

Inlet tube ½" O.D.

To drain or garbage disposer

From dishwasher pump

Figure 19–34
Air gap assembly

Connect the dishwasher pump and the ½-inch outside diameter inlet tube of the air gap assembly with the hose and clamp that comes with the dishwasher. Install the drain air gap assembly as close to the dishwasher as possible so the dishwasher hose can be used. Try to keep the dishwasher within 5 feet of the sink waste connection. (Some codes allow a separation of more than 5 feet.)

If a food disposal unit is installed in a sink, the waste from the dishwasher must connect to the opening provided in the body of the food disposer. If there's no food disposal unit, waste from the dishwasher must connect to a fitting called a "dishwasher Y branch" as shown in Figures 19–35 and 19–36.

Water Heaters

There are two types of water heaters commonly used in single family or small dwelling units: the electric water heater and the gas-fired water heater. The electric heater is the more popular.

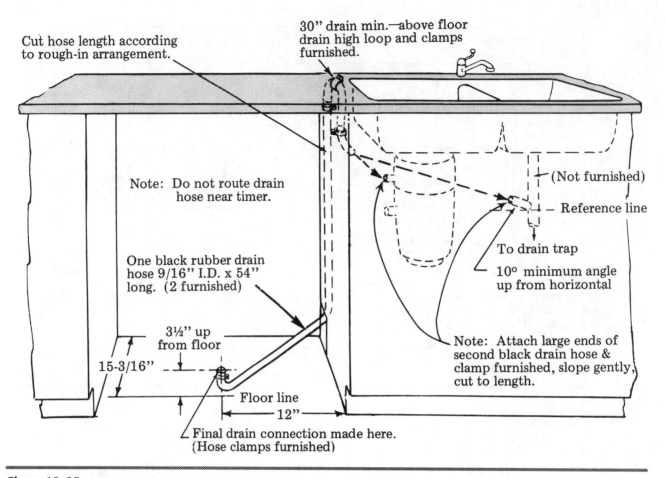

Cut hose length according to rough-in arrangement.

30" drain min.—above floor drain high loop and clamps furnished.

Note: Do not route drain hose near timer.

One black rubber drain hose 9/16" I.D. x 54" long. (2 furnished)

3½" up from floor

15-3/16"

Floor line

12"

Final drain connection made here. (Hose clamps furnished)

(Not furnished)

Reference line

To drain trap

10° minimum angle up from horizontal

Note: Attach large ends of second black drain hose & clamp furnished, slope gently, cut to length.

Figure 19–35
Drain hose connections

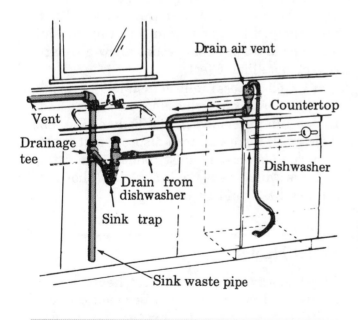

Drain air vent

Countertop

Vent

Drainage tee

Dishwasher

Drain from dishwasher

Sink trap

Sink waste pipe

Figure 19–36
Typical sink with dishwasher

There are generally three designs of electric water heaters: the upright model (Figure 19–37), the utility model (Figure 19–38), and the under-counter or table-top model (Figure 19–39).

■ Water Heater Installation

Select a level location accessible to water lines and the power supply. Don't locate the heater where water lines could be subject to freezing. To minimize heat loss through the pipes, try to install the heater in the center of the area with the greatest hot water use. Locate the heater so that access panels and drain valves are accessible.

Before starting the installation, close the main water supply valve, open a water faucet

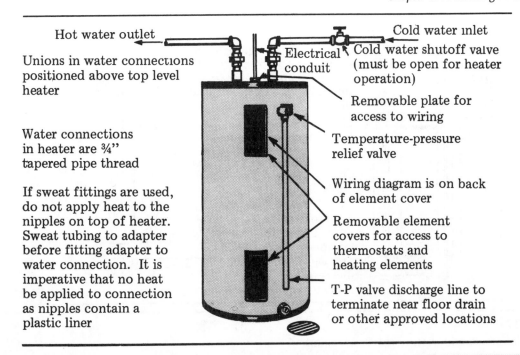

Figure 19–37
Typical upright electric water heater connections

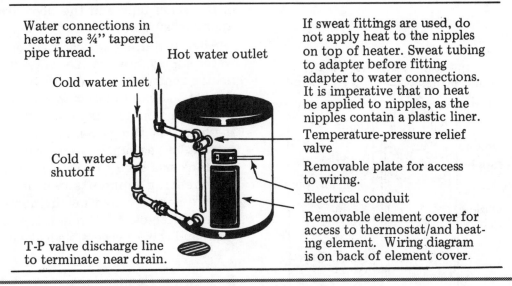

Figure 19–38
Typical utility electric water heater connections

to relieve the house pressure, and then close the faucet.

If your code requires a check valve in the cold water inlet line to the heater, a pressure relief valve must be installed between the check valve and the heater. However, installing a pressure relief valve in this location does not eliminate the requirement for a temperature-pressure relief valve in one of the locations shown in Figure 14-4 back in Chapter 14.

The relief pipe should terminate near the floor drain, where permitted by code, or other suitable location not subject to blocking or freezing. Don't thread, plug, cap, or use the

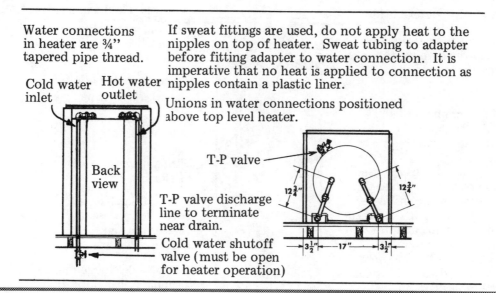

Water connections in heater are ¾" tapered pipe thread.

Cold water inlet

Hot water outlet

Back view

If sweat fittings are used, do not apply heat to the nipples on top of heater. Sweat tubing to adapter before fitting adapter to water connection. It is imperative that no heat is applied to connection as nipples contain a plastic liner.

Unions in water connections positioned above top level heater.

T-P valve

T-P valve discharge line to terminate near drain.

Cold water shutoff valve (must be open for heater operation)

12¾"   12¾"

3½"   17"   3½"

Figure 19–39
Typical table-top electric water heater connections

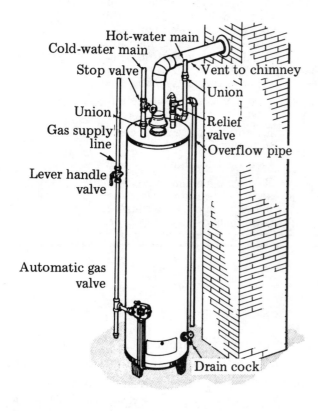

Hot-water main
Cold-water main
Stop valve
Union
Gas supply line
Lever handle valve
Automatic gas valve
Vent to chimney
Union
Relief valve
Overflow pipe
Drain cock

Figure 19–40
Typical gas water heater connection

relief pipe for any other purpose. Leave an air gap of approximately 6 inches between the end of the relief pipe and the floor drain or other approved surface.

After installing water lines, open the main water supply valve to the house and the water heater control valve and fill the heater. Open several hot water faucets to allow trapped air in the heater tank to escape from the system while the heater is filling. Don't turn on the electrical current (or gas supply) to the water heater until the tank has been completely filled with water. The heating elements can be destroyed if they're heated when not immersed in water. When water passes through the faucets, close them and check for possible leaks in the system.

■ Electrical Connections

Before any electrical connections are made, or before the gas is turned on and lighted, be certain that the heater is full of water and that the valve in the cold water supply line is open. An electrical contractor should make the final electrical connections. Otherwise, the warranty for the water heater may be voided. For a gas heater, a gas supplier must light the heater and make final adjustments.

See Figures 19–37, 19–38, 19–39 and 19–40 as well as Chapter 14 for further information on piping electric and gas water heaters.

Questions

1. Why have well-recognized plumbing fixture standards been developed over the years?

2. Plumbing fixtures must be free of what?

3. Plumbing fixtures must be manufactured of what type of material?

4. When fixtures are constructed of pervious materials, what specific condition is not permitted?

5. What potentially dangerous situation exists when bathrooms do not have adequate lighting or ventilation?

6. What are the minimum fixture requirements for a single-family residence?

7. What determines the minimum number and type of fixtures required by code?

8. Besides the usual kitchen sink, water closet, lavatory, and bathtub (or shower), what additional fixture is required in most living units?

9. What determines the number of toilet facilities in light commercial buildings?

10. Besides the regular toilet facilities, what additional plumbing fixtures must be provided for a small office building with 100 employees?

11. To ensure decency and privacy, what is required in a toilet room (that has one water closet, one urinal and one lavatory) which is connected to a public use area?

12. How must toilet bowls be designed when they are for public use?

13. What special plumbing fixture must be provided in a place employing more than five males?

14. What determines the minimum toilet facilities in small restaurants?

15. How are the minimum toilet facilities for fast food drive-in restaurants calculated?

16. What material must be used for floors and walls of public toilet rooms?

17. What type of seat is required for toilet bowls serving the public?

18. In what type of eating establishment must a dishwashing machine or suitable three-compartment sink be used?

19. What must be provided for employees' use in establishments where food is prepared and served to the public?

20. What two fixtures are generally provided with overflows?

21. What two purposes are served when plumbing fixtures are designed with overflows?

22. The overflow passageway from a fixture must be connected to what portion of the plumbing system?

23. What determines the size of a fixture strainer?

24. Name two materials of which modern bathtubs are made.

25. With what must bathtubs be provided when recessed into the finished walls?

26. What is the minimum bathtub waste and overflow size?

27. Name three of the common types of bathtub waste and overflow used today

28. What type of shower is found in most homes?

29. How is water from a tub transferred to the shower head when there is a diverter spout?

30. What are the main advantages of having a shower stall?

31. What are the two general types of shower stalls used in today's plumbing?

32. Why is accuracy very important in roughing-in the waste opening for a shower stall?

33. Give two advantages of a tiled shower over a stall shower.

34. What is the most important requirement for a shower?

35. What is the minimum size waste outlet for a shower compartment?

36. What is the minimum floor space required for any shower compartment?

37. What is the minimum weight per square foot required by most codes for lead shower pans?

38. What must be provided by the carpenter before a shower pan can be installed?

39. What are the design requirements for shower strainers?

40. What protection do lead or copper shower pans installed on concrete floors need?

41. In securing the shower pan, what is the maximum distance from the top of it that nails or screws may be used?

42. How high should the sides of a shower pan extend above the finished curb?

43. At what point in a building's construction should a shower pan be prepared for inspection?

44. When may shower pans be omitted?

45. Shower rods are generally installed how high above the finished floor?

46. Shower doors are generally manufactured from what materials?

47. What two common materials are used in manufacturing lavatories?

48. What two types of lavatories are most often installed?

49. What is the standard height from the finished floor to the overflow rim of a lavatory?

50. On a lavatory, what determines whether you'll use a *combination faucet* or one with *separate hot and cold water valves*?

51. What is the disadvantage of having a single hot and cold water faucet on a lavatory?

52. Why are self-closing faucets sometimes used on commercial lavatories?

53. What are the two types of drain assemblies used on lavatories?

54. Why should lavatory faucets and waste assemblies be installed before the lavatory is in place?

55. How should wall-hung lavatories be finished at the wall contact point?

56. Name three types of flushing action for today's water closets.

57. What part of a combination water closet is installed first?

58. What type of closet flange is required in a plastic drainage system?

59. What is used to make a tight joint between a closet bowl outlet and the building waste pipe opening?

60. Why should a level be used when installing a closet bowl and tank?

61. Why are tank-type water closets seldom used in public toilet rooms?

62. What are the two types of low closet tanks?

63. What does a refill tube accomplish?

64. What is the purpose of an overflow tube in a water closet tank?

65. Why are flushometers generally required in toilet rooms serving large numbers of people?

66. With which other commercial fixture is the flushometer type valve used?

67. What should be done after a water closet is completely installed?

68. Why is it necessary to adjust the float rod in a water closet tank?

69. Toilet seats must be made of what kind of material?

70. What are the two most common urinal designs?

71. What is the minimum size waste pipe for urinals?

72. Why do most wall-hung urinals not require a separate trap?

73. Why must stall urinals be recessed slightly below the finished floor?

74. At what location in the waste pipe must a trap be installed to serve a stall urinal?

75. Why must a flushometer for a urinal deliver water at a certain rate?

76. How may the two compartments of a laundry sink be connected so that one outlet and one trap may be used?

77. What size trap must be used on a kitchen sink waste pipe?

78. What feature of a sink faucet is optional?

79. What is the minimum size waste opening for a domestic sink?

80. How is the drainage system affected when a food waste disposer is used according to the manufacturer's instructions?

81. What do some codes require when a waste disposer is newly-installed in a two-compartment sink?

82. What fitting must be used when a waste disposer is installed on an existing two-compartment sink?

83. What size trap must be used to connect the waste pipe to a food disposer?

84. On which side of the kitchen sink may an undercounter dishwasher be installed?

85. Is a dishwasher served by both hot and cold water?

86. Why should a rubber hose not be used to connect the water supply to a dishwasher?

87. What water piping materials are generally preferred to supply water to a dishwasher?

88. Why should you *not* use a solder fitting on the inlet valve of a dishwasher?

89. Why is it essential to install the high loop fitting on the drain hose?

90. Where must the high loop fitting be installed?

91. When a plumbing code requires that an air gap fitting be used, where must it be installed?

92. By most code standards, at what maximum distance may a dishwasher be installed from the sink waste connection?

93. Where a food disposal unit is installed in a sink, to what must the waste from the dishwasher be connected?

94. What are the two most common types of water heaters installed in residential units?

95. Name three important criteria used in selecting the location for a water heater.

Maintenance of Plumbing Systems

The most common plumbing maintenance problem is clogged drains. In some cases, the cause may be a defective drainage system. But the most common cause of clogged drains is foreign objects in the drainage system.

Foreign matter which finds its way into the drainage system usually enters through the kitchen sink or the bathroom water closet. Grease, cooking fats, butter, gravy, and coffee grounds can clog any kitchen sink drain. These materials shouldn't be put down kitchen drains. Instead, they should go out with the remainder of the household garbage.

Pencils, toys, baby diapers, paper tissues, toothbrushes, sanitary napkins, over-rim bowl deodorants, bath oils, and hair can clog bathroom drains.

Stoppages occur, of course, even with normal use. But a plumbing system that isn't abused should give many years of carefree service.

Identifying Types and Locations of Stoppages

It's unusual to have stoppages in straight horizontal runs of piping, vertical drops from fixtures, or in fixture traps (except those in water

closets). Stoppages usually occur where two pipes are joined together with a fitting or where pipe changes direction. Under the code, all fixture traps must be self-cleaning. (Interceptor traps are the only traps which are not self-cleaning.)

If several fixtures are stopped up and the drain lines are connected to a public sewage system, check first to be sure the public system is functioning properly. Sometimes sewer systems *surcharge* (overload) at peak periods. Maybe a lift station pump has broken down or the power has failed. When that happens, the public sewage system can back up into private sewer lines.

Check with the next-door neighbors to see if they're experiencing any drainage difficulties. If not, you can assume that there is a complete stoppage in the private line, not a malfunction in the public system.

Figure 20–1 shows a typical two-bath floor plan with kitchen and utility room. Circles in the next five figures show likely stoppage points in the isometric drawings based on this plan. In Figure 20–2, a cleanout is located on the building sewer immediately outside the building wall. Most codes require that an accessible two-way cleanout be located within

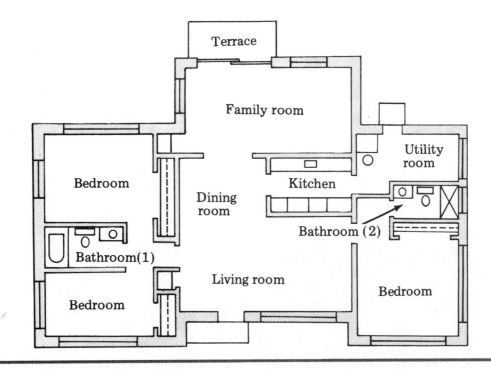

Figure 20–1
Floor plan

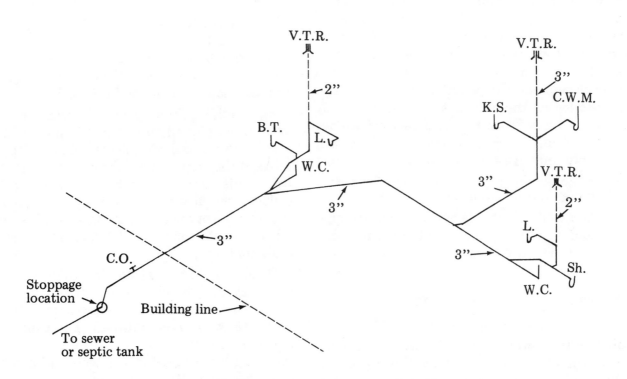

Figure 20–2
Locating a stoppage in building sewer

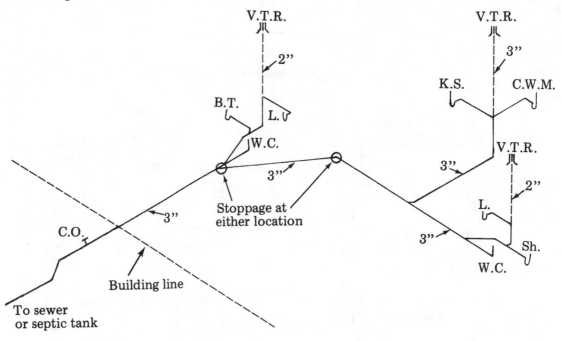

Figure 20–3
Locating a stoppage in building drain

5 feet of the building and another at the property line. The minimum size of this cleanout is usually 4 inches. Use a 14-inch pipe wrench to remove the cleanout plug. This plug may be "frozen" if it hasn't been used previously. If it is, place as much pressure as possible on the wrench and rap the handle hard with a hammer to loosen the plug. A soldering torch may be used on a *brass* plug if it still does not release from the cleanout body. The heat will probably unfreeze the threads. The last resort is to cut the plug out of the cleanout body with a hammer and sharp chisel.

When replacing the cleanout plug, use pipe compound on the threads and tighten the plug snugly into the cleanout body.

■ Using Cables to Free Stoppages

The best tool for cleaning clogged drain lines is an electric-powered sewer cable. If you don't have that type of equipment, there are two choices: The first is a flat 100-foot steel tape approximately ¾-inch wide with a ball on one end. The second is a ½-inch flexible 75 or 100 foot steel spring cable in a cylinder. The cable

can be turned by using a handle attached to the cylinder.

Where the sewer pipe is installed fairly straight and there is plenty of space to work in, use the flat steel tape cable. The strength of the flat steel tape should easily break up the stoppage so that it can be flushed away.

Figure 20–2 shows a stoppage in drainage system connected to a septic tank. To help determine its location, use the flexible spring cable. If the stoppage is in one of the offset fittings, this cable should be strong enough to dislodge the obstruction. If the stoppage is in the inlet tee of the septic tank, the cable's flexibility will let it make the 90-degree turn in the tank inlet tee.

Insert the cable far enough to go through the inlet tee. Then remove it. If the stoppage still exists, assume that the problem is in the septic tank or the drainfield. The tank may need cleaning or the drainfield may need replacing, or both. If this is true, call a septic tank specialist.

Figure 20–3 shows a partial building drain stoppage. The bathroom on the left is not affected. The bathroom on the right, the kitchen

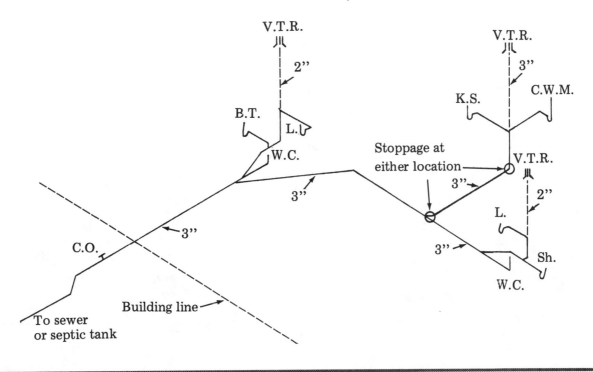

Figure 20–4
Locating a stoppage in waste pipe

sink and the utility room are affected. The location of the stoppage would have to be either in the ⅛ bend or where the building drain pipe connects with a wye fitting into the main drain pipe. Some codes do not require that a cleanout be extended if it is accessible from the outside. The vent pipes extending above the roof are considered adequate for the entry of a cleaning cable.

If cable has to enter through the vent pipes in Figure 20–3, the vent serving the sink and washing machine would be the right entry point. The cable has to pass through two 90-degree offsets and possibly two 45-degree offsets. An electric sewer cable would be best for clearing this type of stoppage, if one is available. Otherwise, use a flexible cable in a cylinder. The cable should move fairly easily through the pipe and fittings until it reaches the blockage.

When you make contact with the obstruction, tighten the thumb screw on the cylinder outlet to grip the cable firmly. Leave about a foot of excess cable above the vent pipe. Do the same if a cleanout fitting is available at ground level. Keep firm pressure on the cable,

as the cylinder has to be turned with a lot of force. This should break up the blockage.

Before retrieving the cable, fill the sink with hot water and flush the line thoroughly. Do this several times. If the water drains away freely, remove the cable.

Figure 20-4 shows a common partial building waste pipe stoppage. Here the kitchen sink and clothes washing machine are affected. The blockage may be located in the sweep at the base of the stack or in the horizontal waste pipe where it connects with a combination into the building drain.

Figure 20–5 shows a blockage that affects only the shower and lavatory in the bathroom on the right. If no ground-level cleanout is available, use the vent pipe opening above the roof to locate the stoppage. Again, the first preference would be an electric sewer cable. If none is available, use a cylinder sewer cable. To clear the blockage in Figure 20–5, use a ⅜-inch diameter cable. That's the best choice when pipes are smaller and the blockage lighter. The procedure would be the same as described above for cylinder-type cables. When the obstruction is broken up, flush the waste pipes

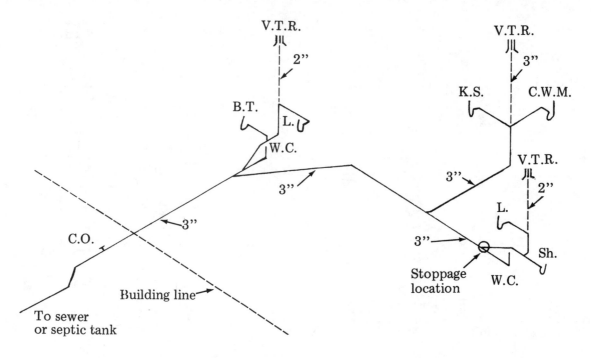

Figure 20–5
Locating a stoppage in fixture branch

with hot water before retrieving the cable. If the water flows away freely, retrieve the cable.

■ **Clogged Sinks**

Figures 20–6 and 20–7 show the stoppages most common in fixture drains. The sink will simply no longer drain. The stoppage usually occurs at the junction of the fixture drain pipe and the waste and vent pipe.

The first practical step is to use a suction cup or "plumber's helper," a flat rubber force cup with a handle. See Figure 20–7. First, remove the basket strainer on the sink drain waste outlet. If the sink has two compartments, plug one waste outlet tightly with a rag to prevent loss of pressure. Place the force cup directly over the drain opening. Fill the sink with approximately 4 inches of standing water so the seal of the force cup can take hold. Take a firm grip with both hands on the force cup handle and push down with a slow, even pressure. Then pull it up quickly and sharply several times. This will usually unstop most fixture drains when the stoppage affects only one fixture.

If this doesn't clear the stoppage, then try the flexible spring cable. First, place a container directly under the trap, loosen the trap's two nuts and remove the "J" bend portion of the trap. Avoid using a cleanout in the "J" bend portion of a trap. This may damage the threads and cause the cleanout plug to leak when replaced. Use a small auger with a ¼-inch spring cable (if an electric auger is not available). Feed the cable slowly into the drain pipe. Rotate the cable by hand as you feed. Turn the handle until the obstruction is broken up. See Figure 20–8. Remove the cable, replace the "J" bend, tighten the trap nuts, flush the fixture drain with hot water, and check for leaks.

■ **Clogged Showers**

The most frequent cause of stoppages in shower drains is hair accumulated in the trap or drain pipe. To unclog a shower, first remove the shower strainer. Some strainers are held in place with two screws. Others are of the snap-in type. Use the "plumber's helper" as shown in Figure 20–7 for the kitchen sink. The obstruction can usually be forced out and into

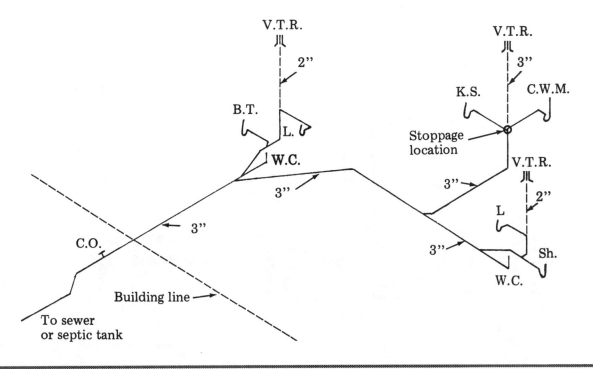

Figure 20–6
Locating a stoppage in fixture drain

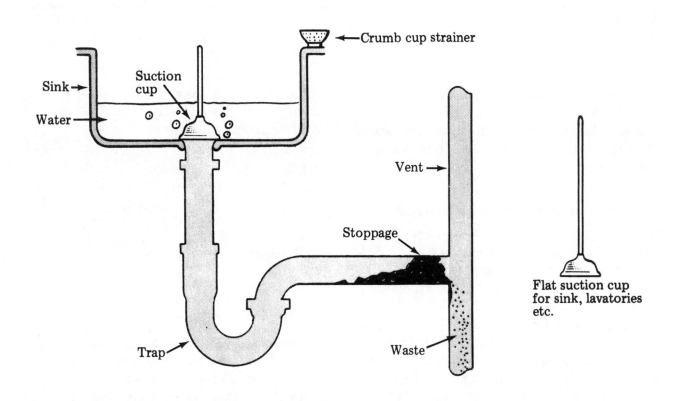

Figure 20–7
Using a force cup to unstop a sink

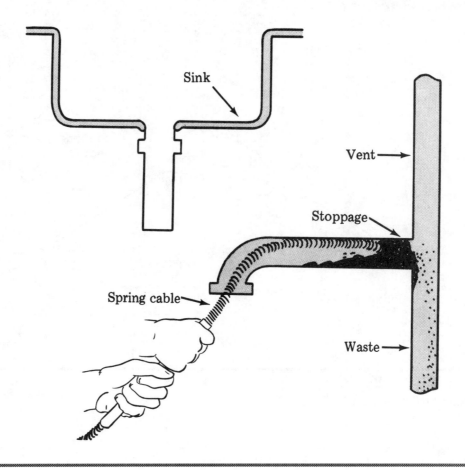

Figure 20–8
Using a spring cable

the larger building drain, where it will be washed away.

If this doesn't work, try using a small auger the same way as described for unstopping a kitchen sink.

■ Clogged Lavatory and Bathtub Drains

If the lavatory or tub has a pop-up type drain stopper, the first probable cause of stoppage is hair accumulated in the stopper. First, remove the lavatory and bathtub drain stopper for cleaning. See Figure 20–9.

When accumulations on the drain stopper are not the cause, the next most probable cause is a stoppage in the trap or waste line. Use a "plumber's helper" as shown in Figure 20–7 to force the obstruction out. Be sure to tightly seal the overflow of the lavatory or tub with a rag before using the force cup. Pumping with the force cup may reduce the water level

in the bathtub. If so, add more water to keep the force cup covered until the flow through the drain returns to normal.

In tubs with a drum trap, the entire cover screws out of the body. A fine penetrating oil applied all around the cover edge will help loosen frozen threads. You may have to use a cold chisel and hammer to free the cover. See Figure 20–10. If the trap isn't clogged, use a small auger as for unstopping a kitchen sink drain line.

■ Clogged Water Closets

Water closets also have their share of stoppages. The water closet trap is a single unit with the bowl. The passageway of the trap is designed to pass "acceptable" materials no larger than 2 inches. A stoppage will be the result when someone puts something in the fixture larger than the trap is designed to

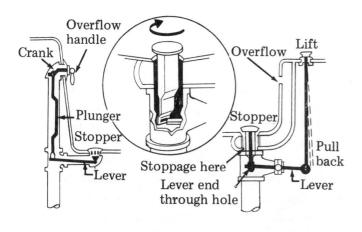

Figure 20–9
Typical tub and lavatory stoppers

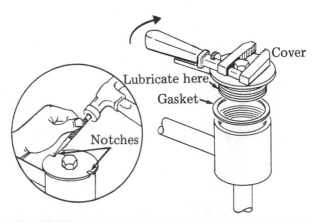

Figure 20–10
Removing a drum trap cover

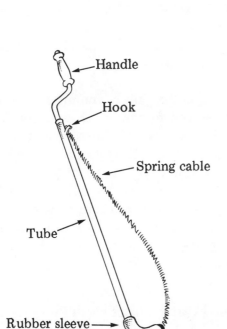

Figure 20–11
Closet auger

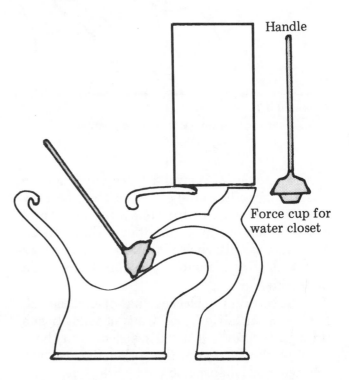

Figure 20–12
Using a water closet force cup

accept. Find the cause of this type of stoppage before you try to unclog it.

Two tools are normally used to unstop a water closet bowl: the closet auger (Figure 20–11) and the force cup especially designed for a closet bowl (Figure 20–12). Start by making a guess about what caused the stoppage. Forcing a large object further down the drainage system may cause more problems

than you've solved. The blockage may be caused by a cloth or some other material that should be retrieved and not forced into the drainage system.

To retrieve an object out of the trap, use a closet auger as shown in Figure 20–13. If the blockage is caused by a comb, pencil, toothbrush or other object that can't be retrieved, try forcing it through the trap into the

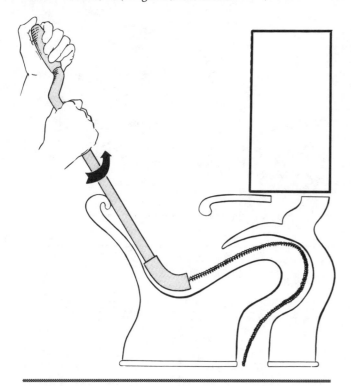

Figure 20–13
Using a closet auger on a water closet stoppage

dle on one end for turning and a spring steel cable with an attached hook on the other end. See Figure 20–11. The curved end of the tube has a rubber sleeve or other soft protective covering to prevent the metal from scratching the bowl's surface. To prepare the auger for use, pull the rod all the way up through the tube until the hook reaches the curved portion of the tube. To use the auger, insert the curved end of the tube into the trap opening and force the spring cable back through the tube until the obstruction is reached. Turn the handle, pressing down on the spring cable as shown in Figure 20–13. Try to hook and retrieve any objects trapped in the bowl.

Hard Water Problems

"Hard" water has high amounts of iron and other dissolved minerals. These minerals collect and settle to the bottom of the water heater and water closet tanks. They remain soft like wet mud as long as they're covered by water. But when they dry out—sometimes at intervals—a residue of tiny hard pieces is formed that can clog up small openings. Following are some solutions to the problems that result.

■ Clogged Water Closet Flush Rim

Around the top of the water closet bowl is the flushing rim. The seat rests on this rim. The purpose of the flushing rim is to both scour the bowl surface and provide enough water volume to give a clean flush.

If the water closet isn't flushing the entire contents of the bowl, raise the seat and check the rim of the bowl. There's a row of small holes spaced at about 1 inch around the entire circumference of the bowl. If some of these holes are partly or completely closed, they need to be cleaned. A 6- to 8-inch length of clothes hanger wire makes an ideal tool for this. Push one end into and through each hole. Then flush the water closet to expel any small particles. You may have to repeat this several times.

drainage system, using a force cup and the closet auger alternately. The obstruction may flow with sewage to the treatment plant and cause no further problem. Should these methods fail, the only choice is to lift off the water closet bowl and remove the object from the underside.

To be certain the blockage has been removed, flush several fairly heavy loads of toilet tissue through the bowl. If it flushes clean and there is no further evidence that other fixtures are affected, you can assume the blockage has been broken up and washed away.

Water Closet Force Cup
The closet force cup has a specially-designed end that fits the opening of the water closet trap discharge side. Refer to Figure 20–12. This type of force cup works almost the same way as the flat-based cup.

The Closet Auger
A closet auger is a hollow metal tube approximately 3 feet long and curved at one end. A solid metal rod passes through this tube and extends to the curved end. The auger has a han-

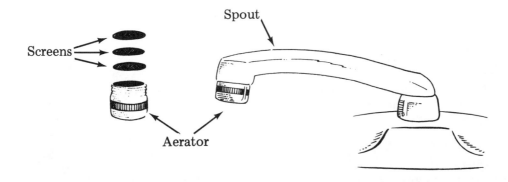

Figure 20–14
Typical aerator

Clogged Shower Heads

Another victim of this type of clogging is the shower head. If some of the holes in the shower head do not allow water to pass through, or if small low volume sprays at a right angle to the main spray, the head may need cleaning.

It's easy to remove and disassemble the head for cleaning. The portion of the head from which the water sprays is called the face plate. It's held in place by a small knob or screw in its center. Remove the face plate and soak it in vinegar. Vinegar loosens the mineral deposits. Then scrub thoroughly with a brush and replace the face plate.

Water Heater Maintenance

Whenever a water heater must be drained, the electric power or the gas supply source must be shut off first. *Only then should the water supply be turned off.* Take the following two general precautions to ensure efficient operation and long tank life:

1. Drain the heater through the drain valve until the water runs clear. Do this at least once a year and more often if the sediment content warrants it. Failure to do this may result in noisy operation and lime and sediment buildup in the bottom of the tank. At the same time, check the temperature-pressure relief valve. Raise the test lever at the top of this valve to make certain that the waterways are clear.

2. If the building is left unoccupied during cold weather months, drain the heater completely to prevent the tank from freezing.

Reduced Water Pressure

Loss of water pressure may be caused by clogged aerators or screens, or the wrong washer in a compression-type faucet. We'll take the possible causes one at a time.

Clogged Aerators

Nearly all combination lavatory and sink faucets today have an aerator at the end of the spout. See Figure 20–14. The aerator blends air with the water, thus conserving water and minimizing water splatter. The aerator has several fine screens inside it. After several years, these will usually become clogged. Remove the aerator to clean the screens. First wrap the outside with adhesive tape to keep from marring the chrome finish. Then loosen by turning the aerator counterclockwise with pliers. Take out the screens and thoroughly clean them until all debris is removed. Then reassemble exactly as before. If the aerator is the female type, place a small amount of pipe compound on the outside threads of the spout only. If it's the male type, place some pipe compound on the outside threads of the aerator only. Tighten the aerator enough to prevent water from seeping out around the threads. Then remove the tape.

■ Clogged Screens

The hot and cold water hoses that supply water to clothes washers are equipped with filter screens. Each hose has a cone-shaped screen to prevent water-formed scale and other sediment from entering the washer tub.

When water enters the washing machine tub under reduced pressure, the screens are probably partially clogged and need to be cleaned. The screens can be removed with your fingers after disconnecting each hose from its faucet. Clean each screen thoroughly and replace as before.

Valves serving a clothes washing machine are usually left "on" at all times. When the house is to be vacant for any length of time, the faucets should be turned off. This prevents flooding if one of the hoses should rupture.

■ The Wrong Washer

Gradual reduction in water pressure from a compression-type faucet (hot water side only) may be caused by the wrong type of washer installed in the faucet. Heat expands washers that are not designed for hot water use. When that happens, water flow may be reduced to a trickle. To check if a washer is obstructing water flow, open the valve about one-quarter turn and let the water flow for about a minute. If water pressure falls as the temperature of the water rises, probably the wrong type of washer has been installed. Replace the washer with a proper one. (Replacement procedures are discussed in Chapter 21.)

Here's another possible cause of reduced water pressure in a compression-type faucet. A flat washer may have been replaced with a beveled washer. Worn washers which have given good service should always be replaced with new washers of the same type.

Pressure Fluctuation

There are two probable causes of pressure fluctuation on either the hot or cold water side of a compression-type faucet.

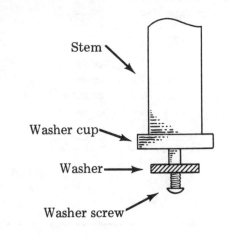

Figure 20–15
Loose washer

1. Alternating strong and weak water flow accompanied by a chattering or whistling noise in the pipe is usually caused by a loose screw that holds the washer tightly in place at the bottom of the faucet stem. See Figure 20–15. The washer floats up and down on the loose screw, causing the pressure to fluctuate. Usually you'll hear an irritating chatter as the pressure changes. To correct this, remove the stem (as shown in Chapter 21) and tighten the screw.

2. The second cause of pressure fluctuation is rare but still possible. If the screw holding the washer in place is not tightened, it can loosen and fall into the supply tube under the fixture. See Figure 20–16. The washer may stick to the washer cup at the end of the stem and remain firmly in place, even without the screw. The faucet continues to work normally to control the flow of water, but there may be a fluctuation in the water flow.

Shut off the water control valve to the fixture and remove the faucet stem assembly as if you were replacing the washer. Close the waste opening on the fixture. Cup your hand over the opening in the faucet body where the faucet stem was removed. This will deflect the water down into the fixture bowl. Have someone open the control valve and then close it immediately. Your hand will deflect the surge of

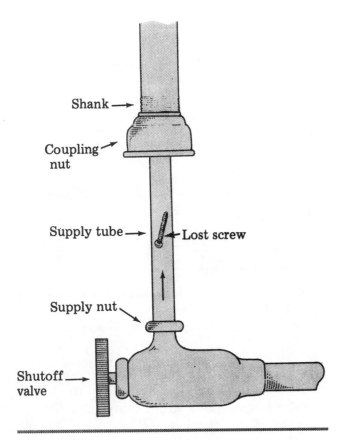

Figure 20–16
Lost bibb screw

water downward and should cause the screw to fall into the fixture bowl as it's blasted out of the washer cup.

If this doesn't solve the problem, disconnect the supply tube under the fixture. There's a nut at each end of the supply tube. Loosen the coupling nut with a basin wrench; loosen the compression nut with a small smooth-jawed adjustable wrench. Remove the supply tube. The screw should then fall out. If the screw is good, reinstall it. If not, replace it with a new one. Replace the supply tube, covering the threads with a little pipe compound. Tighten the nuts snugly, turn the water on, and check for leaks.

Complete Loss of Water

There are four common causes of total water loss to a building.

1. The house control valve may have been closed by someone. This valve is readily accessible to anyone from the building exterior. Open the valve by turning it counterclockwise.

2. The utility company supplying water to the community may have shut off the water to make emergency repairs. Call the utility company. They should be able to confirm the cause and estimate the time when service will be resumed.

3. In colder climates, suspect frozen pipes as the cause of a blockage in the main service pipe. If this is the problem, see the procedure for thawing frozen water pipes later in this chapter.

4. If only the hot water is lost, it may be that someone (possibly a child) has been "playing plumber," closing valves at random. Check the shutoff valve controlling cold water flow to the water heater. It may have been closed.

Water Hammer

Water hammer, a banging in the pipes when a faucet is closed quickly, is caused by air lost from the air chambers. It may affect the entire system, but is generally limited to either the hot or the cold water line or to one bathroom. To correct water hammer, take these steps.

1. Shut off the main house valve and open one or more outside faucets. Then open all faucets, hot and cold, inside the building for a few minutes. This lets water in the pipes drain from the system. Air will refill the pipes. After the water has stopped draining through the outside faucets, close them and all inside faucets. Then open the house valve. As water refills the pipes, air will be pushed back into the air chambers again. This works very well when the piping system is installed high enough above the outside faucets to permit complete draining.

2. If the water system pipes won't empty and refill with air, try this method. Shut off the main house valve and open one or

more outside faucets as before. Then open the faucet on one fixture. Remove the aerator if the faucet has one. Place your mouth over the faucet spout opening, take a deep breath, and blow with all your might. The objective is to blow trapped water out and through the hose faucet.

Repeat with the other hot and cold water faucets in each fixture. This can force enough air into the pipes to replace lost air in air chambers and should eliminate the banging noise.

3. If the two methods above are not successful, install exposed air chambers in both hot and cold water pipes. The easiest and most likely place for these would be near the water heater. An air chamber should be one pipe size larger than the pipe to which it connects, and a minimum of 12 inches long. (Look back to Figure 12–2 in Chapter 12 and Figure 16–7 in Chapter 16.)

Thawing Frozen Pipes

In cold climates, water pipes can freeze in cold weather. The old saying that "an ounce of prevention is worth a pound of cure" applies here. All exposed piping should be insulated well and should be checked before winter. Replace any unsound insulation.

But if pipes do freeze, only heating the pipes can thaw them. There are several ways to apply heat.

1. If there is no danger of fire, use a blowtorch or a burning twist of newspaper. To thaw a pipe using an open flame, open the nearest faucet to indicate when the flow of water starts. Apply heat at one end of the pipe and work the flame along the entire length of the pipe. Continue heating the pipe until the water flows from the faucet.

2. If there is danger of fire, wrap the affected section of pipe with rags and pour boiling water over the rags. This also gives satisfactory results. Leave a faucet open so you know when water is flowing again.

3. Frozen water supply pipes located where heat cannot be readily applied may be

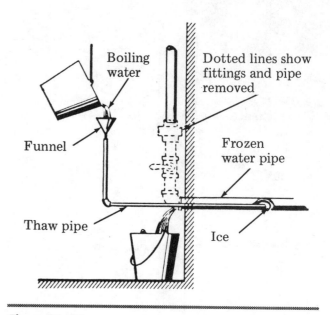

Figure 20–17
Thawing an underground or otherwise inaccessible pipe

thawed by the following procedure. Close the house valve and remove the fitting that closes the exposed end of the pipe. Then insert a smaller pipe or rubber hose with a funnel into the pipe until it contacts the ice blockage. See Figure 20–17. Pour boiling water into the funnel. The hot water circulates to the ice blockage through the smaller pipe. It then comes back out the opening where the smaller pipe enters. Catch this spillage in a pail if water damage is likely. As the ice melts, push the "thaw pipe" deeper into the building pipe. Then withdraw the pipe quickly when the flow starts and allow the flow to continue until the thaw pipe has been completely withdrawn and the building pipe is cleared of ice.

4. Inaccessible pipes may be thawed with a low voltage electrical thawing unit which may be attached to any alternating current electric outlet. The unit has a transformer that takes a relatively small amount of current at house voltage but sends a large amount of current at low voltage through the pipes. Two heavy cables may be clamped to each side of the frozen section of pipe. The transformer produces heat,

which is transmitted to all portions of the pipe between the clamps.

To thaw frozen underground water service, connect one clamp where the service pipe is accessible within the building. Connect the other clamp to the street valve. Heat transmitted to the frozen section of pipe will usually give quick, satisfactory results.

Cleaning Discolored Fixtures

No matter how dingy the fixtures look, never attempt to remove stains with scouring powder or pads. These destroy the glaze. Instead, clean fixtures with any fine-grained cleaning powder or with a liquid cleaner such as Soft Scrub.

Whiten a stained porcelain tub or sink by filling the fixture with warm water and adding some chlorine bleach. For best results, let this stand for several hours.

If the water source is a well, I recommend a water softener. The softener will improve water taste and color and remove most staining agents from the water, especially the iron.

Iron stains porcelain fixtures a deep reddish brown, which is practically impossible to remove without damage to the fixture finish.

Vinegar is usually effective on stubborn mineral stains. Soak a pad in vinegar and place it directly over the stain for an hour or longer. Rust stains sometimes respond if you rub the area with a fresh slice of lemon. If stains do not respond to these efforts and heavy-duty cleaners must be used, follow the instructions on the label carefully to avoid damaging the porcelain finish.

■ Cleaning Toilet Seats, Chrome and Tile

Don't use anything stronger than a soft cloth and mild soap when cleaning toilet seats, chrome, or tile. To keep mildew from forming on tile walls, shower unit enclosures, and bathtubs, always wipe excess water from the tile or bath enclosure with a towel after use. Also wipe the chrome shower or tub trim dry at the same time. Both will remain bright and new-looking for years if properly cared for.

Questions

1. What is the major cause of clogged drains?

2. In what portions of a drainage system do stoppages rarely occur?

3. At what point is a stoppage most likely to occur in a drainage system?

4. Besides a local blockage on private property, what are other possible causes of a complete stoppage of a drainage system?

5. In order to make stoppages accessible for unclogging, what do most codes require on a private sewer?

6. If a brass cleanout plug is frozen and will not loosen, what other procedure should be tried?

7. What are some of the factors determining the right kind of sewer cable to select for a particular type of stoppage?

8. When an accessible cleanout is not available, what portion of a sanitary system is considered adequate for cleaning purposes?

9. Which plumbing fixture is most subject to clogging?

10. What is a "plumber's helper"?

11. When using a "plumber's helper" how should a two-compartment sink be blocked up to prevent loss of pressure?

12. What is the most likely cause of a clogged shower drain?

13. What tool should be tried first to unstop a shower drain?

14. When the drain stopper of a lavatory or bathtub is not causing the stoppage, what tool should be tried first to clear the trap or drain line?

15. What is the passageway of a water closet trap designed to do?

16. Why is it important to determine the cause of a stoppage in a water closet trap before proceeding to unclog it?

17. What two tools are usually used to unstop a water closet bowl?

18. What tool should be used to clear a closet bowl trap in a stoppage caused by cloth?

19. How should you check to be certain a closet bowl blockage has been removed?

20. What substances are contained in "hard" water?

21. Where is the flushing rim of a closet bowl?

22. What is one of the first checks you should make when a water closet is not flushing properly?

23. What procedure should be used to clear mineral deposits from a shower head's face plate?

24. What should be the first step before shutting off the water supply to an improperly-working water heater?

25. What yearly check should be made on a water heater to find out whether it is working properly?

26. What two things are accomplished when an aerator is used on the end of a faucet spout?

27. If an aerator has male threads, where should the pipe compound be applied when reassembling?

28. What is the main cause of reduced water pressure to a clothes washing machine?

29. What should be done to a washing machine to protect the property from possible water damage when the house is to be vacant for any length of time?

30. What is the most likely cause of a complete loss of hot water in a building?

31. What would be the most probable cause of a complete loss of water to a residence in Chicago during January?

32. What would be the most probable cause of a complete loss of water to a residence in Key West, Florida, in January?

33. What is the most probable cause of a gradual pressure reduction in a compression faucet on the hot water side?

34. What is another possible cause for reduced pressure on a compression faucet?

35. What usually causes a chattering or whistling noise in a compression-type faucet when the water is turned on?

36. How do you recover a lost bibb screw that has dropped into the supply tube?

37. What is the major cause of water hammer?

38. What three steps should be tried to eliminate water hammer in a water distribution system?

39. "An ounce of prevention is worth a pound of cure." How does this apply to protecting a water piping system from freezing?

40. What is used to thaw water frozen in pipes?

41. Where there is no danger of fire, what procedure should be used to thaw a frozen section of pipe?

42. Where danger of fire exists, what procedure should be used to thaw a section of frozen water pipe?

43. What method is used to thaw pipe that is inaccessible?

44. What should you not use to rid plumbing fixtures of heavy stains?

45. How may bathtub or sink stains be removed safely?

46. How can fixture stains be avoided when the source of water is a well?

47. What substance may be used to remove stubborn mineral stains without damage to the fixture finish?

48. To avoid damaging the finish, what is recommended for cleaning toilet seats, tile, and chrome?

21

General Plumbing Repairs

Leaks account for much of the water loss in any system. For example, a drop of water that takes 2½ seconds to form and drip to the sink wastes about 365 gallons of water per year. If the drip is from the hot water side, the leak wastes both 30 gallons of water a month and 65 kilowatt hours of electricity costing about $2.50. Obviously, leaks can be expensive.

Leaks in Valves

General plumbing repair work includes changing worn-out washers and packings as well as repairing other types of leaks in the system. Every plumber should be able to find, diagnose, and eliminate the most common types of leaks in all the common types of valves. Valve leaks are generally caused by faulty or worn washers or leaking packing nuts.

Figures 21–1, 21–2, and 21–3 show valves used in residential construction. Valves come in many styles and may differ somewhat in general design from those shown. Gate and globe valves are the most common types of shutoff valves used to control water flow to fixtures and appliances. In Figure 21–1, you can see the large water passages in the globe

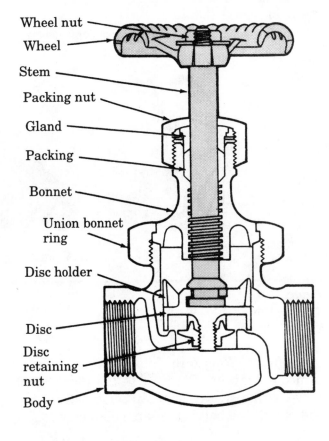

Figure 21–1
Details of globe valve

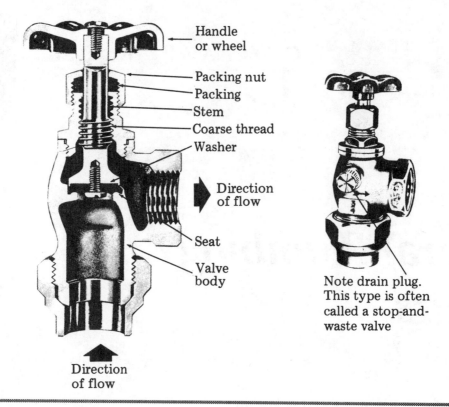

Figure 21–2
Globe angle valve (Faucets are similar in construction)

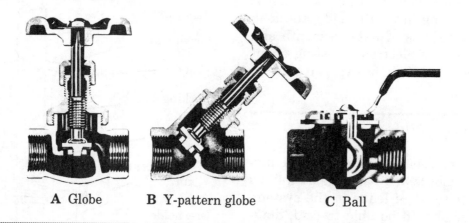

Figure 21–3
Types of valves

valve. In the Y-pattern globe valve, the flow is almost straight. Figure 21–3 shows a ball valve with a straight flow. In some ball valves, the port of the ball is the same diameter as the pipe.

Globe valves and gate valves are very similar. The repair instructions below will apply to both types.

■ Globe or Gate Valve Repair

When a globe valve (also known as a *compression valve*) leaks, the washer on the end of the stem is probably worn and needs to be replaced. Shut off the water at the building main and drain the line. Then disassemble the valve in this order: remove the handle or wheel, the packing nut, the packing, and the stem. Fit the handle or wheel back on the stem and use it to unscrew and remove the stem from the valve body. Do not use a wrench to remove the stem. Figures 21–1 and 21–2 show valve parts in detail.

Remove the screw and the worn washer from the stem. Clean the washer cup and install a new washer of the proper size and type. Reassemble the parts in reverse order: the stem, the packing, the packing nut, and then the handle.

If the washer in a globe valve requires frequent replacement, the replacement washer is probably the wrong type for the seat. Or the seat may be pitted and rough, scoring and wearing away the washer prematurely. Most globe valves have seats which may be replaced when they become worn or pitted. Replaceable seats have either square or hexagonal water passages, and the seat removal tool must be either square or hexagonal to fit the passage. Seat dressing tools are available for both replaceable and nonreplaceable seats.

If water leaks around the stem, either the packing nut has loosened and needs tightening or the packing is worn and needs replacing. Tighten the packing nut slightly with a suitable wrench such as an adjustable crescent wrench. If the leak continues, remove the handle and then the packing nut. Wrap a few strands of graphite packing string or lamp wick string dipped in pipe joint compound around the stem. Slide the packing nut down the stem and tighten it with a wrench as before.

Gate valve repair usually requires only replacing the stem packing. The procedure is the same as for globe valves.

■ Repairing Compression Faucets

Dripping faucets are the most common plumbing problem. Normally, a new washer in the faucet is all that is required.

Most plumbing fixtures today have ordinary compression faucets: two separate valve and water supply units with a common spout. The washer in each side of the faucet body (one for hot and one for cold) eventually wears down and must be replaced. Each unit must be repaired independently.

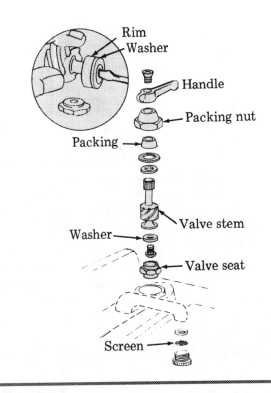

Figure 21–4
Typical faucet assembly

To replace the washer in a compression faucet, first shut off the water supply to the faucet. Most newer buildings have shutoff valves directly below the fixture, one for the hot water supply and one for the cold. Turn the valve handle clockwise to shut off the water. In older buildings where no such valve is available, shut off the main house control valve. This valve is usually located on the water service pipe close to the point of entry to the building or in some other accessible place. (Look back in Chapter 12, Figure 12–1.)

Figure 21–4 lists the stem assembly parts for a common compression faucet. The appearance of these parts varies from brand to brand, but their function remains the same regardless of the manufacturer, the type of faucet, and its particular use. Here are some of the most common problems plumbers face in servicing compression faucets:

Water Leaking Around the Handle

When water leaks from under the faucet handle, the stem packing (packing washer) almost certainly needs replacing. The original packing washer is preformed to fit the stem firmly, and the packing nut holds the washer in place. After thousands of turns, the stem wears away the packing, or the packing nut loosens, allowing water to seep out around the stem.

This is a very simple repair. First remove the handle screw and handle. (Some handle screws are located under a snap-on button.) Use a smooth-jawed adjustable wrench. Adjust the wrench to fit the packing nut and tighten the nut slightly. Often this adjustment takes care of the leak. If the leak continues, remove the packing nut completely. Remove the old packing washer and replace it with a new packing washer of the same size if one is available. If none is available, wrap a few strands of graphite packing string or lamp wick string dipped in pipe joint compound around the stem. Slide the packing nut down the stem and tighten it.

Some compression faucets have one or two rubber "O" rings secured to the stem in shallow grooves instead of a packing washer. The "O" rings serve the same purpose as the packing washers and prevent water from seeping out around the stem. Here the replacement procedure is to completely remove the stem from the faucet body and replace the old "O" rings with new ones of the same size (not thicker or thinner).

Water Leaking from Spout

When water drips from a faucet spout, a tub spout, or a shower head when the handle is fully closed, the cause is almost always a worn washer.

To replace the washer, shut off the water to the faucet. Then open the faucet to relieve the last of the water pressure. Remove the handle screw and the handle and loosen the packing nut. Then unscrew the stem from the valve body. The bibb screw holding the damaged washer in place may be brittle from age and hard to remove with a screwdriver. If so, grip the head of the screw with pliers and remove it by turning counterclockwise. Clean the washer cup and replace the old washer and screw with the same size and type. Remember that flat washers are used on seats with a crown or a round ridge. Tapered or rounded washers are used with tapered seats. A perfect fit is essential. A near fit is *not* good enough.

Reassemble the faucet in reverse order: Turn the handle to the closed position, tighten the packing nut, and reset the handle to the proper position. Finally, replace the handle screw.

If a compression washer requires frequent replacement, it may be that the washer has become scored by a pitted or rough seat. Use a seat dressing tool to smooth the surface of the seat if the roughness is slight. If the seat is badly pitted, remove the seat and replace it with one of the same type and size.

On old faucets which have not had their washers replaced regularly, the handle may have frozen to the stem over the years. If you can't remove the handle, try a tool called a *handle puller.* If this doesn't work, rap the underside of the faucet handle with the plastic end of a screwdriver. This usually loosens the handle without damaging its chrome finish.

A Noisy Faucet

Occasionally a compression faucet becomes very noisy when water is flowing through it. This may be caused by a loose washer or by worn threads on the stem and worn receiver threads in the faucet body. Either condition can cause the stem to vibrate or chatter. Determine the exact cause by pressing down on the handle. If the stem vibration does not stop, the washer is probably loose.

Replacement stems are available. However, if the receiving threads are worn excessively in the faucet body, a new stem would not eliminate the problem completely. In some faucets, you can replace the stem receiver, the stem, and the seat, restoring all normal wearing parts in the faucet. If the faucet is old and the finish is worn, replace the entire faucet.

Some types of faucets are designed with no washers and should provide easier operation and a long, drip-free service life: The American Standard Aquaseal faucet shown in Figure 21–5 is such a faucet. Instead of conventional "washer-grinding" principle, the Aquaseal works on the diaphragm principle of water control. All moving parts above the Aquaseal are outside the flow area, so lubrication on the stem threads is effective for the

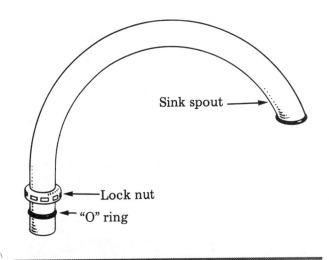

Figure 21–6
Compression faucet spout with "O" ring

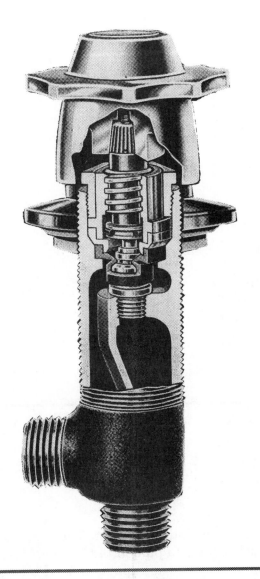

Figure 21–5
American Standard Aquaseal faucet

life of the fitting. There is no seat washer to wear out and cause leaks.

Leaks Around the Locknut

Compression-type sink faucets tend to develop leaks around the locknut that secures the swing spout to the faucet body. Continuous side-to-side movement of the spout wears down the rubber "O" ring or rings located at the base of the spout. See Figure 21–6.

To repair this leak, first loosen the locknut at the spout base. Some locknuts are designed with square corners so a smooth-jawed wrench can be used. Others are round, requiring the use of channel lock pliers or a similar tool. Wrap round nuts with adhesive tape to avoid

marring the chrome finish. Once the locknut is loosened, the spout lifts out. Replace the "O" ring washer with a new one of the same size.

Some types of older faucets may have a split graphite washer. Replace this with a washer of like type or use an adequate amount of graphite string. Reassemble the faucet and tighten the locknut. Remove the adhesive tape and test for leaks. Note that the water to the faucet does not have to be shut off for this repair.

■ Single-Handle Mixing Faucets

This type of faucet can be expected to give many years of maintenance-free use before repairs are necessary. Single-handle mixing faucets are reliable because moving parts control the flow without grinding the washer against a metal seat. These faucets are popular on sinks, lavatories, bathtubs, and showers. Every plumbing professional should know how to maintain and service them.

Figure 21–7 shows parts for the Delta mixing faucets and valves. Other manufacturers' mixing faucets and valves may differ somewhat, but these illustrations will be quite helpful if you have to disassemble and reassemble similar valves.

Repairing the Delta Faucet

Delta faucets and valves are designed to control water temperature and flow with a single ball that rests against a seat assembly. Figure 21–7

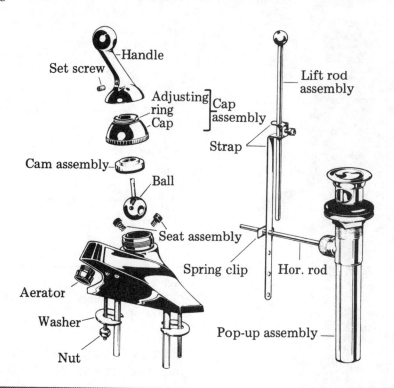

Figure 21–7
Delta lavatory faucet

shows a Delta lavatory faucet, Figure 21–8 a Delta deck-type sink faucet, and Figure 21–9 a Delta tub and shower valve. The routine maintenance instructions for each type of faucet are shown in each figure. Each of these faucets has identical parts to control the water flow. When you learn to repair one Delta faucet, you can repair all faucets in the Delta line. Follow these steps to make proper repairs on a Delta faucet. Remember that the Delta sink faucet has a swing spout. That means that you use *two* "O" ring washers instead of one. This is the only difference between repairing this faucet and repairing the other Delta faucets.

1. If the faucet leaks from under the handle, loosen the set screw and pull off the handle. Tighten the adjusting ring until no water leaks around the stem when the faucet is on and pressure is exerted on the handle to force the ball into the socket.

2. If the faucet drips at the spout outlet, the tub spout, or the shower head, shut off both hot and cold water. Then loosen the set screw and pull off the handle. Unscrew and remove the cap assembly. Pull up on the ball stem to remove the cam and ball assembly. Replace the rubber seats and springs. Check the ball and replace it if there is a sharp edge or any roughness around either of the two small holes.

Reassemble the faucet in the reverse order. Make sure that the slot inside the ball and *lug* on the side of the cam is *inserted into the slot* on the side of the body. Screw the cap down until it's tight.

3. If the faucet leaks from either the bottom or the top of the spout (Figure 21–8), shut off both hot and cold water, loosen the set screw, and pull off the handle. Unscrew and remove the cap assembly. Turn the spout slowly and pull it up to remove. Replace the two body "O" rings. Reassemble in the reverse order. Tighten the cap until it's tight.

Delta faucets have earned the reputation of superiority in design, engineering, performace and durability in millions of home installations. Routine faucet maintenance to compensate for foreign materials in varying water conditions, will restore like new performance and extend the life of the faucet. Ease and simplicity of service and maintenance is another advantage of having Delta faucets throughout the home.

A. If you should have a leak under handle—tighten adjusting ring following steps 1 and 9. Reassemble as in step 10.

B. If you should have a leak from spout—shut off water supply, and follow steps 1, 2, 3, 4 and 5. Reassemble as in steps 6, 7 and 8. Set adjusting ring as in 9. Replace handle as in 10.

1. Loosen set screw and lift off handle.

2. Unscrew cap assembly and lift off.

3. Remove cam assembly and ball by lifting up on ball stem.

4. Remove seats.

5. Place new seats over springs and insert into sockets in body.

6. Place ball into body over seats.

7. Place cam assembly over stem of ball and engage tab with slot in body. Push down.

8. Partially unscrew adjusting ring and then place cap assembly over ball stem and screw down tight onto body.

9. Tighten adjusting ring until no water will leak around stem when faucet is on and pressure is exerted on handle to force ball into socket.

10. Replace handle. Tighten handle screw—tight.

Use only genuine Delta replacement parts.
—All Delta Lever Single Handle Valves.
Kit 3614—"O" Rings (2)
       Cam Assembly
       Seat Assembly (2)
       Wrench

 **Delta Faucet Company**
P.O. Box 31 · Greensburg, Indiana 47240
A Division of Masco Corporation of Indiana

Figure 21–7a
Routine maintenance instructions for Delta series 500, 520, and 530 single control lavatory faucets

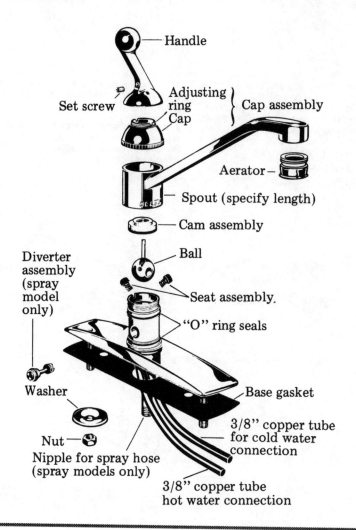

Figure 21–8
Delta deck-type sink faucet

Repairing the Moen Faucet

Moen faucets and valves are designed to control water temperature and flow with a single cartridge. Figure 21–10 shows a Moen lavatory faucet, Figure 21–11 a Moen deck-type sink faucet, and Figure 21–12 is a Moen tub valve. Each of these has identical parts to control water flow. As with Delta, the swing spout on the sink faucet requires the *two* "O" rings instead of one. The following steps explain how to make repairs on Moen faucets:

1. If the faucet drips at the spout outlet, tub spout, or shower head, first shut off both hot and cold water. Then remove the handle cover, handle screw, handle, and stop tube. Lift the retainer clip and pull the cartridge out by the stem from the faucet or valve body.

To reassemble, reinsert the replacement cartridge by pushing it all the way into the faucet or valve body *until the front of the ears on the cartridge shell are flush and aligned with the body*. See Figures 21–10 and 21–12. Replace the retainer clip so that its legs straddle the cartridge ears and slide down into the bottom slot in the body. This prevents the cartridge from rotating and locks it into the body.

Reinstall the stop tube, handle, handle screw, and handle cover. When mounting the handle, be sure the *red* flat on the stem is pointing *up*.

2. If the sink faucet drips at the spout outlet or leaks from either the top or the bottom of the spout, follow the steps in Figure

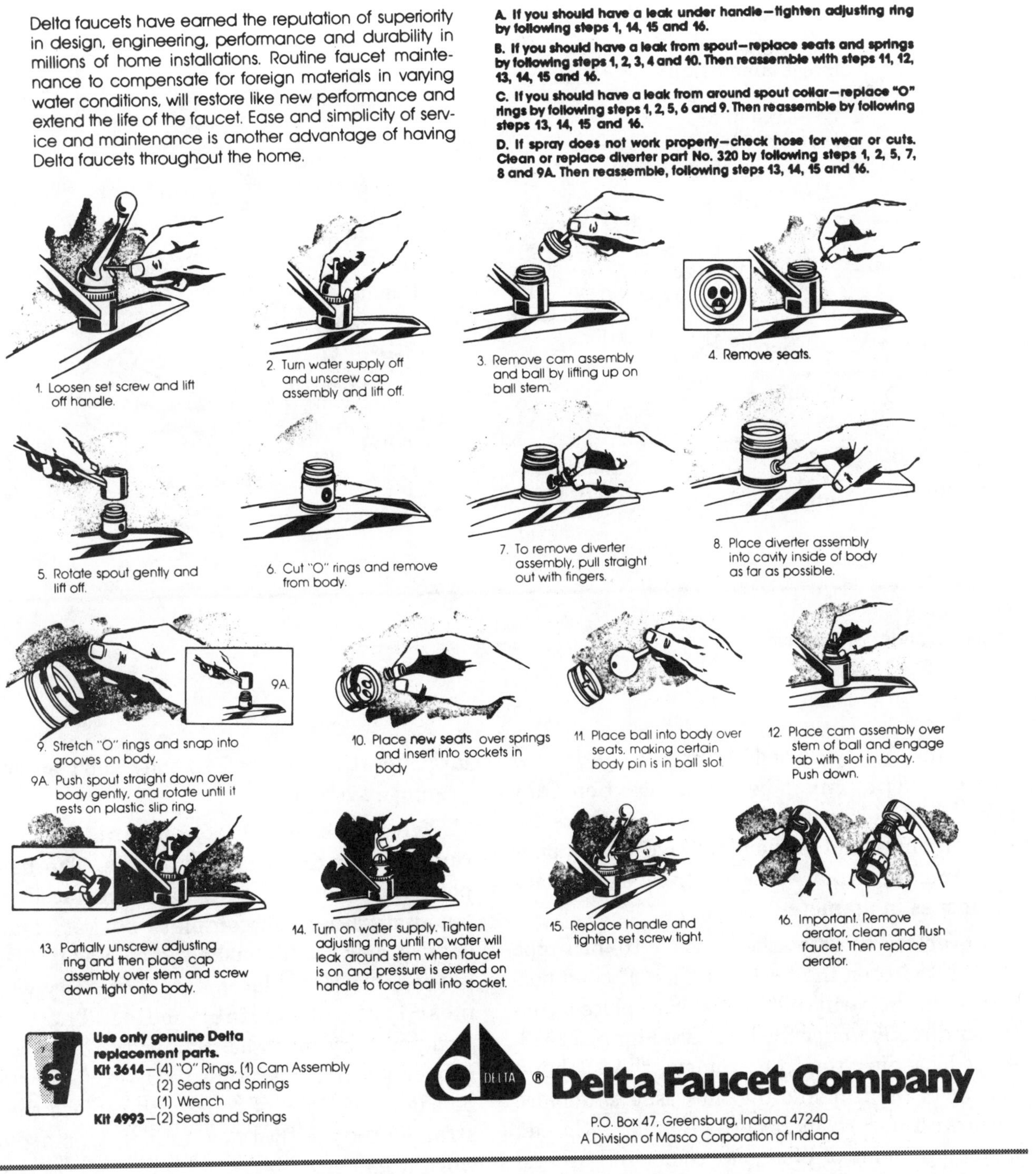

Delta faucets have earned the reputation of superiority in design, engineering, performance and durability in millions of home installations. Routine faucet maintenance to compensate for foreign materials in varying water conditions, will restore like new performance and extend the life of the faucet. Ease and simplicity of service and maintenance is another advantage of having Delta faucets throughout the home.

A. If you should have a leak under handle—tighten adjusting ring by following steps 1, 14, 15 and 16.

B. If you should have a leak from spout—replace seats and springs by following steps 1, 2, 3, 4 and 10. Then reassemble with steps 11, 12, 13, 14, 15 and 16.

C. If you should have a leak from around spout collar—replace "O" rings by following steps 1, 2, 5, 6 and 9. Then reassemble by following steps 13, 14, 15 and 16.

D. If spray does not work properly—check hose for wear or cuts. Clean or replace diverter part No. 320 by following steps 1, 2, 5, 7, 8 and 9A. Then reassemble, following steps 13, 14, 15 and 16.

1. Loosen set screw and lift off handle.

2. Turn water supply off and unscrew cap assembly and lift off.

3. Remove cam assembly and ball by lifting up on ball stem.

4. Remove seats.

5. Rotate spout gently and lift off.

6. Cut "O" rings and remove from body.

7. To remove diverter assembly, pull straight out with fingers.

8. Place diverter assembly into cavity inside of body as far as possible.

9. Stretch "O" rings and snap into grooves on body.

9A. Push spout straight down over body gently, and rotate until it rests on plastic slip ring.

10. Place **new seats** over springs and insert into sockets in body

11. Place ball into body over seats, making certain body pin is in ball slot.

12. Place cam assembly over stem of ball and engage tab with slot in body. Push down.

13. Partially unscrew adjusting ring and then place cap assembly over stem and screw down tight onto body.

14. Turn on water supply. Tighten adjusting ring until no water will leak around stem when faucet is on and pressure is exerted on handle to force ball into socket.

15. Replace handle and tighten set screw tight.

16. Important. Remove aerator, clean and flush faucet. Then replace aerator.

Use only genuine Delta replacement parts.
Kit 3614—(4) "O" Rings, (1) Cam Assembly
(2) Seats and Springs
(1) Wrench
Kit 4993—(2) Seats and Springs

Delta Faucet Company

P.O. Box 47, Greensburg, Indiana 47240
A Division of Masco Corporation of Indiana

Figure 21–8a
Routine maintenance instructions for Delta series 100, 300, 350, 400, and 450 single lever kitchen faucets

21–11 to replace the cartridge or the two body "O" rings.

■ Other Types of Washers and Packings

Other parts of the plumbing system contain washers and packings that may eventually need replacing.

Trap Washers

The fixture trap design most commonly used under lavatories, kitchen sinks, and laundry trays is called a *P-trap*. P-traps are usually constructed of chromed brass but may be of plastic where plastic systems are permitted. The trap is two-piece and is held together by

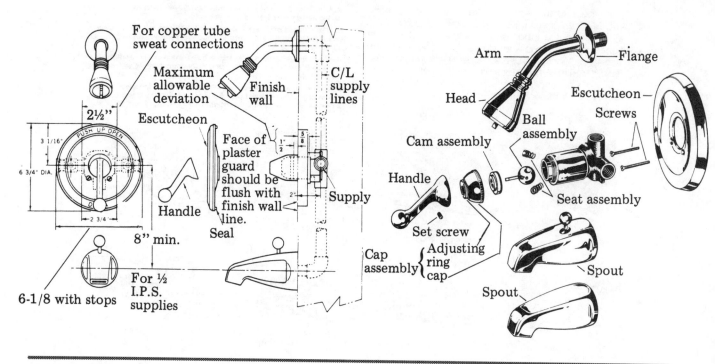

Figure 21–9
Delta tub and shower valve

two nuts and two washers called *square cut washers*. They must be replaced when they begin to deteriorate. Square cut washers for kitchen sinks and laundry trays are 1½ inches in diameter, while washers for lavatories are 1¼ inches in diameter.

To replace these washers, use a 10-inch pipe wrench to loosen the two nuts. The "J" bend portion of the trap will contain water, so place a container directly under the trap. See Figure 21–13. After loosening the nuts, remove the "J" bend portion of the trap and the old washers. Replace them with new washers of the right size. Spread a small amount of pipe joint compound around and over each washer and tighten the nuts snugly. Fill the fixture with water and run water through the trap to check for leaks.

Kitchen Sink Basket Strainer

Crumb basket strainers are used in most kitchen sinks. The basket, when fully seated, keeps water from flowing down the drain. When the basket is not fully seated, water passes down the drain but the basket catches larger particles of food so they don't enter the drainage system.

Two parts of a kitchen sink may need replacement due to wear or age. The basket portion of the strainer is easily replaced without disturbing the complete basket strainer body. A new crumb basket, known as a "fit-all," is available for replacement. It fits very well in most types of strainer bodies. The entire strainer body must be removed to replace the rubber washer. See Figure 21–14. This would be the ideal time, of course, to replace the old strainer body with a new unit if it's marred or worn from age.

To remove the strainer body or install a new washer or a complete unit, first loosen the 1½-inch tail piece nut. See Figure 21–15. If the proper tool (a *combination locknut wrench*) is not available to loosen the strainer locknut, you can use a 14-inch pipe wrench. The jaws should open wide enough to grip the locknut. But it's hard to keep a grip on the nut with a pipe wrench. You may need help from a second person when doing this work.

Delta Faucets have earned the reputation of superiority in design, engineering, performance and durability in millions of home installations. Routine faucet maintenance to compensate for foreign materials in varying water conditions can restore like new and extend the life and performance of the faucet. And, ease and simplicity of service and maintenance is another advantage of having Delta faucets throughout the home.

A. If you should have leak under handle—tighten adjusting ring by following steps 1 and 9 leaving water supply and faucet turned on. Replace handle as in step 14.

B. If you should have leak from spout—replace seats and springs by following steps 1, 2, 3, and 4. Reassemble following steps 5, 6, 7 and 8. Turn water supply on and follow steps 9 and 14.

C. If diverter fails to operate properly follow steps 1, 10, 11, and 12. Reassemble with steps 13 and 14.

Shut off water supply

1. Loosen set screw and lift off handle.

2. Unscrew cap assembly and lift off.

3. Remove cam assembly and ball by lifting up on ball stem.

4. Remove seats and Springs

5. Place new seats over springs and insert into sockets in body.

6. Place ball into body over seats.

7. Place cam assembly over stem of ball and engage tab with slot in body. Push down.

8. Partially unscrew adjusting ring and then place cap assembly over ball stem and screw down tight onto body.

Turn on water supply

9. Tighten ring until no water will leak around stem when faucet is on and pressure is exerted on handle to force ball into socket.

10. Remove escutcheon.

11. Unscrew diverter. Check for sediment, clean off, and flush cavity. If flapper is nicked, or plunger sticks, replace with new diverter.

12. Screw diverter back into body.

13. Replace escutcheon.

14. Replace handle. Tighten handle screw—tight.

PRESSURE BALANCED VALVES

If unable to maintain constant temperature, clean balancing spool as follows:

1. Remove handle and escutcheon.

2. Close stops.

3. Unscrew balancing spool assembly (part No. 574) and remove—slowly—so as not to damage "O" Ring Seats. NOTE: On shower only installations water trapped in shower riser will drain out when balancing spool assembly is removed.

4. Examine valve hole for chips or any other foreign matter.

5. Remove any chips from spool sleeve before attempting to remove spool from inside of sleeve.

6. Remove spool from sleeve and clean all deposits from both sleeve and spool.

7. When spool will slide freely in sleeve, replace assembly in original position.

8. Open stops and check cap of spool assembly and stems of check stops for leaks before reinstalling escutcheon.

Use only genuine Delta replacement parts.
Kit 3614—"O" Rings (4)
Cam Assembly (1)
Seats (2)
Wrench (1)
Kit 4993—Seats and Springs (2)

Figure 21–9a
Routine maintenance instructions for Delta series 600 single lever bath valves, with or without pressure balance

Whichever tool is used, place the handle end of a pair of pliers between the crossbars in the strainer body. This keeps the body portion of the strainer from turning. Use a medium size screwdriver as in Figure 21–16 for additional leverage. Then unscrew the locknut.

After the complete unit has been removed (and if the old unit is to be reused), remove the old putty with a knife or the blade of a screwdriver. Whether the old strainer unit is to be reused or a new unit installed, place a generous amount of putty around the flange portion

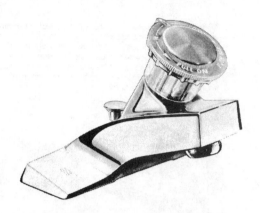

Cartridge is the only part that requires service

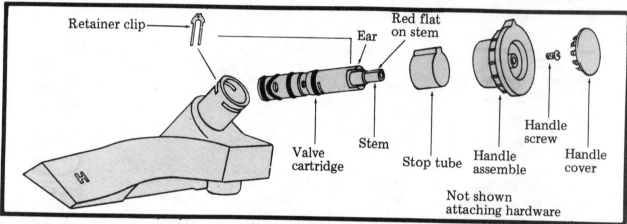

To Disassemble: Remove handle cover, handle, and stop tube. Lift out retainer clip and pull the cartridge out of the body by the stem.

Correct position of retainer clip

To Re-Assemble: Reinsert cartridge by pushing it all the way into the body *until the front of the ears on the cartridge shell are flush and aligned with the body.* Replace the retainer clip so that the legs straddle the cartridge ears and slide down into the bottom slot in the body. This prevents the cartridge from rotating and locks it in the body. Reinstall stop tube, handle, and handle cover. The red flat on the stem must point up when mounting the handle.

Figure 21–10
Moen lavatory faucet

of the strainer. Reassemble the drain and tighten it into the original position. Fill the sink with water and check for leaks.

Sink Tail Piece Washer

The tail piece for a kitchen sink, shown in Figures 21–15 and 21–17, fits between the strainer unit and the upper (first) slip joint nut on the trap. The top of the tail piece has a flat surface that rests against the base of the

strainer unit. It requires a flat 1½-inch washer. To replace this washer, loosen the nut on the strainer unit and also the top nut on the trap. The tail piece should drop downward into the swedged portion of the trap a quarter of an inch or so. This should be sufficient to permit removal of the worn washer and the replacement of a new one without disturbing the rest of the trap. Coat both sides of the washer with pipe compound, reassemble, and tighten both

TO DISASSEMBLE

(Need Pliers, Screwdriver,
Flat-Jawed Wrench)

1. Turn off both hot and cold water supplies. Pry off handle cap. Remove handle screw and pull handle off.

2. Remove wire clip by hooking one end of wire with finger, pulling it out of groove on slotted stop guide, and lifting it over cartridge stem.

3. Remove handle stop. Unscrew and remove retainer nut. Lift off slotted stop guide.

4. Lift and twist spout off.

5. With screwdriver, pry up both legs of cartridge clip and remove it. See illustration below.

6. Grasping stem of cartridge with pliers, pull out cartridge as shown below.

If necessary to Flush Supply Lines, Turn On Both Hot and Cold Water Supplies Slowly.

TO REASSEMBLE

7. With stem of cartridge pulled out, put in cartridge by pushing down on its ears. See illustration below.

8. Turn cartridge so its ears face front and back of faucet as shown below.

EARS

9. Replace cartridge clip by pushing down on both sides so it locks on both sides. See illustration below.

10. Turn red (notched) flat of cartridge stem so it faces rear of faucet. Replace spout, pushing down and twisting until it nearly touches escutcheon.

NOTE: For cross piping installations where supply piping is reversed, red (notched) flat faces front of faucet.

11. Replace slotted stop guide so it straddles cartridge ears and lies flat against cartridge clip. Screw on retainer nut with fingers. Do not cross thread. Tighten with wrench.

12. Put on white handle stop. Push cartridge stem all the way down. Replace wire clip making sure both sides catch in slots of stop guide. See illustration at right.

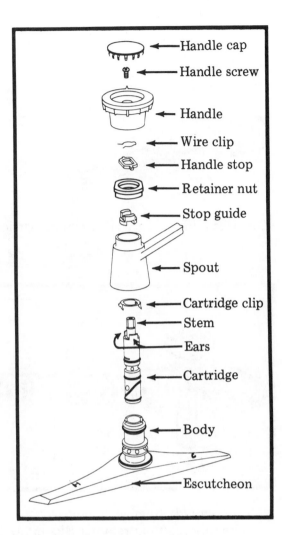

- Handle cap
- Handle screw
- Handle
- Wire clip
- Handle stop
- Retainer nut
- Stop guide
- Spout
- Cartridge clip
- Stem
- Ears
- Cartridge
- Body
- Escutcheon

13. Push handle onto cartridge stem with handle's pointer facing rear of faucet.

14. Replace handle screw. Tighten securely. Replace handle cap.

NOTE: If hot and cold water reversed, red (notched) flat of cartridge stem is just opposite of correct position. (See Step 10.)

NOTE: For proper water flow, aerator must be free of foreign particles. If flow is weak or irregular, unscrew aerator, clean and replace.

Figure 21–11
Moen deck-type sink faucet

Cartridge is the only part that requires service

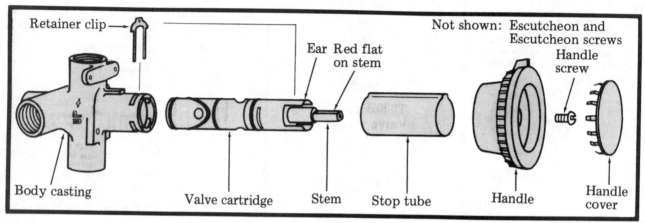

To Disassemble: Remove handle cover, handle, and stop tube. Lift out retainer clip and pull the cartridge out of the body by the stem.

Correct position
of retainer clip

To Re-Assemble: Reinsert cartridge by pushing it all the way into the body *until the front of the ears on the cartridge shell are flush and aligned with the body*. Replace the retainer clip so that the legs straddle the cartridge ears and slide down into the bottom slot in the body. This prevents the cartridge from rotating and locks it in the body. Reinstall stop tube, handle, and handle cover. The red flat on the stem must point up when mounting the handle.

Figure 21–12
Moen tub and shower valve

nuts firmly. Fill the sink with water and release it through the drain to check for leaks.

Lavatory Pop-Up Assembly and Washers
The lavatory waste assembly has two washers. The *mack washer* seals the opening on the bottom side of the lavatory. The second washer, or *seal*, prevents leakage around the pop-up rod pivot ball. Parts of a Delta lavatory pop-up assembly are shown in Figure 21–18. Figure 19–16 in Chapter 19 shows another pop-up assembly.

If water drips from the bottom of the lavatory, the mack washer is not holding and should be replaced. (Remember to first place a

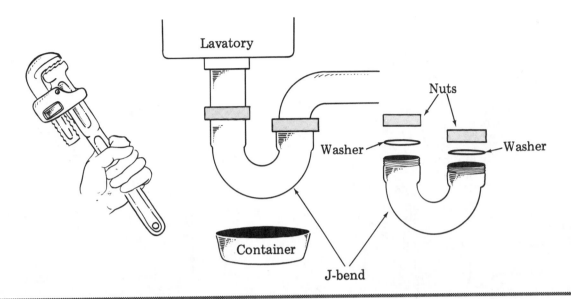

Figure 21–13
Replacing trap washers

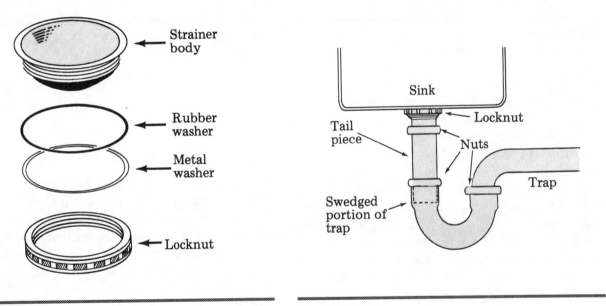

Figure 21–14
Sink strainer parts

Figure 21–15
Assembled tail piece

container under the lavatory trap to catch the water in the trap.) Remove the "J" bend portion of the trap. Loosen the compression nut on the pop-up rod and slide the rod free from the pop-up body. Unscrew the pop-up body from the drain assembly flange. Lift the flange out and remove the old putty. Place a small ring of new putty on the underside of the flange and press it firmly down into its original position.

Place a new mack washer, beveled side up, over the threaded part of the pop-up body. Tighten the pop-up body into the flange. Apply pipe compound on the threaded parts and on each side of the mack washer. Lift the metal washer to the bottom side of the mack washer and tighten the locknut. Replace the "J" bend of the trap. Then connect the pop-up rod to the pop-up body as before and tighten the

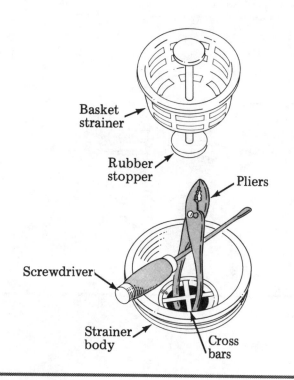

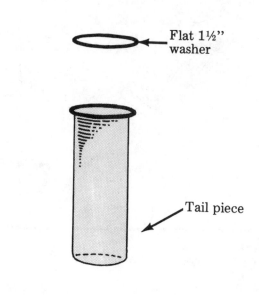

Figure 21–16
One way to remove strainer body

Figure 21–17
Tail piece and washer

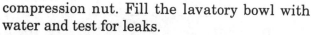

compression nut. Fill the lavatory bowl with water and test for leaks.

The second washer or seal is located on the pop-up rod assembly. This seal is not difficult to replace. The only tool necessary is a pair of pliers. Most lift rods are secured to the pop-up rod with either a set screw or a spring clip. The complete pop-up assembly can be taken out by freeing the lift rod from the pop-up rod and unscrewing the compression nut. A new seal can then be added. Replace the assembly in reverse order and check for leaks.

Water Closet Tank Washers

Close-coupled closet tanks are secured to the closet bowl by two bolts with washers. If there's too much pressure on the tank, the bolt washers may loosen and create a leak. If water drips from one or both tank bolts, try tightening the bolts. In most cases, this stops the leak. The top of the bolt has a slot for a screwdriver head. Use a screwdriver large enough so the head fits snugly into the slot. Tighten the nut under the tank with a pair of pliers or a socket wrench. Do not overtighten, as this may crack the tank. If the leak continues, first shut off the water

and drain the tank. Then remove the tank bolts and install new washers.

Wall-Mounted Tank Leaks

A wall-mounted water closet tank supplies water to the bowl through a 2-inch flush elbow. See Figure 21–19 and Figure 19–24 back in Chapter 19. Two slip nuts and rubber washers secure the flush ell to the closet tank and bowl. If water drips from one or both nuts, the washers should be replaced. If a spud wrench is not available, use a 14-inch pipe wrench to loosen and remove the nuts. Install new 2-inch square cut washers. Coat both sides of the washers with pipe compound and retighten the nuts firmly. (See water closet elbow parts detail in Figure 21–20.) It isn't necessary to shut off the water for this repair.

Water Closet Spud Washer

The spud washer fits over the end of the flush valve. It provides a watertight seal between the closet tank and bowl of close-coupled water closets. If water is leaking from under the tank, replace the spud washer.

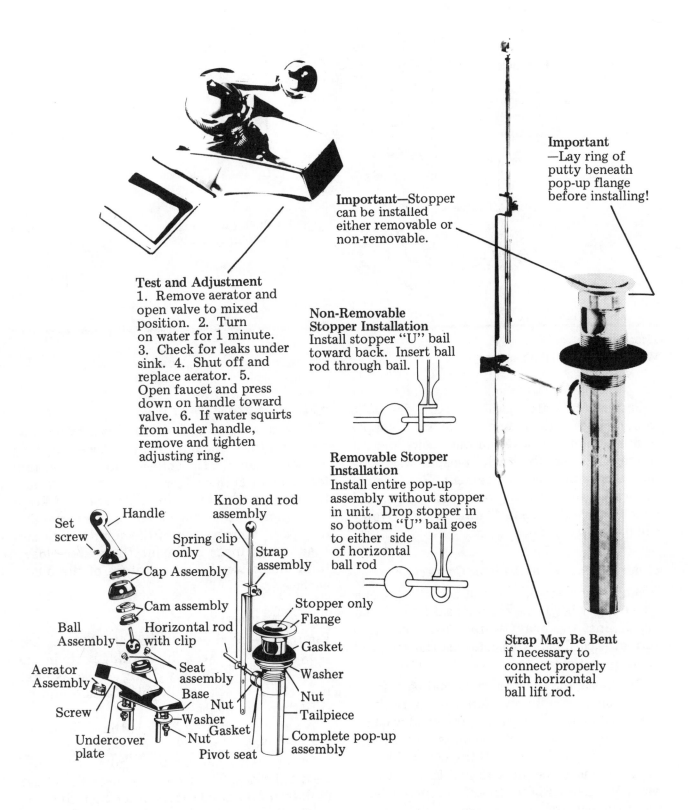

Important—Stopper can be installed either removable or non-removable.

Important—Lay ring of putty beneath pop-up flange before installing!

Test and Adjustment
1. Remove aerator and open valve to mixed position. 2. Turn on water for 1 minute. 3. Check for leaks under sink. 4. Shut off and replace aerator. 5. Open faucet and press down on handle toward valve. 6. If water squirts from under handle, remove and tighten adjusting ring.

Non-Removable Stopper Installation
Install stopper "U" bail toward back. Insert ball rod through bail.

Removable Stopper Installation
Install entire pop-up assembly without stopper in unit. Drop stopper in so bottom "U" bail goes to either side of horizontal ball rod

Strap May Be Bent if necessary to connect properly with horizontal ball lift rod.

Set screw

Handle

Knob and rod assembly

Spring clip only

Strap assembly

Cap Assembly

Cam assembly

Ball Assembly

Horizontal rod with clip

Stopper only
Flange

Gasket

Aerator Assembly

Seat assembly

Washer

Screw

Base

Nut

Nut

Tailpiece

Undercover plate

Washer
Gasket

Nut

Pivot seat

Complete pop-up assembly

Figure 21–18
Delta series 500, 510, 520, and 530 single control lever lavatory faucet and pop-up

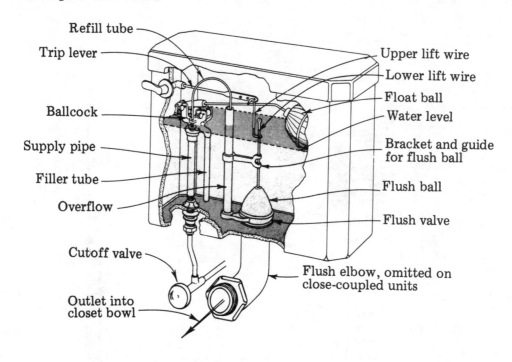

Refill tube
Trip lever
Ballcock
Supply pipe
Filler tube
Overflow
Cutoff valve
Outlet into closet bowl

Upper lift wire
Lower lift wire
Float ball
Water level
Bracket and guide for flush ball
Flush ball
Flush valve
Flush elbow, omitted on close-coupled units

Figure 21–19
Water closet (toilet) flush tank

Shut off the water to the tank and drain the tank of all water. Disconnect the water supply from the tank. Remove the tank bolts and lift the tank up from the bowl. The spud washer is now accessible for replacement. Reassemble the water closet, fill the tank with water, and check for leaks. (See "Tank Installation," Chapter 19.)

Water Closet Leaks at the Floor Connection

When water seeps out from under the base of a water closet bowl, the setting seal is not holding against the heavy water discharge pressure of each flush. This could be the result of building settlement.

To replace the seal, the tank and bowl must be removed completely. Remove the tank as described above for replacing the spud washer. Then remove the closet bolt nuts and lift the bowl up from the floor flange. See Figure 21–21. Remove the old seal and install a new wax seal, as in Figure 21–22. Reinstall both the bowl and the tank as before. (See "Bowl and Tank Installation," Chapter 19.) Fill the tank with water and flush several times to check for leaks.

"Sweating" Water Closet Tanks

Cold water entering a water closet tank may chill the tank enough to cause "sweating" or condensation of atmospheric moisture on the tank's outer surface. Prevent this by insulating the tank with a tank liner. This should keep the temperature of the outer surface above the dew point temperature of the surrounding air. Insulating liners fit inside the water closet tank and raise the temperature of the outer surface.

Repairing Water Closet Tanks

Figure 21–19 shows a typical water closet tank. It's easy to keep the working parts in the tank in good repair by doing an occasional inspection. Check the flushing action and be sure there are no leaks.

A valve called a *ballcock* is the main working part. See Figures 21–23, 21–24 and 21–25 for components and parts of three commonly-used ballcocks. The ballcock causes the closet tank to refill automatically after each flush. It shuts off the water when the closet tank is full. The ballcock is attached to a *float rod* which is attached to a *float ball*. As the tank refills after a flush, the float ball rises with the water

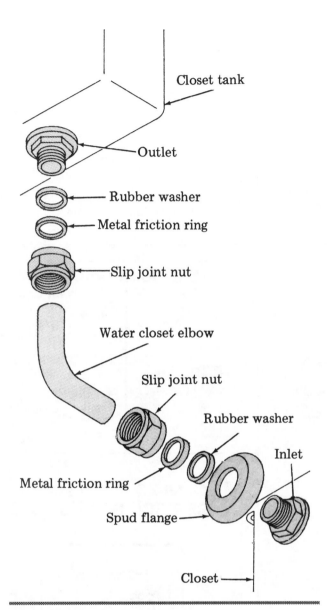

Figure 21–20
Detail of water closet elbow

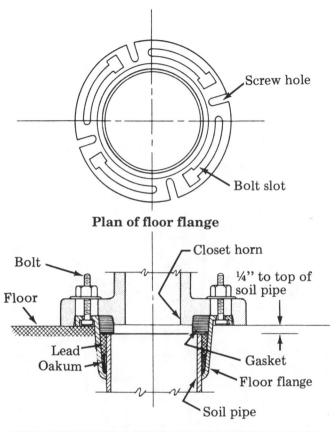

Plan of floor flange

Figure 21–21
Connection of water closet to floor and soil pipe

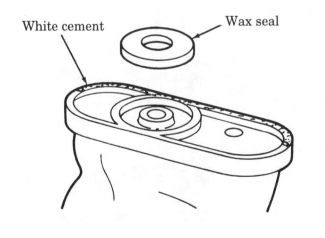

Figure 21–22
Putting wax seal on base of water closet

level in the closet tank. The rod attached to the float ball closes the ballcock valve when the proper level is reached.

A flush valve forms the outlet connection between the closet tank and the water closet bowl. This valve may be of brass, copper, or plastic with a machined seat. A rubber *flush ball* or *flapper* keeps the flush valve opening closed except during actual flushing. The rubber flush ball is connected to the *trip* lever by a lower and upper *lift wire*. This wire raises the rubber flush ball to begin the flushing cycle when the operating handle is pushed down.

The *flapper type* uses an adjustable beaded chain attached to the operating handle in lieu of the lower and upper lift wires. The flapper does not require a guide assembly as does the rubber flush ball.

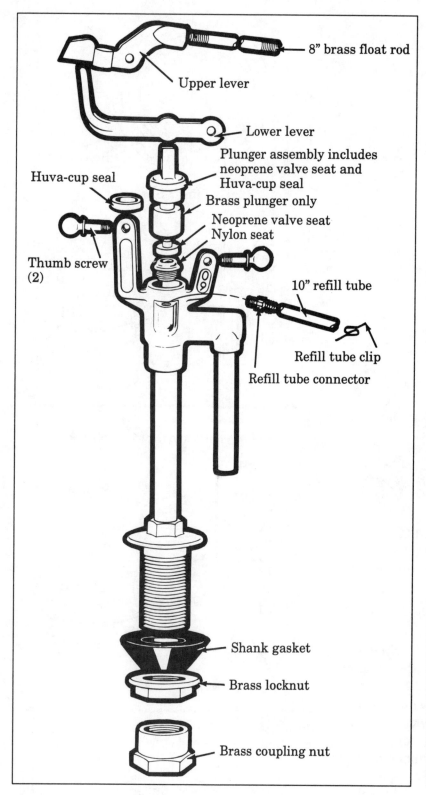

8" brass float rod

Upper lever

Lower lever

Plunger assembly includes neoprene valve seat and Huva-cup seal

Brass plunger only

Neoprene valve seat

Nylon seat

Huva-cup seal

Thumb screw (2)

10" refill tube

Refill tube clip

Refill tube connector

Shank gasket

Brass locknut

Brass coupling nut

Shank Extender

Extends all regular ballcock shanks to 2¾" length. Eliminates need for repair shank ballcocks.

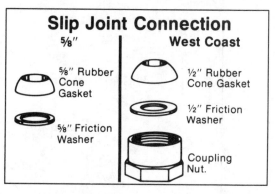

Slip Joint Connection

5/8"	West Coast
5/8" Rubber Cone Gasket	½" Rubber Cone Gasket
5/8" Friction Washer	½" Friction Washer
	Coupling Nut.

03-09-XQ16 BALLCOCK SERVICE PAK No. 7066

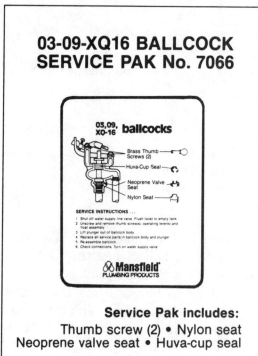

Service Pak includes:
Thumb screw (2) • Nylon seat
Neoprene valve seat • Huva-cup seal

Figure 21–23
Mansfield 09 ballcock

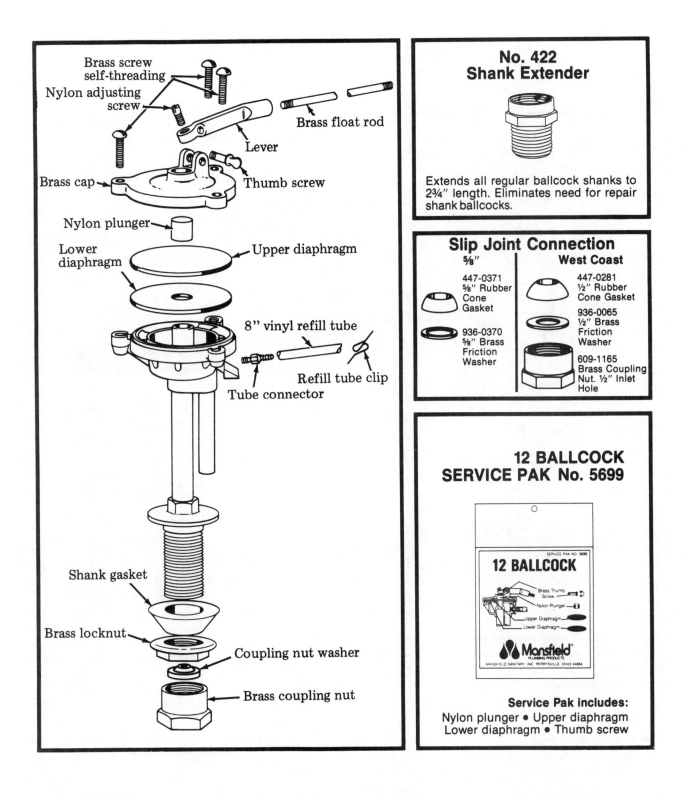

Figure 21–24
Mansfield 12 ballcock

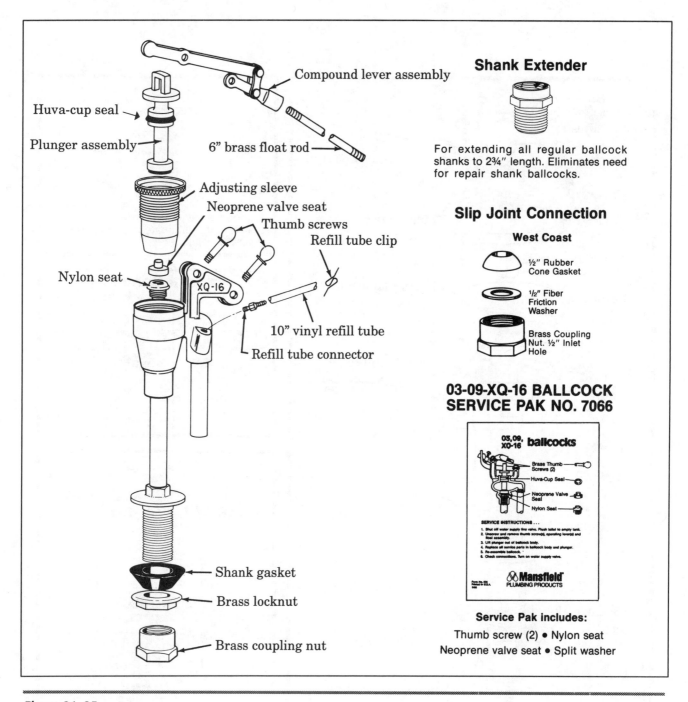

Shank Extender

For extending all regular ballcock shanks to 2¾" length. Eliminates need for repair shank ballcocks.

Slip Joint Connection

West Coast

½" Rubber Cone Gasket

½" Fiber Friction Washer

Brass Coupling Nut. ½" Inlet Hole

03-09-XQ-16 BALLCOCK SERVICE PAK NO. 7066

Service Pak includes:
Thumb screw (2) • Nylon seat
Neoprene valve seat • Split washer

Figure 21–25
Mansfield XQ-16 ballcock

The float ball is full of air and remains in the raised position, floating on the water's surface until the toilet is flushed and the water level in the tank begins to fall. As it descends within the draining tank, the weight of the float ball opens the valve assembly in the ballcock. Fresh water begins to enter the tank through the valve. Gravity causes the flush ball to return to its seat as the last of the water passes through the flush valve and into the closet bowl.

Flushing mechanisms and water closet tanks vary in design from one manufacturer to another. But they are all similar enough that these general repair instructions fit nearly any design.

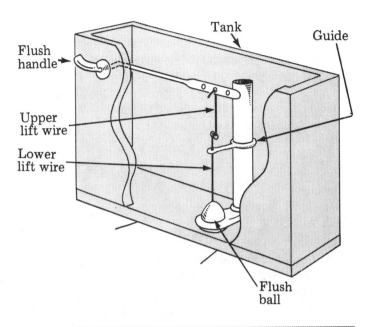

Figure 21–26
Water dripping into bowl

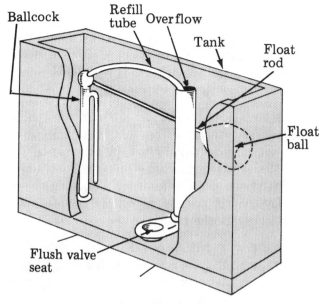

Figure 21–27
Continual running water

Water Dripping into Bowl

When a trickle of water continues to flow from the tank into the bowl after the flushing cycle is complete, the rubber flush ball is probably not seating properly. See Figure 21–26. There are three checks that should be made in this case; one of them should correct this condition.

1. Shut off the water at the valve below the water closet tank. (Codes require a valve at this location.) Remove the tank lid. Flush the toilet to empty the tank. This eliminates any resistance to the up and down movement of the flush ball. Now trip the flush handle several times to see if the guide is lined up directly over the center of the flush valve seat. If it is, the flush ball will fall smoothly and freely onto its seat. If not, adjust the guide assembly until the guide is directly over the center of the flush valve seat.

2. If the water is corrosive, copper or brass lift wires and the guide assembly can deteriorate in a few years. If these parts are corroded, they won't work smoothly. The flush ball won't seat properly. If this is the case, disassemble the mechanism

and clean off all the loose corrosion, or replace with new parts.

3. If steps 1 and 2 don't stop the leaking, raise the flush ball as high as it will go. Or better, unscrew the flush ball from the lower lift wire and remove it completely. Use fine steel wool and clean the valve seat to remove any irregularities that might prevent the flush ball from making a tight seal. Inspect the flush ball while it is out of the tank. If it is soft or out of shape, replace it with a new flush ball.

Continual Running of Water

Sometimes the water keeps running after the tank is full even though there's no sign of water entering the bowl. Very likely the ballcock is not closing completely to shut off the water to the tank. See Figure 21–27. Water rises higher in the tank than its normal shut-off level and leaves the tank through the overflow tube. There are three possible remedies for this problem:

1. If the float ball is riding too high on the surface of the water, bend the float rod down slightly. Use both hands and work carefully

to avoid placing too much strain on the ball-cock assembly. Then flush the tank. The water should not rise as high this time. (The normal water level is approximately 1 inch below the overflow tube.)

2. If the float ball rides low in the tank water, it has probably lost some buoyancy (become waterlogged) and should be replaced. Unscrew the float ball from the float rod and replace it with one of copper, plastic, or foam. One material works as well as another.

3. If the float ball is not the cause, repair or replace the ballcock. The valve of the ball-cock is made of brass, copper, or plastic and has a machined plunger. The plunger has two washers. One is a split round leather or rubber washer which encircles the plunger to keep water from squirting out around the top of the ballcock. The other washer is a round washer in the bottom of the plunger. This washer closes off the flow of water to the closet tank when the float ball reaches the proper water level. Leaks at either of these two washers are the likely source of the running water sound.

To replace the washers, shut off the water and drain the tank. Two thumbscrews hold the plunger in place. Unscrew both of these and lift out the plunger. Unscrew the cup on the bottom and insert a new washer of rubber or leather. Remove the split leather washer and replace it with a new one. If the ballcock assembly is badly corroded or worn out, install a new unit rather than replace the washers.

To replace the ballcock, first shut off the water to the closet tank. Flush the tank. This leaves about an inch of water in the bottom of the tank that won't drain away. Place a shallow pan directly under the shutoff valve to catch this water. Unscrew the float ball, the float rod, and refill tube and remove these from the tank. Use a pair of channel lock pliers to loosen the supply nut and ballcock shank nut. Lift the ballcock up and out. Replace it with a new ballcock. Be sure the beveled part of

the rubber washer is facing down toward the bottom of the tank. Coat the washer and the threads with pipe compound. Retighten the ballcock shank nut snugly. Don't overtighten, as this may crack the tank. Reconnect the supply nut and replace the float rod, the float, and the refill tube. Open the shutoff valve to refill the closet tank. Check the supply and ballcock shank nuts for leaks.

Flush Valve

The flush valve (Figure 21–28) consists of a machined seat and an overflow tube. It's seldom the source of leaks. The machined seat is not subject to wear since the rubber flush ball which closes the opening is soft. The valve is made of brass, copper, or plastic. It is not usually subject to ordinary corrosion and should last for the life of the closet tank. If repair is necessary, the flush valve must be replaced as a unit.

Flush Ball Guide

To install this unit, swing the refill tube to one side and loosely place the guide holder over the overflow. Insert the lower lift wire through the ring at the bottom of the upper lift wire and the opening in the guide holder. Screw the threaded portion of the lower lift wire into the rubber flush ball. Center the flush ball over the flush valve opening. Adjust the loose guide holder forward or backward until the ball seats properly. Securely tighten the guide holder to the overflow tube while it is in that position. See Figure 21–28.

Water Closet Seats

The water closet seat is usually easy to remove and replace. The only tool necessary is a pair of pliers or a socket wrench.

Two brass or plastic bolts pass through two holes in the water closet bowl. The seat is held firmly in place with two nuts. See Figure 21–29. For some one-piece (special) water closets, the seat fastens directly to the tank.

On some older seats, one or both nuts may be corroded so they will not loosen. If time isn't critical, place some penetrating oil around each bolt and wait an hour or so. Otherwise, cut through the bolts with a new fine tooth hack-

Figure 21–28
Flush valve components

saw blade without the frame. Insert the blade between the top of the bowl and the hinge portion of the seat. Push the blade slowly and easily back and forth until you cut through the bolt. Then replace the seat. Tighten the nuts snugly so the seat will not wobble.

NOTE: **Never sit or stand on a seat cover (lid). This places a strain on the hinge and may cause the seat cover to split. The hinge may break or the hinge screws may pull out.**

Use only a mild soap, a soft cloth and warm water when cleaning a seat. Abrasive cleaners can destroy the finish.

Using Valves to Locate the Leak

In some cases you may have trouble determining exactly which pipe is leaking. By closing valves one at a time, you can usually find the source of the leak. Follow the steps below to isolate the leak to a given section of piping. This method can save considerable time and can help you decide whether to repair the leak or to replace the pipe with a new material.

First, close the house valve. Remove the cover of the water meter and watch for movement of the meter's hands. If hands on the meter are moving, there's a leak in the water service pipe between the meter and the house valve. You'll have to excavate along that line until the leak is found. Pipes that are split because of freezing or that are weakened by corrosion must be replaced.

In an emergency, a small hole in a pipe can often be repaired temporarily with a pipe clamp. A pipe clamp consists of a rubber patch and a hinged frame that's pulled together tightly by two bolts. The rubber patch seals the leak.

If a pipe clamp is not available, a sharpened wooden stick can be forced tightly into some holes. As the wood swells, it seals the leak without materially affecting the flow of water through the pipe. Remember, these measures are emergency repairs only. Plan to make permanent repairs as soon as practicable.

If a small section of piping is split or corroded through, cut out the bad section with a hacksaw. Then join the two sections of piping with a *dresser coupling* to make a watertight connection. (See Figure 21–30.) A dresser coupling can provide a permanent repair if the rest of the piping is sound.

Sometimes the leak will be at a threaded connection. Try unscrewing the fitting and applying pipe compound to the threads. This seals the joint when the fitting is screwed onto the pipe again.

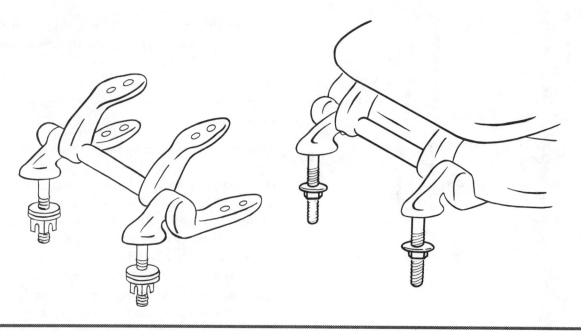

Figure 21–29
Typical water closet seat hinge

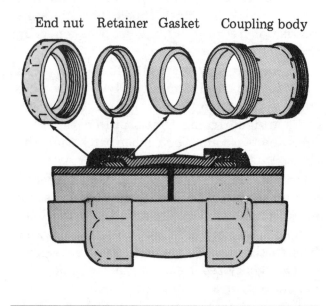

End nut   Retainer   Gasket   Coupling body

Figure 21–30
Dresser couplings

If you're sure that the building water service piping isn't leaking, open the house valve and close the control valve to the water heater. Recheck the water meter hands for movement. If the hands are moving, the leak is located in the cold water distributing pipes under the floor. Rule out a slow leak in the cold water distributing pipes by placing an ear against one of the inside fixture faucets. If you don't hear running water, the leak is in the hot water distributing pipes.

Leaks in cold water distribution systems under concrete floors are very hard to locate. If there's a leak in the cold water system, usually you'll advise abandoning the old pipes and installing new piping overhead or around the outside of the building.

Leaks in hot water piping under a concrete slab are much easier to locate. Just find the hot spots by walking barefoot on the slab. Break a hole in the floor and repair the leak. Leave the hole open for a few days to be sure the repair is successful.

If the hot water pipe is heavily corroded, install new piping overhead in the attic.

Leaking Tanks

Leaks in tanks are usually caused by corrosion. Occasionally a safety valve may fail to open, causing a rupture in the tank.

If you detect a corrosion leak in the tank wall, suspect leaks from the same cause in other parts

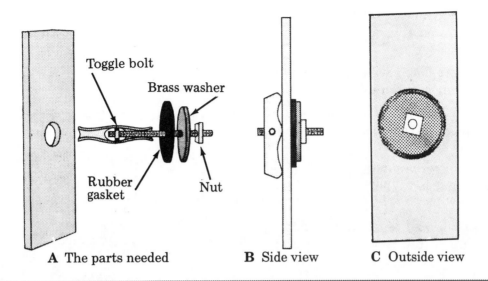

Figure 21–31
Patching a hole in the tank

of the tank wall. Repair to leaking tanks should be considered temporary. Usually the tank will have to be replaced as soon as possible.

A leak in a tank can be temporarily repaired with a toggle bolt, a rubber gasket, and a metal washer. See Figure 21–31. Drill a hole in the tank large enough to install the toggle bolt. Draw the bolt up tight to compress the rubber gasket against the tank wall and seal the leak.

Questions

1. What types of valves are normally used to control the water to or in a building?

2. Valve repairs are generally limited to _____ or _____.

3. When a globe valve does not control the water, what must be done to correct it?

4. What is the *first* step in disassembling a control valve?

5. What may be the problem when a washer requires frequent replacement?

6. How are replaceable valve seats identified?

7. What is the name of the tool used to remove pits or rough surfaces from valve or faucet seats?

8. What should be done first to repair a leak around the stem of a control valve?

9. What type of faucet is now most often used on plumbing fixtures?

10. What is the most common repair required on fixture faucets?

11. In which direction must a valve handle be turned to shut off the water?

12. Where is a house control valve usually located?

13. When water leaks from under the handle, what is usually needed to repair it?

14. What do newer type compression faucets use to prevent leaks from under the handle?

15. What repair is needed when water drips from the spout?

16. What is the procedure for removing a frozen faucet handle when a "handle puller" is not available?

17. When water flowing through a faucet is "noisy," what, other than a loose washer, may be the cause?

18. What must be replaced when water leaks around the swing spout of a faucet?

19. What precautions should be taken when pliers are needed to loosen a chrome swing spout locknut?

20. Why do single handle mixing faucets last longer than other types before repairs are needed?

21. How do the Delta and Moen faucets differ in controlling water temperature and flow?

22. Describe and size the two washers in a sink trap.

23. What is the U-shaped portion of a P-trap called?

24. Name the two parts of a sink basket strainer that usually must be replaced because of age.

25. What distinguishes a washer designed for a sink tailpiece from one designed for a P-trap?

26. What is the name of the washer that seals the bottom of a lavatory bowl?

27. What should be tried first to stop a leak at the water closet tank bolt?

28. Why is it important not to overtighten water closet tank bolts?

29. What is the device that provides the waterway between a wall-mounted tank and a closet bowl?

30. Where is a water closet spud washer located?

31. What would be the probable cause of a water closet leak at the floor connection?

32. Why does a water closet tank sometimes "sweat"?

33. What may be used to stop tanks from sweating?

34. What is the valve called that controls the water to a water closet tank?

35. What three checks should you make when a trickle of water continues to flow from the tank into the bowl?

36. What three checks should you make when there is the continual sound of running water coming from a water closet but no visible water entering the bowl?

37. What must one be familiar with to repair a ballcock?

38. What should be done first to repair or replace a ballcock?

39. What does the guide in a water closet tank do?

40. When the nuts are frozen to the bolts of a water closet seat and will not come loose, how can you remove the old seat?

Simple Gas Installations

Matter occurs in three physical states: solid, liquid, and gas. Although gas is much lighter per cubic measure than the other two forms of matter, it does have weight. It can be forced through very small spaces, a valuable characteristic. Gas has neither a fixed shape nor a fixed volume. It is made of constantly moving atoms. When these atoms are forced into a container, they take on the container's shape but occupy only about one-thousandth of the container's interior space.

Gas liquefies when cooled to below its boiling point. When this temperature is reached, the gas particles are pulled together to form a liquid. This principle is used in making liquid oxygen.

Kinds of Gas

Natural gas, *or methane*, contains chemical impurities which are valuable for uses other than as a fuel. These impurities are removed before the gas is piped to the customer. What we know as natural gas is known as dry or sweet gas to chemists. Natural gas is not poisonous but can cause suffocation in a closed space because it drives out the oxygen. It's also explosive when well-mixed with oxygen. Since natural gas is clean, dry, and has no odor, a gas leak in a pipeline or in pipes within a building might go undetected until an explosion occurs. A chemical scent is added to gas before it enters the pipelines to warn anyone in the area of escaping gas before the concentration can reach danger level.

Manufactured gas is usually made from coal. It burns with a blue flame and is generally added to other fuels to increase its heating capacity. Manufactured gas is also used by consumers as fuel in homes and industries. It can be poisonous, since it contains carbon monoxide. It's also explosive under certain conditions.

Liquefied petroleum gas is also known as LP or bottled gas. It's produced by plants that process natural gas. LP gas consists primarily of butane or propane, or a mixture of both. LP gas liquefies under moderate pressure. That makes it easy to transport and store in special tanks. When LPG enters a building's gas piping system, the liquid drops to normal atmospheric pressure and temperature. This turns it back to a gas.

LP gas is heavier than air, colorless, and nonpoisonous. Since it's easily containerized and transported, it makes a convenient fuel for homes and businesses in remote areas.

Sizing Gas Service Piping

The *gas service pipe* is sized and installed under the direct control of the gas supplier. This pipe extends from its connection at the gas supplier's pipe (located on public property) to the gas meter (usually located outside the building). You'll be responsible for sizing and installing the gas supply piping only within the building.

Plumbing codes seldom, if ever, govern the sizing or installation of gas supply systems. The plumbing code usually refers the installer to the local gas code. The gas code is like the plumbing code in one respect: it can be hard to interpret correctly. Trying to learn and comply with these codes can be both frustrating and discouraging, especially to someone installing something as simple as a gas line in a home.

To make your task even more difficult, the information you need is scattered throughout the gas code. This chapter you're reading now provides the basic information from the gas code, but *it doesn't replace the gas code.* If you follow the instructions here, you should find all you need for simple residential installations. Commercial installations and less common materials used with gas systems are examined in the more advanced *Plumber's Handbook Revised* by this author.

■ What You Need to Know

You can size the gas main and branch lines once you know:

- The maximum gas demand at each appliance outlet

- The length of piping required to reach the most remote outlet

Figure 22–1 takes into consideration key sizing factors such as pressure loss, specific gravity, and diversity.

Gas appliance manufacturers always attach a metal data plate in a visible location on each appliance. This plate shows the maximum input rate in Btu. (Btu is the abbreviation for British thermal units. One Btu is the quantity of heat required to raise the temperature of 1 pound of water 1 degree Fahrenheit.)

Length	Nominal Iron Pipe Size (inches)*		
(feet)	1/2	3/4	1
10	176	361	681
20	121	251	466
30	98	201	376
40	83	171	321
50	74	152	286
60	67	139	261	Residential
70	62	126	241
80	58	119	221
90	54	111	206

*More complete sizes are found in your local gas code

Figure 22–1

Maximum capacity of pipe in cubic feet of gas per hour, based on natural gas at 1,000 Btu per cubic foot

The tables in the code prescribe the sizing of gas piping in cubic feet of gas rather than in Btu. So you have to convert each Btu input rating to cubic feet of gas before sizing the distribution piping.

You can assume that each cubic foot of natural gas releases 1,000 Btu per hour. The Btu rating of some gas varies from this figure, but using 1,000 Btu per cubic foot is a safe assumption.

Assume you're sizing pipe for a cooking range with a maximum demand of 68,000 Btu per hour. Divide the value in Btu by 1,000 to find the demand in cubic feet per hour. Thus, 68,000 Btu divided by 1,000 is 68 cubic feet per hour.

On used appliances, the Btu rating on the data plate may not be legible or the plate itself may be missing. In such a case, it's a safe practice to make the appliance inlet pipe no smaller than the supply pipe serving the appliance.

A larger size supply pipe could be installed without violating the code, but the appliance wouldn't work any better. The supply pipe should never be smaller than the appliance's inlet pipe, and under no circumstances can it be smaller than 1/2 inch.

Figure 22–1 can make your sizing chores quick and easy. *Use it only to help understand how to use the table in your local gas code.* This low pressure gas table is the one most

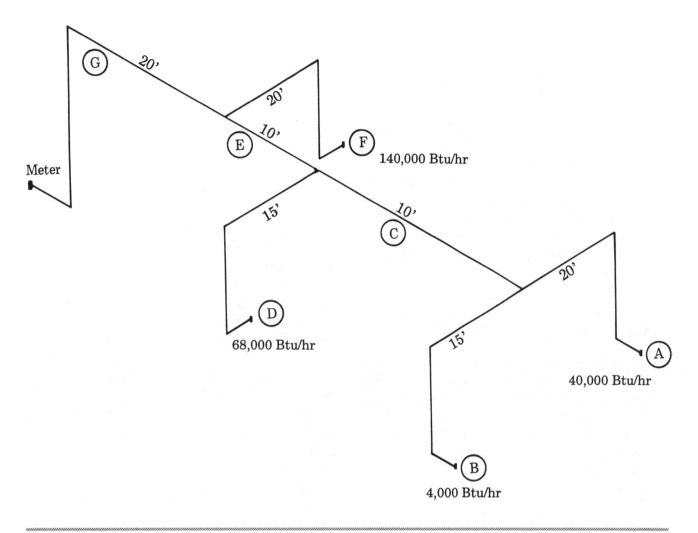

Figure 22–2
Natural gas residential installation

commonly used by professionals. Low pressure gas is used in millions of home and business installations.

Sizing a Sample System

Figure 22–2 shows a simple gas piping system similar to what you might find in most single family residences. Note that each section of piping must be sized to serve the Btu input rating of the appliance being served. Studying this illustration and the explanation in the rest of this chapter will help you size nearly any residential gas line correctly. We'll use Figure 22–1 to size the gas piping system in Figure 22–2.

Assume that the total developed length of gas piping in Figure 22–2 is 60 feet from the meter to outlet A. Find that distance in Figure 22–1. Underline all figures in each vertical demand column opposite 60. (This row is shaded in Figure 22–1.) The pipe size listed at the top of each demand column is the size that carries the volumes listed in that column.

Notice that the maximum gas demand of outlet A is 40,000 Btu per hour, or 40 cubic feet per hour (cfh). The correct pipe size is ½ inch because ½-inch pipe will support up to 67 cubic feet per hour.

The maximum demand of outlet B is 4,000 Btu, or 4 cfh. The correct pipe size is still ½ inch. The combined maximum demand for A and B is 44 cfh. The correct pipe size for section C is ½ inch. Maximum demand of outlet D is 68,000 Btu per hour, or 68 cfh, and the correct pipe size is ¾ inch. The combined maximum gas demand for A, B, C, and D is 112 cfh. The correct pipe size for section E is ¾ inch. The maximum demand of outlet F is 140,000 Btu per hour, or 140 cfh, and the correct pipe size is 1 inch. The combined maximum gas demand for A, B, C, D, E, and F is 252 cfh. This is the maximum gas demand for the entire residence. The correct pipe size for section G is 1 inch.

When you've sized the gas pipe in Figure 22–2, size each section of piping in Figures 22–3 and 22–4. Then have someone knowledgeable in plumbing check your work. The more practice you get, the more proficient you'll become.

Materials for Gas Installations

Your local gas code regulates the materials and installation methods for gas systems. All gas codes limit the use of certain materials and prohibit the use of others. This section explains how to install either of the two more common gas piping systems on a simple residential job.

Before ordering gas supply pipes and fittings, consider characteristics of gas that will flow through the system. For example, natural gas in some areas is classified as corrosive. Any gas that has more than 0.3 grains of hydrogen sulfide per 100 cubic feet is corrosive. If the gas supply in your community is corrosive, some types of piping will be prohibited. If you don't know anything about the character of the gas to be used, call your gas supplier before ordering materials.

▨ Galvanized Steel Pipe

Galvanized steel pipe is the most common gas piping material. Because of its versatility and strength, it can be installed both underground and above ground and is approved for use where gas is corrosive. Galvanized steel pipe in simple gas installations is threaded. Fittings should be of the same material as the pipe. Threaded joints must be sealed tight with an approved pipe compound. Galvanized steel pipe cannot be installed under a concrete slab.

▨ Copper Seamless Pipe

Copper seamless pipe or *copper tubing, Type K or L,* can be used for interior gas piping if first approved by the local authority or the area gas supplier. You can't use copper pipe or copper tubing if the gas is corrosive. Copper pipe or copper tubing can be used underground outside of buildings but can't be installed under a concrete slab. Fittings must be of the same material as the pipe or tubing.

Where copper pipe is selected, joints between fittings and pipe should be made with a hard solder, usually a silver solder. Joints between fittings and pipe can also be brazed. This usually means that a filler of brass is used to join the metals. This type of gas piping generally requires special equipment and a knowledgeable plumbing professional. Don't try brazing joints in gas piping unless you have expert assistance.

Copper tubing may be used with approved gas flare fittings. Copper tubing may be concealed within partitions if the tubing is continuous and of one piece. Flare fittings used for gas openings or for joining two sections of pipe must not be concealed. Copper tubing may be used to advantage where the system can be exposed or is accessible, as in a building crawl area or basement.

Installing Gas Piping

Metallic gas piping installed underground in exterior locations shall have a minimum of 12 inches of earth covering. Pipe installed in corrosive soils should be protected with an approved wrapping or one or two coats of asphaltum paint. Where freezing is common in winter months, the trench bottom should be below the frost line. That keeps the moisture in the gas from condensing, freezing and then rupturing the pipe. Insulate pipe where it enters a building above ground, in crawl spaces, and anywhere it's not protected from the cold. Otherwise, moisture in the gas can freeze and block free flow of gas in the pipe.

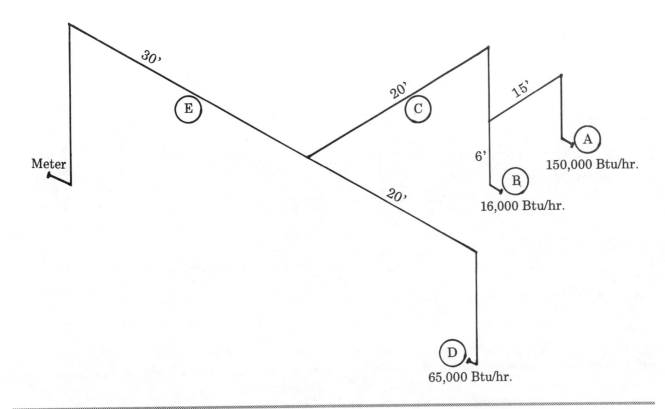

Figure 22–3
Residential diagram for sizing practice

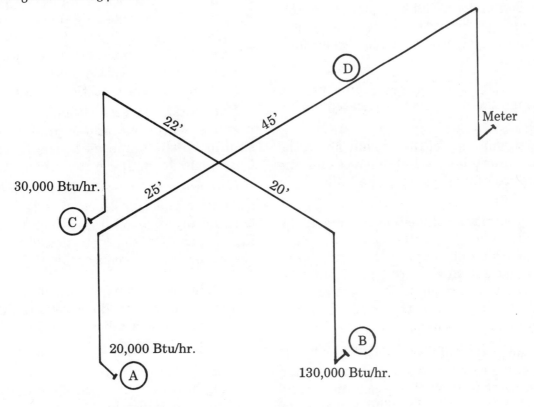

Figure 22–4
Light commercial diagram for sizing practice

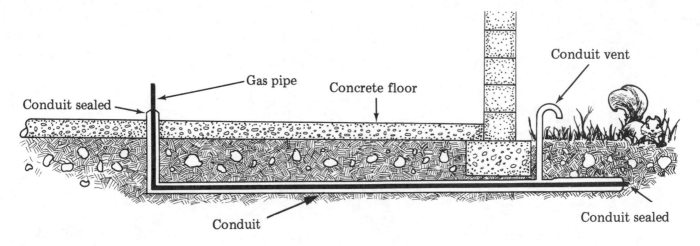

Figure 22–5
Gas piping installation in protective conduit

Lay underground gas piping in open trenches on a firm bed of earth. The pipe should be securely supported to prevent sagging and excessive stress during backfill. Use only fine material for backfilling.

Occasionally gas piping must be installed underground under a building slab. When this is unavoidable, the code permits this type of installation only if the following three conditions are met:

1. The entire length of gas piping up and through the floor must be encased in conduit.

2. The termination of the conduit above the floor must be sealed to prevent the entrance of any gas into the building in case of a leak.

3. The termination of the conduit outside the building must be tightly sealed to prevent water from entering the conduit. A vent must be extended above grade and be secured to the conduit. This vent conveys any leaking gas to the outside of the building. See Figure 22–5.

Installing gas appliances can be a problem if they're located on a concrete slab in the center of a room away from adjoining partitions. The best solution is to install the piping in an open channel cut into the concrete floor. The channel must be as deep as the outside diameter of the pipe. A removable grille or cover must be installed over the channel to provide access for repair or replacement of the gas piping. See Figure 22–6.

When gas piping must pass through suspended concrete slabs (as in a basement ceiling) or through masonry walls, the pipe must be properly protected against corrosion by sleeving or by painting with asphaltum paint. Where gas piping must be installed in vertical masonry walls, adequate chases must be provided to protect the pipe. See Figure 22–7.

Horizontal and vertical supports for galvanized gas piping, copper pipe and tubing should be the same as those described in Chapter 16 for similar water piping.

Here are some rules to follow when installing galvanized piping or copper pipe or copper tube horizontally in wood-framed partitions walls. Provide the following protection to the building structure and to the pipe or tubing:

1. Install short runs of horizontal gas piping or tubing which don't require additional joints through a hole drilled in the center of partition studs.

2. Install longer runs of horizontal gas piping or tubing in notches cut deep enough into the wall stud to conceal the pipe or tubing. (To avoid weakening the stud, don't cut deeper than 40 percent of its total width).

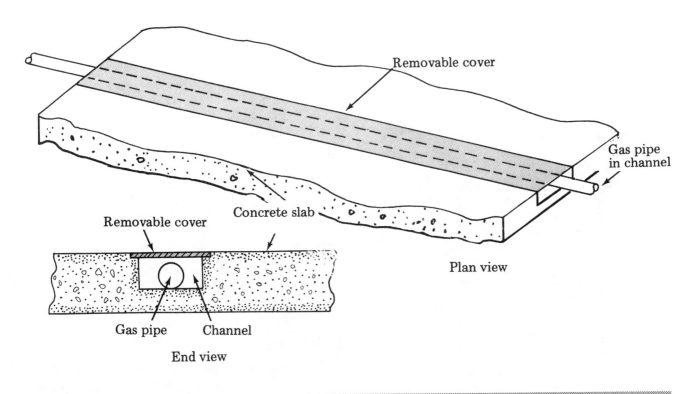

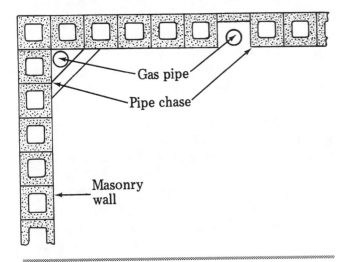

Figure 22–6
Gas piping installation in open channel

Figure 22–7
Gas piping installed in vertical masonry wall
Plan view

3. Protect copper pipe or tubing in a notched partition with a metal stud guard to avoid penetration by lath nails. See Figure 16–6 back in Chapter 16.

Metal stud partitions are replacing wood partitions in many new buildings. Metal studs are hollow rather than solid. Gas pipe or tubing can be fed through holes already cut in the center of studs. The pipe or tubing must be wrapped with an approved material to prevent contact with the studs. Secure the pipe to the studs with tie wire.

▇ The Drip Pipe

The gas main must be installed to drain dry, and the pitch or grade must be toward the gas meter. Many gases contain moisture. Most codes require that drip pipes be provided to receive condensation that forms within the pipe. The drip pipe must also be accessible for emptying. This drip pipe is usually assembled from a tee, nipple and cap, as shown in Figure 22–8. This drip pipe should not be smaller than the pipe it serves. The drip pipe, if exposed, should be protected from freezing in colder climates.

Connect gas branch pipes to other horizontal pipes at the *top* or the *side* of the feeder pipe, never from the bottom. This keeps condensate from filling and obstructing the flow of gas through the branch lines. See Figure 22–9.

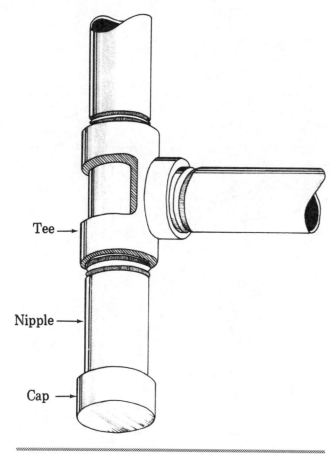

Figure 22–8
Gas drip pipe

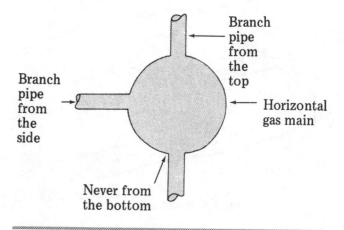

Figure 22–9
Correct connection of branch pipes to horizontal gas main

Shutoff Valves

A shutoff valve must be provided near the gas meter for each building. It should be located on the outside of the building or at some other accessible location. Its primary purpose is to shut off gas to the building when making repairs or additions to the existing gas piping. To help prevent accidental or malicious tampering, the valve usually has a square nut head which can be turned only with a special tool.

Each gas appliance within a building must have an accessible, manually-operated shutoff valve. This valve must have a lever handle that does not require a tool, so the occupants can shut off the gas in an emergency. This valve should be located as close as possible to the gas outlet pipe serving each appliance.

Shutoff valves are manufactured in two types: a straight pattern and an angle pattern. The shutoff valve must be installed not more than 6 feet from the appliance it serves.

Other Requirements

Galvanized or copper pipe installed in a concealed location must not have unions, right or left couplings, running threads, bushings, or swing joints. Copper tubing must not have flare fittings.

Use only a ground joint union to make a new connection to an existing line in a concealed threaded gas piping system. Punch the center nut on the joint to keep it from working loose under vibration. No new connection is permitted on a concealed copper tubing gas installation.

Threads for gas pipe must conform to the standards adopted by the American Standards Association. Look back to Figure 16–12 in Chapter 16 for the number and length of standard pipe threads. Don't use pipe with chipped or torn threads. Cutting, threading, and reaming gas pipe is done the same way as for water pipe.

Do not conceal the completed gas installation until it is pressure tested. Cap each outlet and install a pressure gauge on one of the outlets. Pressure test at 10 to 20 pounds. The system must remain airtight (no loss of pressure) until it is inspected. Leave the caps in place until the appliances are ready to be connected. If a leak occurs during the test of the rough piping, find the leak by brushing a liquid soap around each joint. The leaking joint will blow bubbles.

It's the plumber's job to connect the gas from the wall outlet to the appliance. This connection can be made with either a rigid pipe or an approved flexible connector. When the appliances have been set in place and connected, the gas supplier purges the lines of air, checks for leaks at the joints that connect the appliance to the gas system, lights the pilot lights, and adjusts each appliance.

▓ Venting Requirements

Gas appliances such as water heaters and clothes dryers can be installed on the floor of a residential garage under certain circumstances: the floor of the garage must be higher than the driveway or adjacent ground and the combustion chamber must be a minimum of 18 inches above the floor or adjacent ground. The appliance may be installed in a separate room off the garage if the walls, ceiling, and door of the room have a 1-hour fire rating. Ventilation must be provided through permanent openings with a total free area of 1 square inch for each 5,000 Btu per hour of input rating. One vent opening must be a minimum of 12 inches above the floor. The second opening should be a minimum of 12 inches below the ceiling. This permits air circulation for combustion and dilution of the flue gases. See Figure 14–1 in Chapter 14.

Gas appliances should be installed so there's access to the appliance for repairs, replacement, and cleaning, as well as for the intended use. Water heaters must not be installed in any living area which may be closed, such as bedrooms or bathrooms.

Some gas appliances don't have to be vented. An appliance rated at 30 Btu or less per hour per cubic foot of room space needs no vent. For example, suppose a water heater with an input rating not over 5,000 Btu is installed in a room (not normally closed) 6 feet long by 4 feet wide with a ceiling height of 7 feet. Is a vent required? Here's the calculation:

$$6' \times 4' \times 7' = 168 \text{ CF}$$
$$5,000 \div 168 = 29.76 \text{ Btu per hour}$$

No vent would be required. That's why small, freestanding gas space heaters and gas logs for fireplaces don't need a vent. However, water heaters with an input rating of 5,000 Btu or less are rarely used.

Gas water heaters with an insulated jacket must not be installed closer than 2 inches from any combustible material. Water heaters should not be installed closer than 1 inch from enclosures constructed of 1-hour fire-rated materials.

Appliances that must be vented should be installed as close to the vent pipe as possible. If a draft hood is required, the vent pipe should never be smaller than the opening of the draft hood. Gas appliances and their vents should be installed so there is sufficient clearance from combustible materials to avoid a fire hazard.

The most common vent pipe is *double wall metal pipe*. The clearance distance specified by the manufacturer will be stamped in the metal. At least a 1-inch clearance from any combustible material is recommended.

Vent pipes installed in partitions constructed of combustible material must have an approved metal spacing device. This spacer keeps the surface temperature of the flue pipe below 160 degrees F. See Figure 22–10.

Single wall metal vent piping can be used for exposed vent pipes installed in a room built of noncombustible material. All horizontal vent pipes must be supported to prevent sagging or misalignment. Straps or hangers should be constructed of at least 20 gauge sheet metal. The horizontal vent section length must not exceed 75 percent of the vertical vent length.

All gas vent pipes above the roof of a building must terminate in a UL-approved cap. Figure 22–11 shows several examples of acceptable vent pipe terminations outside a building.

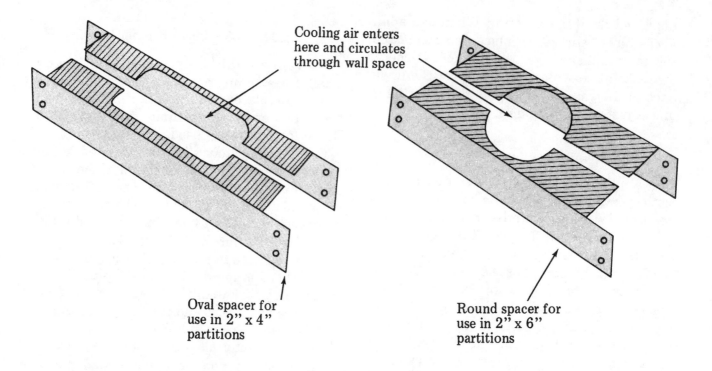

Cooling air enters here and circulates through wall space

Oval spacer for use in 2" x 4" partitions

Round spacer for use in 2" x 6" partitions

Figure 22–10
Vent spacers

Questions

1. Matter occurs in what three physical states?

2. One of the very valuable characteristics of gas is that it can be _____ through very small spaces.

3. Gas has neither a fixed shape nor a fixed volume, but is made of constantly moving _____.

4. Gas particles are liquefied only when they are cooled below what point?

5. What is added to the natural gas system before it enters the pipelines so that gas leaks can be detected?

6. What are some of the other names for natural gas?

7. How can natural gas cause death even though it is not poisonous?

8. From what substance is manufactured gas chiefly produced?

9. What substance is contained in manufactured gas that makes it poisonous?

10. Why is LP gas convenient to use as fuel in remote areas?

11. Who sizes the gas service pipe to a building?

12. Who or what governs the sizing and installation methods of gas supply piping?

13. What does the abbreviation "Btu" stand for?

14. One Btu is the quantity of heat required to do what?

15. How many Btus are assumed to be in each cubic foot of natural gas?

16. If you know the maximum Btu rating for an appliance, how do you convert the Btu into cubic feet?

17. What minimum pipe size should be used when connecting a gas supply pipe to an appliance that has lost its Btu rating plate?

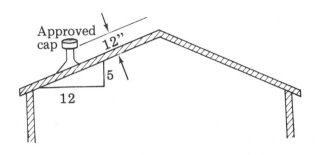

1- Roof pitch 5'/12' (22½°) or less (includes flat roofs). Maintain minimum clearance of 12'' as illustrated.

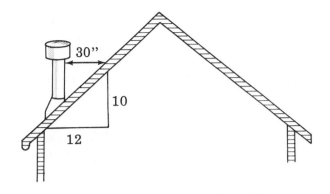

2- Roof pitch 5½'/12' to 12'/12' (45°). Maintain 30'' horizontal distance as illustrated.

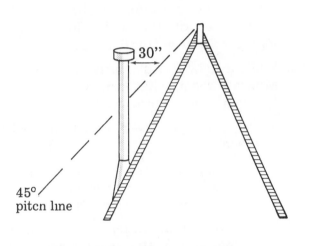

3- Roof pitch greater than 12'/12', or vertical wall. Maintain 30'' horizontal distance from 45° pitch (12'/12') line.

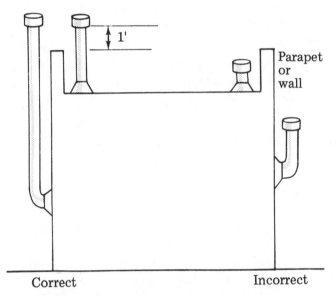

4- Vent top should be located one foot above parapet or wall when within 30''

Figure 22–11
Acceptable vent pipe terminations

18. What is the minimum size gas pipe outlet that can be used under any circumstances?

19. What two things must you know before sizing any gas main or branch lines?

20. What regulates the materials and installation methods for a gas system?

21. What is the most common material used in residential gas piping?

22. What condition might prevent the use of copper piping in a gas system?

23. What two weights of copper pipe and tubing are required for interior gas piping?

24. Copper pipe or tubing may be used outside underground except for what condition?

25. When joints are necessary in a copper gas piping system, what type of solder must be used?

26. Flare fittings may be used in a copper gas piping system only if they are _____ and _____.

27. Name two things that should be done to protect gas piping when it is installed underground in trenches.

28. When backfilling a trench containing gas piping, what type of backfill should be used?

29. Describe the three steps that must be taken when gas piping must be installed under a slab.

30. How should gas piping be installed to serve an appliance located in the center of a room?

31. What must be provided to protect gas piping installed in vertical masonry walls?

32. When might it be necessary to drill a hole in the center of the partition stud for horizontal gas piping?

33. Why must you *not* notch a partition deeper than 40 percent the width of the stud?

34. What is generally used to secure gas piping installed in metal stud partitions?

35. A gas main must be _____ to drain dry.

36. What must be installed for certain gases containing moisture?

37. Why should gas branch pipes be taken only from the top or side of a gas feeder pipe?

38. What is the purpose of installing a shutoff valve near the gas meter?

39. What type of shutoff valve should each gas appliance in a building have?

40. What two types of shutoff valves are manufactured for appliances?

41. What is the maximum distance allowed from a shutoff valve to the appliance it serves?

42. When are left and right threaded couplings in a concealed gas piping system permitted?

43. What must be done to prevent a union in an existing concealed gas line from working loose?

44. When is a new connection in an existing concealed copper tubing gas line allowed?

45. To what standards must the threads for gas piping conform?

46. The procedure for preparing threads for gas piping is the same as for _____.

47. What last major step must be taken before gas piping can be concealed?

48. What is the best and safest way to check gas piping for leaks?

49. What is the minimum height above the garage floor that the combustion chamber for a gas water heater may be set?

50. When a gas water heater is installed in a separate room, what type of ventilation openings must be provided?

51. Where must gas water heaters never be installed?

52. If a gas appliance input rating does not exceed 30 Btu per hour per cubic foot of room space, what may be omitted?

53. What is the minimum separation between any combustible material and a gas water heater with an insulated jacket?

54. What size vent pipe would be required for a 30 gallon gas water heater with a 4-inch draft hood?

55. What are the two types of vent piping materials acceptable for installations?

56. What must be provided for vent pipes installed in partitions constructed of combustible material?

57. What gauge should metal straps or hangers be to support horizontal gas vent piping?

58. All gas vent pipes terminating above a roof must be equipped with what?

23

Private Swimming Pools

Approximately 75,000 new pools are installed in the United States each year. Most of these are considered permanent pools by plumbing codes because they're constructed in the ground and can't be disassembled. Others are considered nonpermanent because they can be disassembled and reassembled to their original condition. These are considered seasonal and aren't intended for real swimming.

The only common type of swimming pool today is the recirculating type. This pool is equipped with a pump to *recirculate* water from the pool through a filter system. Some pools at private residences have a device that adds chlorine or fluorine automatically. Most owners of private pools add their own chemicals or use a professional pool service to maintain the water quality.

The recirculating swimming pool can accommodate heavy use without using very much water. Fresh water is added as needed to replace water lost by evaporation, splashing, or backwashing. Good filtration equipment assures pool users of water that is clear of organic matter and safe from harmful bacteria.

Water Supply and Water Disposal

Water for the pool can come from a well or from the public water system. Most owners of private pools use the public water supply to refill the pool by simply attaching a garden hose to a hose bibb. When this is done, the hose bibb must have a vacuum breaker to prevent a cross-connection. Pools may also have a direct connection to the public water system if a fill spout with an air gap above the overflow rim of the pool is installed. See Figure 23–1.

If pool water comes from a well, the water supplied must be clean and must meet the bacterial requirements for a domestic water supply system. If well water doesn't meet this requirement, a filtration system is needed to bring the water up to standard.

Occasionally the pool must be emptied or the filter flushed. This requires disposal of waste water. The most common waste water disposal methods for residential pools are listed here. Any one of these is acceptable.

- Waste water may be expelled into an adequately sized drainfield.

- Waste water may be disposed of through a sprinkler system used for irrigation. The

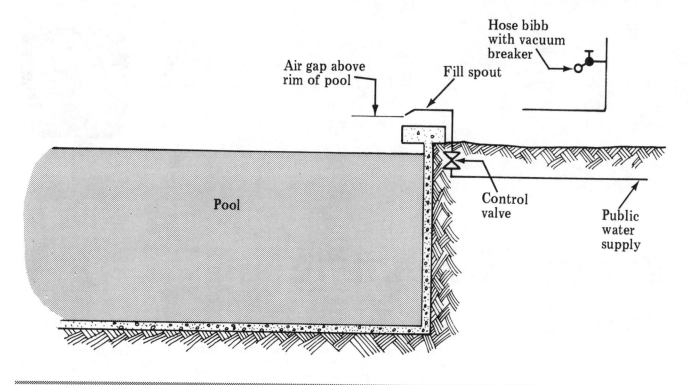

Figure 23–1
Water supply layout

water must be confined to the property from which it originates. It must not flow on to or across any adjoining property, public or private. However, backwash waste water must not be discharged through a sprinkler system.

- Both pool waste and backwash water may be puddled on private property if the disposal area is big enough and is properly graded to retain the waste water within the property. Standing water cannot remain pooled on the property for more than one hour after discharge. The disposal area must be at least 50 feet from any supply well.

Minimum Pool Equipment

A certain minimum of equipment is required by code for every pool. See the piping diagram (Figure 23–2) and the plan view (Figure 23–3) for a typical installation. Each pool must have a main drain (outlet) located at the pool's lowest point. This main drain must have a grate with an unobstructed open area at least 4 times that of the pipe it serves. The grate must be securely fastened so that tools are required for its removal. (The suction from the drain pipe under the grate could hold a small person under water.) The main drain must be located at the pool's lowest point so that it can drain dry for cleaning, painting, and general repairs. See the main pool drain detail in Figure 23–4 and the main drain installation in Figure 23–5.

A hair, lint, and sediment interceptor equipped with an easily-removable screen must be installed in the suction line ahead of the pump. The screen must be sized to have a free area 5 times the cross-section area of the suction pipe. See the interceptor detail in Figure 23–6.

Recirculation inlets (supply fittings) must be sized and spaced to produce uniform circulation of incoming water throughout the pool. One inlet is required for each 350 square feet of pool water surface or fraction thereof. See the supply fitting detail in Figure 23–7. When the main drain is used for a return, it must be

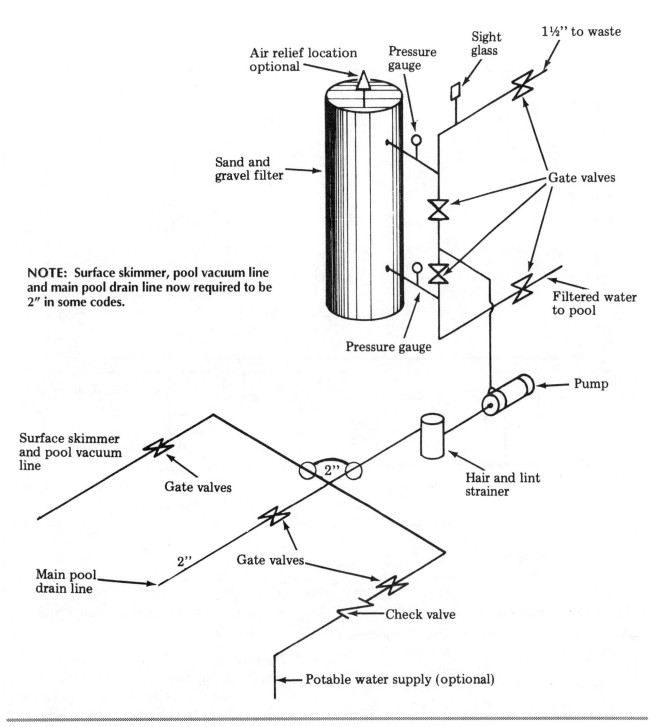

Figure 23–2
Pool equipment piping diagram

considered an inlet (supply) and sized at a minimum of 2 inches.

A vacuum fitting a minimum of 2 inches in diameter must be provided on all pools. This fitting is connected to piping which connects to the suction side of a pump. This fitting has to be installed in an accessible location a maxi-mum of 18 inches below the water line. See the vacuum fitting detail in Figure 23–8.

A valve must be installed on the main drain (outlet or suction) line. Valves, pumps, filters, and other equipment must be installed so that they are readily accessible for operation, main-tenance, and inspection.

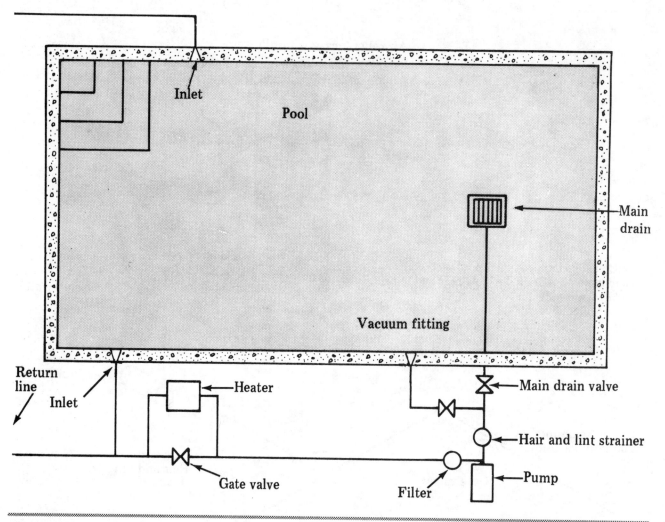

Figure 23–3
Plan view of swimming pool

▓ Filtration Equipment

Below are code requirements for common types of pool filtration equipment. A filter of a type other than these may be installed if a test shows it to be as efficient as sand and gravel filters, and if it is approved by the local authority.

Sand Filters

- Pressure sand filters must have a filtration rate not over 5 gallons per minute per square foot of filter area.

- Pressure sand filters must have a backwash rate of not less than 12 gallons per minute per square foot of filter area.

- Sand filters must contain a minimum of 19 inches of suitable grades of screened,

sharp silica sand properly supported on a graded silica gravel bed.

- A sufficient freeboard must be provided above the surface of the sand and below the overflow troughs or filter pipes to permit 50 percent expansion of the sand during the backwash cycle. There should be no loss of sand.

- The inflow and effluent lines must be provided with pressure gauges. See Figure 23–2.

- The backwash line must have a sight glass installed so that backwash water can be visibly checked for clarity. See Figure 23–2.

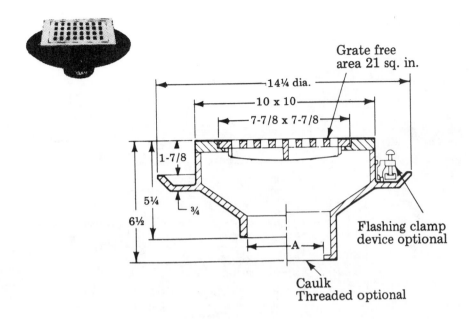

Figure 23–4
Pit and pool drain detail

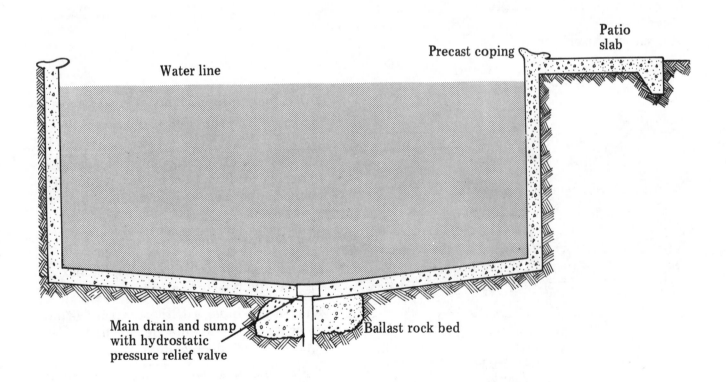

Figure 23–5
Main drain installation

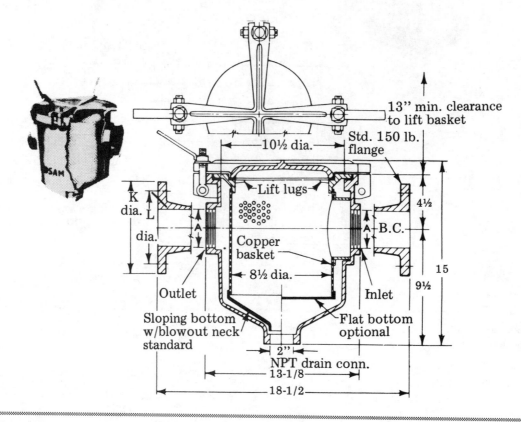

Figure 23–6
Sediment interceptor

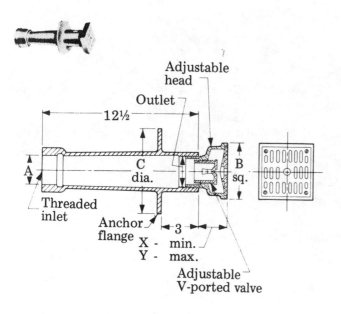

Figure 23–7
Swimming pool supply fitting

- Tanks longer than 24 inches must have an access hole measuring a minimum of 11 inches by 15 inches.

- All pressure filter tanks must be designed to remove air from the tank by an approved method or device. See Figure 23–2.

Diatomaceous Earth Filters

- Diatomaceous earth filters may be either the vacuum type or the pressure type.

- They must have a filtration rate of no more than 2 gallons per minute per square foot of effective filter area.

- A filter aid must be able to be introduced into the filter tank so that the filter septum or element is evenly precoated before it is put back in operation.

- The filter piping must be designed and installed so that during the precoating operation the filter aid recirculates or

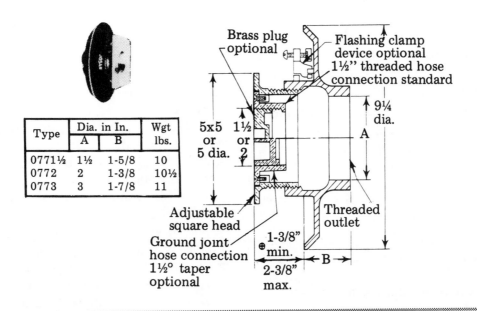

Figure 23–8
Swimming pool vacuum fitting

discharges through the waste pipe and does not return to the pool water.

- Provision must be made for removing the caked diatomite by either backwash or disassembly.
- The filter must be designed and installed so that the filter elements can be removed easily.
- Pressure or vacuum gauges must be installed on these filters to determine the pressure differential across the filter and the need for cleaning.
- All pressure filter tanks must be tested to a minimum of 50 psi.
- Filter elements must be spaced a minimum of 1 inch from face to face.
- An air relief device must be installed on each pressure filter tank at its high point.

Cartridge Filters
- Cartridge filters may be either the pressure type or the vacuum type.
- The filtration rate cannot exceed 1 gallon per minute per square foot of actual effective filter area.
- The cartridge must be manufactured of materials suitable for use in potable water.

- Filters must be designed and installed to permit ready disassembly and removal of the cartridge for cleaning.
- Filters must be equipped with a pressure or vacuum gauge. This gauge must be located to indicate when the filter needs cleaning.
- Filter tanks must be hydrostatically tested to a minimum pressure of 50 psi.

■ Surface Skimming Device

Swimming pools must have some form of surface skimming device to remove floating debris from the surface of a pool. One skimming device must be provided for each 600 square feet of pool surface or fraction thereof. The skimming device should have a volume control which will divert through the skimmer up to 50 percent of the filter capacity. Two types of skimming procedures are acceptable under most codes:

1. Skimmers built into the pool wall. These must have skimmer weirs that are automatically adjustable to variations in water level over a range of 3 inches. The skimmer weirs must be a minimum of 5 inches wide. A removable basket with a minimum surface of 75 cubic inches must

be provided. The basket must be accessible for cleaning through an opening in the deck and must have a minimum diameter of 5 inches.

2. An overflow gutter at the end of the pool. Gutters should have drainage grates and should be properly connected to the filter system.

Pumps

The pump selected for the pool must have enough capacity to filter and backwash the pool water at the proper pressure and rate required for the filter and piping system used. The pump must sit on a base made of materials suitable for outdoor use. The bottom of the pump motor must be elevated at least 4 inches above the surface level.

Pool Heaters

Gas-fired swimming pool heaters must comply with AGA (American Gas Association) standards. Pool heaters that have been tested and design-certified by the AGA will, of course, meet the energy requirements of most states.

Pool heaters must be provided with a thermostatic or high temperature control which maintains a maximum temperature differential of 15 degrees F, or with some other acceptable overheat protection device.

Pool water heating equipment installation is the same as for domestic water heaters. See Chapter 14.

A plumbing permit is required for the installation and repairing of pool piping and equipment. The entire pool pressure piping system, including the main drain, must be water-tested at 40 psi and prove tight before the installation is concealed.

Materials and Installation Methods

The piping materials and fittings of a swimming pool may be the same as those for potable water systems. This section reviews the two most commonly-used piping materials for residential pools. Installation methods of

other material types can be found in *Plumber's Handbook Revised,* by this author.

Thermoplastic pipe must be continuously marked on opposite sides by the manufacturer with the size, type, schedule, U.S. Commercial Standard and National Sanitation Foundation seal of approval.

PVC plastic pipe and fittings must be Schedule 40. The assembly and installation method is the same as for potable water piping of the same material.

Polyethylene plastic pipe is flexible and uses insert-type fittings. Stainless steel clamps must be used to secure each connection. Polyethylene plastic pipe that's approved for a building water service line may be used for swimming pool piping on pressure lines only. The assembly and installation method is the same as for sprinkler piping of the same material. See Chapter 17.

Polybutylene plastic pipe comes in rigid 20-foot pipe lengths or in flexible coils of varying lengths.

PE plastic is the newest plastic accepted by most codes. Installation methods can be found in Chapter 15.

Unless supported directly on existing ground, pool piping must be secured to the pool structure with pipe hangers or heavy-duty strap iron. Hangers must be spaced a maximum of 4 feet apart, center to center. Backfill must be fine material free of rocks larger than ¾ inch in diameter. Ells installed on any suction line underground must be of the long radius type. Short radius 90-degree ells (generally used on water piping) are prohibited.

Dielectric fittings are required where dissimilar metals are installed in pool and filter piping.

Spas and Hot Tubs

Spas and hot tubs have become very popular for recreation and therapeutic use in recent years. Most spas are permanent construction, while hot tubs may be disassembled and reassembled. Water capacities for spas and hot tubs can range from approximately 300 gallons to 700 gallons. Neither is used for swimming purposes and both are generally classified as special pools.

Most codes require that spas and hot tubs be of the recirculating type. The water supply, waste water disposal, and minimum equipment requirements are the same as for swimming pools (see Figure 23–2). The pool heater must heat the water to a temperature high enough to serve the facility's intended purpose.

Energy Conservation Law

The 1979 national energy conservation law created quite a stir for several years. All manufacturers of heaters and equipment using electric or fossil fuel (gas) had to meet rigid standards. Water heaters and pool heaters (as well as automobiles and home appliances) had to improve their energy efficiency, yet retain good performance.

Pool and spa heating have tapped into several acceptable energy sources.

- Natural gas (where available) or LP gas. This is the most popular type. Gas-fueled heaters are manufactured for both outdoor and indoor installation. They're highly energy efficient. They have built-in safety features and perform well, even in harsh climates and tough weather conditions.

- Heat pumps. Their energy source is electricity. Heat pumps come with either single or dual thermostat controls. They're more energy-efficient than a standard electric pool heater. They generally meet energy conservation law requirements.

- Solar. During the 1970s energy crunch, solar heating for pools found new life, especially in the Sun Belt states. Once the cost of installation was met, the owner had an unlimited, "free" energy source. Solar systems have proven to be practical and very efficient. They will generally meet the heating demands of most homes, including pool heating.

Today, most authorities will issue permits for pools, spas and hot tubs. Before proceeding with any one of these installations, however, you should check with your local code authority. Use only those that have been approved for your area.

Questions

1. A pool that is considered permanent can not be _____.

2. What is the most common type of residential pool?

3. How does a good filtration system protect pool users?

4. What must be provided for a hose bibb when it is used for adding water to a swimming pool?

5. How is cross-connection prevented when water is added to a swimming pool by means of a direct water supply connected to public water?

6. What are the special problems when well water is used to supply a swimming pool?

7. What are two acceptable ways of disposing of swimming pool waste water?

8. In what part of a pool is the main drain located?

9. Other than recirculating pool water, what other purpose does the main drain serve?

10. Why must the grate over a main pool drain be securely fastened so that tools are required for its removal?

11. Where must an interceptor be located in a pool piping system?

12. Why is it important to size and space pool inlets properly?

13. How many square feet of water surface may one pool inlet adequately serve?

14. What is the minimum pipe size connected to a main drain?

15. What is the minimum diameter size of a pool vacuum fitting?

16. What is the maximum depth below the waterline to which a vacuum fitting may be installed?

17. When valves, pumps, filters, and other pool equipment are installed, what provision must be made for their continued operation and maintenance?

18. Name the types of filtration equipment accepted by most codes.

19. Describe the two methods commonly used for surface skimming a pool.

20. What is the minimum height above the surface level for installing a pump motor?

21. What are the most commonly used materials for swimming pool piping?

22. If pool piping is not supported directly on the ground, how should it be supported?

23. What type of ell should be used underground on any swimming pool suction line?

24. The installation of pool water heating equipment is the same as for what other type of heaters?

25. At what pressure must pool piping be water-tested to prove the system is tight?

26. What is required before installing and repairing pool piping?

27. How do most spas and hot tubs differ in the method of construction?

28. What is the general use of a spa?

29. Gas-fired swimming pool heaters must meet whose standards?

30. What is the most popular energy source for swimming pool heaters?

Answers

Chapter 2

1. The three basic components of a plumbing system:
   1) the drainage and vent system
   2) the water service and distributing pipes
   3) the plumbing fixtures.
   See page 7

2. The three major classes of a drainage system:
   1) the drainage pipes
   2) the fixture traps
   3) the vent pipes.
   See page 7

3. A DWV system:
   the drain, waste and vent pipes of a building.
   See page 8

4. The component usually first to be installed and inspected:
   the drainage and vent piping.
   See page 8

5. The reason it's important to know how to read isometrics:
   most construction blueprints include locally approved isometric drawings for DWV systems.
   See page 9

6. The two basic components of a water supply system:
   the water service pipe and the distributing pipes.
   See page 9

7. The two separate systems of a water supply system:
   the hot and cold water distributing pipes.
   See page 9

8. A private water supply system is:
   a domestic well.
   See page 9

9. In diagramming the identification of hot and cold water lines:
   cold water piping: a solid line broken with a single dot or dash;
   hot water piping: a solid line broken with two dots or dashes.
   See page 9, Figure 2-2

10. Placement of hot water outlet (as you face the fixture):
    on the left side.
    See page 10

11. Common abbreviations:
    water closet = W.C.
    shower = SH
    water heater = W.H.
    See page 10

You can't lay out and install a plumbing system unless you know the location and source of water, and location and type of sewage disposal facility.
See Figures 3-1, 3-2 and 3-3

Chapter 3

1. Approved plans are called:
   "as-built" plans.
   See page 11

2. Steps taken if changes are made in approved plans:
   revised plans must be submitted for reapproval.
   See page 11

3. Included in a plot plan:
   1) shape and size of the lot
   2) location, size of building
   3) setback from property lines
   4) all permanent outside construction above ground (e.g., driveway, walkway)
   5) electric service
   6) easements (if any) and streets.
   See page 11

4. Below-ground information in plot plan:
   1) location and size of the gas service;
   2) source, location and size of water facility;
   3) storm drainage and disposal facility, if any;
   4) type, location and size of sewage disposal facility.
   See page 11

5. When it's acceptable to install facilities as shown on plot plans:
   after approval by plans examiner(s).
   See page 11

6. Plot plans approval for septic systems comes from plans examiner, plus:
   The Department of Environmental Resource Management (DERM) or, in some cases, local health department.
   See page 15

7. Necessity of plumber's knowing how to interpret plot plans:

Chapter 4

1. Purpose of isometric drawings:
   Professionals use isometric drawings to communicate their ideas about plumbing systems to others.
   See page 16

2. In an isometric drawing, pipes meet at angles of:
   120 (not 90) degrees
   See page 16

3. Type of triangle needed to do isometric drawing:
   a 30-60 right triangle.
   See page 16

4. The lines on isometric drawings represent:
   pipe and fittings.
   See page 17

5. Importance of knowing names of pipes:
   in order to understand how to design, lay out and install pipes and fittings.
   See page 18

6. Maintenance of a public sewer:
   usually done by the local water and sewage authority via taxation.
   See pages 18, 22

7. Three pipes which are not actually a part of a building drain:
   fixture drains, wet vents, waste pipes.
   See page 23

8. Three residential fixtures which a waste pipe (fixture branch) may serve:
   kitchen sinks, lavatories, bathtubs. (Any fixture that does not carry solid fecal matter is an acceptable answer. The only incorrect answer is "water closet.")
   See page 23

9. Three residential fixtures which a soil stack or soil pipe may serve:
   a water closet, a bidet, a shower. (A soil

pipe may carry waste from any fixture.)
See page 23

10. Definition of a stack:
    the vertical pipe or pipes of a building.
    See page 23

11. Number of fixtures at the same level served by a common vent:
    two.
    See page 23

12. Name given to continuation of vertical vent of the waste pipe to which it connects:
    a continuous vent.
    See page 23

13. Primary function of vent stack:
    to provide circulation of air to and from all parts of a drainage system.
    See page 24

14. Composition of vent system:
    all the vent pipes of a building.
    See page 24

Chapter 5

1. The maximum fixture unit load is used to size these sections of the drainage system:
   1) building sewer
   2) building drains
   3) individual waste pipes.
   See page 26

2. The gallons of water flow per minute in one fixture unit:
   7.5 gallons (1 cu. ft.)
   See page 26

3. The pipe sizes for a building drainage system as listed in the code table represent:
   the minimum sizes required by code.
   See page 26

4. The part of a waste pipe considered to be the normal capacity of waste piping:
   one-third full.
   See page 30

5. The reason drainage piping should be the smallest size permitted by code:

to promote scouring action within the pipe and help to prevent stoppages.
See page 30

6. The two primary differences in model plumbing codes in determining waste pipe sizes:
   1) each has its own fixture unit loads for the various plumbing fixtures
   2) each code has a distinct way of applying these fixture units to waste pipe sizes.
   See pages 26-27

7. The minimum size of a building sewer:
   4 inches. (This is uniform in all codes.)
   See page 28

8. A building drain may be considered a drain pipe when connected to a septic tank only under this condition:
   the developed length does not exceed 10 feet.
   See page 28, Figure 5-4

9. The size sewer that must be used to serve an accessory building with one bath, located on the same lot as main building and connected to a single building sewer:
   3 inches. (It's the same as a building drain, but is considered a horizontal branch line, not a building sewer.)
   See pages 28, 30, Figure 5-6

10. It's permissible to install a kitchen sink and clothes washing machine at different levels on a 2-inch waste if:
    a relief vent is used on the clothes washing machine fixture drain.
    See page 30, Figure 5-7

Chapter 6

1. The two major reasons a fixture trap seal is destroyed when not properly vented:
   back pressure or siphoning action
   See page 33

2. When vent pipes are inadequately sized and arranged, the following five problems may occur to a drainage system:
   1) plumbing fixtures drain slowly

2) water closets may have inadequate flush

3) back pressure could force sewer gases into the building

4) negative pressure could siphon the liquid trap seal

5) in very cold climates vent terminals are in danger of frost closure. Increasers are code-required.

See page 33

3. The four things that determine the sizing of vent pipes:

   1) maximum fixture unit load

   2) the developed length of the vent pipe

   3) type of plumbing fixtures vented

   4) the diameter of the soil or waste stack the vent pipe serves.

   See page 33

4. In a building with three vent stacks, the minimum size for at least one of these three:

   not less than 3 or 4 inches in diameter.

   See page 33

5. The minimum size vent stack for an accessory building with a full bath located on the same lot as the main building and sharing one common building sewer:

   2 inches.

   See page 34

6. When it's permissible for the diameter of a vent stack to exceed the diameter of the soil or waste stack to which it connects:

   never. (A vent stack can be the same as or smaller than the connecting soil or waste stack, but never larger.)

   See page 34

7. The smallest vent pipe in inches that can serve any plumbing fixture:

   1¼ inches.

   See page 34

8. In stack venting, the procedure that must be followed to connect each fixture to the stack:

   must connect independently.

   See page 34, Figures 6-2 and 6-3

9. The two major functions fulfilled by stack venting a group of fixtures:

   1) a single pipe receives the liquid waste

2) a single pipe also vents the group of fixtures.

See page 34, Figures 6-2 and 6-3

10. The plumbing fixtures that a vertical combination waste and vent stack may serve:

    low-unit rated plumbing fixtures (like lavatories, bidets, bathtubs, showers and sinks), under certain conditions.

    See page 36

11. Can a combination waste and vent stack connect to a stack vent:

    No, regardless of size.

    As to the reason:

    A pipe which generates waste cannot connect to a stack vent serving other fixtures.

    See page 36

12. A common plumbing fixture that cannot be installed on a 2-inch combination waste and vent stack:

    a kitchen sink.

    See page 36

13. Three plumbing fixtures that may connect to a wet vent system:

    acceptable answers include bathtubs, showers, bidets, and lavatories. (A wet vent can convey only waste from fixtures with low unit ratings.)

    See page 37

14. Two changes in direction a wet vent may assume and still be within the intent of the code:

    a vertical as well as a horizontal position.

    See page 36, Figure 6-5

15. The maximum distance acceptable by code for a horizontal wet vent:

    15 feet in developed length. (Check local code.)

    See page 37, Figure 6-5

16. The maximum height of a vertical wet vent connected to a horizontal wet vent:

    6 feet. (Check local code.)

    See page 37, Figure 6-6

17. The name given to a vent pipe that serves two fixture traps connected at the same level:

    a common vent.

    See pages 37, 38

18. When measuring the developed length of a fixture drain, the measurements that must be included:

crown weir of a fixture trap and any offsets or turns.

See page 39, Figure 6-11

19. The reason most authorities agree that a trap closer to the vent on a minimum slope serves it better:

the trap's seal is less likely to be broken by siphonage or back pressure when this condition is met.

See page 38

Chapter 7

1. To meet code requirements, each fixture must be equipped with:

a water seal trap.

See page 45

2. A trap protects human health by:

providing a liquid seal which prevents drainage system odors, gases and vermin from entering the building at fixture locations.

See page 45

3. The most commonly used trap in the trade:

a P-trap.

See page 45

4. The term used to identify a trap that does not retain foreign substances:

self-cleaning

See page 45

5. The code prohibits a 1½-inch trap from connecting to a 1¼-inch fixture drain because:

the code states that no trap outlet can be larger than the fixture drain to which it is connected.

See page 45

6. Traps that are prohibited by code for use in new construction:

¾ S-trap, full S-trap and crown vented trap, bell trap, running trap, pot trap, and trap with slip-joint nuts and washers on the discharge side of the trap above the water seal.

See pages 45, 47, Figure 7-2

7. The minimum depth permitted by code for a trap water seal:

2 inches

See page 47, Figure 7-3

8. Other than an interceptor type trap, the maximum depth permitted by code of the water seal for a trap:

4 inches

See page 47, Figure 7-3

9. All traps should be installed level:

to prevent self-siphonage.

See page 47

10. When two or more fixtures are permitted to use a single trap, the waste pipe for this installation is:

a continuous waste. (See text for limitation provisions.)

See page 47

11. The maximum center-to-center measurement allowed between two or more fixtures using a single trap:

30 inches. (See text for limitation provisions.)

See page 47

12. When three individual fixtures are permitted by code on a single trap, the fixture on which the trap must be located is:

the center fixture.

See page 47

13. (a) Double-trapping is:

when the liquid waste discharged from a fixture goes through one trap and then a second trap.

(b) It is permitted:

never.

See page 47

14. The maximum vertical length (drop) from a lavatory outlet to its trap water seal:

24 inches. (Some local codes permit 18 inches.)

See page 49, Figure 7-8

15. When a fixture tailpiece (drop) exceeds the maximum length set by code, the possible result:

self-siphonage of the fixture trap water seal.

See page 47

16. The maximum length of the vertical drop from a water closet outlet to the building drain:
    24 inches.
    See page 49, Figure 7-9

17. Description of the trapping function when a water closet is operated:
    the contents are siphoned with each flush. When the flush is activated, the trap seal is lost but is automatically restored as the flush tank refills by means of a refill tube located in the flush tank.
    See page 49

18. In a tank-type water closet, the action which automatically restores the trap seals:
    the ballcock has a refill tube that discharges into the overflow pipe of the flush valve. As the tank fills, water passes through the refill tube into the overflow pipe, restoring the trap seal.
    See page 49, Figures 21-19 and 21-27

19. The thing that's accomplished by using the shortest possible fixture tailpiece to connect to its trap:
    the fixture trap is made more efficient.
    See page 47

20. The thing which most codes prohibit on concealed fixture traps:
    cleanouts. (Concealed trap cleanouts on bathtubs or showers are not accessible and the seal may deteriorate, in time, and leak waste water into the ground beneath a building.)
    See page 49

21. Drum traps can be used in today's plumbing:
    only after approval by local authority and limited in use to special fixtures.
    See page 50

22. All drum traps, when approved for use, must be equipped with:
    a removable screwed cover for cleaning purposes.
    See page 50, Figure 7-10

23. The reason building traps are no longer required or permitted by code:
    today's buildings are properly vented to

protect fixture traps.
See page 50

24. Five ways the trap seal may be broken:
    1) wind effect
    2) evaporation
    3) capillary attraction
    4) trap siphonage
    5) back pressure.
    See pages 51-52

25. Difference between negative and positive pressures within a drainage system:
    Negative pressure within the fixture drain is known as siphonage action, which tends to form a vacuum, pulling the water seal from fixture trap.
    See pages 51-52, Figure 7-13

    Positive pressure within the fixture drain is known as back pressure. This occurs when a higher pressure develops in a drainage system that is improperly designed and vented.
    See page 52, Figure 7-15

26. It's illegal for a trap to have a slip joint nut and washer on the discharge side above the water seal because:
    washers have a tendency to deteriorate with age. This would permit sewer gases to enter a building. (Check local code.)
    See page 54

27. Caution should be used in installing a plastic trap on a kitchen sink waste because:
    sometimes home owners pour "too hot" liquids into the sink. Plastic traps and tailpieces can withstand heat up to 180 degrees F, only. (When plastic parts are installed, home owners should be cautioned to blend boiling water with cold tap water to avoid damaging plastic parts.)
    See page 54

28. S-traps may still be approved for use by local authorities when:
    they're used only to replace an existing S-trap in an older home.
    See page 47

29. The resulting action to a fixture trap when the fixture is installed on a drain pipe

exceeding the critical distance set by code and not properly vented:

can cause self-siphonage of the fixture trap.
See page 52, Figure 7-14

30. The crown weir of a trap is located:

at the highest part of the inside portion of the bottom surface at the crown of a trap.
See page 55, Figure 7-16

31. The portion of the trap that's known as a J-bend:

the part that holds the water which forms the seal of a trap.
See page 55, Figure 7-17

32. The dip of a trap is formed by:

the lowest portion of the inside top surface of the channel through the trap.
See page 55, Figure 7-16

33. At fixture locations a trap seal accomplishes three purposes:

1) prevents drainage system odors from entering the building
2) prevents gases from entering the building
3) prevents vermin from entering the building
See page 45

34. In a two-compartment sink with two different depths, the maximum depth of one compartment connected to a single trap:

not more than 6 inches deeper than the other.
See page 47, Figure 7-6

Chapter 8

1. The code prohibits cutting a hole in a drainage or vent pipe to remove a blockage because:

the hole would have to be patched. Such patch jobs often deteriorate, allowing raw sewage to seep out and into the ground.
See page 57

2. The dual role of a cleanout tee when installed at the end of a building sewer:

1) for testing purposes
2) for cleaning purposes.
See page 57

3. The advantage of a two-way cleanout fitting:

it permits upstream and downstream rodding.
See page 58

4. The required cleanout in a horizontal drain pipe for any change of direction:

when the change in direction is greater than 45 degrees.
See page 58

5. The creation of a dead end:

when the drain pipe is extended 2 feet or more to the outside of a building and is terminated with a cleanout.
See page 59

6. A countersunk plug is required for a cleanout when:

a cleanout terminates in an area where people walk, such as rooms, hallways and walkways.
See page 59, Figure 8-6

7. The maximum height from a finished floor to a cleanout installed in a vertical stack:

4 feet.
See page 59, Figure 8-7

8. The two possible procedures for accessibility to a concealed cleanout in a vertical stack:

1) a covering plate
2) an access door.
See page 59, Figure 8-8

9. The necessary clearance from walls or other obstructions for rodding purposes of a 3-inch cleanout:

18 inches.
See page 59, Figure 8-9

10. In a one-story building, cleanouts may be omitted, provided:

1) the vent stack is vertical throughout and extends up through the roof,
2) the vent stack is the same size as the waste pipe it serves,
3) the drainage system does not have more than one 90-degree change in direction,
4) the vent pipe must not be reduced beyond the stated minimums.
See page 60

11. The minimum size of a cleanout serving a 4-inch horizontal drain pipe:
4 inch.
See page 60

12. The distance between cleanouts in a horizontal straight 4-inch drain line:
should not exceed 75 feet.
See page 58

Chapter 9

1. Plumbing maintenance problems are easier in older homes because:
plumbing installations were simple
See page 63

2. The reason builders allot more space for bathrooms in today's homes:
Homeowners are looking for more than functionality in their plumbing. Beauty and style are important, also.
See page 63

3. The importance of having an isometric layout before beginning work:
to illustrate the type of installation best suited for the job; to show the number, type and size of fittings needed to do the work.
See page 63

4. At least three items must be included by the architect on his floor plans:
1) location and type of plumbing fixtures,
2) floor dimensions,
3) location of walls, doors and windows.
See page 63

5. An isometric drawing should show:
1) a stack system,
2) a flat system,
3) a horizontal wet vent system,
4) a vertical wet vent system.
See page 63

6. Other important steps one must consider before beginning the actual installation are:
1) the spacing of fixtures,
2) rough waste outlets at right locations,
3) avoid door and window openings.
See pages 63-64

Chapter 10

1. A plumbing code is needed:
to ensure that the design, installation and maintenance of plumbing systems will protect the health, welfare and safety of the public.
See page 73

2. The reason the drainage system should be designed carefully:
so one's work will not be turned down by the plumbing inspector.
See page 73

3. The purpose of properly arranging and sizing the vent pipes:
to provide free circulation of air within the drainage system.
See page 73

4. The code has the following provision to cope with stoppages:
it requires that an adequate number of cleanouts be installed to make all portions of drainage system accessible for cleaning.
See page 73

5. Two code-approved fittings that can be used in vertical piping when the flow is from the horizontal: (Any two of the following are acceptable.)
sanitary tee, 1/4 bend, sanitary tapped tee, sweep. (There may be others not listed in this book.)
See page 74, Figure 10-1

6. Two code-approved fittings that can be used in horizontal piping when the flow is from the vertical: (Any two of the following are acceptable.)
combination, sweep, wye and 1/8 bend, reducing closet bend. (There may be others not listed in this book.)
See page 75, Figure 10-2

7. Two code-approved fittings that can be used in horizontal piping when the flow is from the horizontal: (Any two of the following are acceptable.)
wye and 1/8 bend, wye and sweep, combination, sweep
See page 76, Figure 10-3

8. Two reasons that fittings having drainage patterns opposite to flow should not be used:

   1) double hub fitting can restrict flow and cause stoppages.

   2) inverted Y fitting can permit waste to flow into branch drain, causing stoppage or waste drain problems.

   See pages 73, 77, Figures 10-4 and 10-5

9. It's against code to drill or tap drainage and vent pipes because:

   raw sewage could seep onto or into the ground; sewer gases could enter building.

   See page 77

10. It's against the code to install drainage pipes in trenches that are not open because:

    the piping might sag, causing stoppages, and there's no way to properly support piping installed in this manner.

    See page 77

11. The minimum head required to water test a drainage system:

    5 feet.

    See page 77

12. The standards and specifications to which plumbing materials are subject:

    the code regulations in a given area.

    See pages 77-78

13. The two pipes and fittings considered new materials for DWV installations (although they have been in use for some three decades):

    1) no-hub cast iron pipe and fittings,

    2) plastic pipe and fittings.

    See page 78

14. The three most frequently used materials for DWV residential systems:

    1) cast iron hub-and-spigot pipe and fittings,

    2) cast iron no-hub pipe and fittings,

    3) ABS or PVC plastic pipe and fittings

    See page 78

15. The two weights associated with cast iron pipe:

    1) extra heavy

16. The schedule rating for plastic pipe and fittings used in a DWV system:

    schedule 40.

    See page 78

17. The two types of plastic pipe and fittings approved for use in DWV systems:

    1) ABS (acrylonitrile-butadiene-styrene)

    2) PVC (polyvinyl chloride)

    See page 78

18. Two commonly used piping materials for fixture drains in a cast iron system:

    1) copper pipe Type DWV

    2) copper pipe Type M

    See page 78

19. The fitting with which plastic or copper pipe may be used for vent extensions in a cast iron system:

    proper conversion adapters.

    See page 78

20. The two materials the code prohibits to be intermingled in a plastic drainage system:

    ABS and PVC plastic pipe and fittings.

    See page 78

21. Code requirement for screwed, copper or plastic fittings used in a sanitary drainage system:

    must be of the recessed drainage pattern type.

    See page 78

22. The first purpose served by cast iron pipe:

    to replace deteriorated wooden water mains.

    See page 78

23. DWV system joints must accomplish two things:

    1) watertightness

    2) gastightness.

    See page 78

24. The characteristic of oakum in lead and oakum joints which other types of joints do not have:

    its ability to expand after several hours of being wet, to make joints water and gas tight.

    See page 79

25. No-hub joints are made in the following way:

    by joining two pieces of no-hub material (pipe or fittings), by using a one-piece neoprene gasket, a stainless steel shield and retaining clamp.
    See pages 79, 92-93, Figures 10-24 and 10-25

26. Two advantages in using no-hub pipe: (Any two of the following are acceptable.)
    1) can be used in thinner partitions
    2) can be used in limited-access areas
    3) joints are easy to make up
    4) installation is fast and efficient
    5) can be used in combination with a lead and oakum joint pipe to meet almost any specific requirement.
    See pages 79, 81

27. A well-made lead and oakum joint will generally last:

    as long as cast iron soil pipe. In fact, it's likely to last longer than buildings into which it is installed.
    See page 81

28. Three characteristics of cast iron soil pipe which make it superior material for a building sewer:
    1) its strength
    2) its durability
    3) its resistance to trench loads.
    See page 79

29. The pipe barrel can be kept in firm contact with solid ground by:

    excavating for the hub, so the weight will be distributed along the full length of the pipe barrel.
    See page 79

30. The minimum size cast iron soil pipe manufactured for use in a DWV system:

    no smaller than 2 inches (inside diameter).
    See page 79

31. To avoid waste, when a short piece is needed in a hub and spigot cast iron system, the type of pipe it should be cut from:
    a double hub pipe.
    See page 79

32. There is less waste in a no-hub or plastic system because:

    these pipes can be joined together without the necessity of having a hub or hubs on each piece of pipe as is required in a lead and oakum system.
    See page 79

33. Tees or tee branches must be installed only in soil, waste or vent:

    piping which is vertical (stacks).
    See page 81

34. The term used for a tee that enters the straight through section of a fitting on a downward curve:

    a drainage pattern.
    See page 81

35. The section of a DWV system where straight tees are used:

    only in the vent system, as it is designed to carry air and not liquid waste.
    See page 81

36. Sanitary tees are designed to carry:
    waste substances.
    See page 81

37. The two purposes of a test tee:
    1) to test a plumbing system for leaks
    2) to act as a permanent cleanout after testing.
    See page 81

38. A wye (Y) branch is used to make changes in direction of:
    45 degrees.
    See page 81

39. Combination Y and 1/8 bends are used to make changes of direction of:
    90 degrees.
    See page 82

40. The purpose of wyes and combinations in the building's main drain:

    principally to connect horizontal waste pipes and branch drains to the building main drain.
    See page 81

41. The advantage of using a combination instead of a Y and 1/8 bend:

    a joint is saved. (And of course, this saves money.)
    See page 82

42. The side takeoff from a wye or combination may never be larger than:
the through section.
See pages 82, 84

43. The method used in sizing plumbing fittings:
the through section is always sized first, reading from the spigot end to the hub end, and the branch is always sized last: 3 x 3 x 2 inches.
See page 84

44. The way bends are classified:
by the degree of turn and the radius of the curve.
See page 84

45. The degree of turn for:
a 1/16 bend: *22½ degrees.*
a 1/8 bend: *45 degrees.*
a 1/6 bend: *60 degrees.*
See page 84

46. A 1/8 bend offset is generally used:
in a DWV system to bypass an obstruction.
See page 84

47. The difference between a straight cross and a sanitary cross:
a straight cross is used in a vent system; a sanitary cross is used in a drainage system.
See page 84

48. The number of vent lines that may connect to a straight cross installed in a stack vent:
two vent lines are commonly connected to it because it has two openings.
See page 84

49. The two types of stacks which a sanitary cross may use:
either a soil or waste stack.
See page 84

50. An increaser is used:
to increase the size (diameter) of a straight through line. (Permitted in vent systems only.)
See page 84

51. The most frequent use of an increaser is:
to increase the diameter of a vent stack terminal to prevent frost closure in cold climates.
See page 84

52. The place and purpose for using a reducer:
permitted in drainage pipes of a plumbing system to reduce the size (diameter) of a straight through line.
See page 84

53. Kafer fittings are commonly used:
in lead and oakum jointed plumbing systems to replace a section of soil pipe or to install a fitting into an existing soil pipe line. (Used very infrequently today.)
See page 84

54. The thing that makes a kafer fitting unique:
its hub is threaded onto its body and can be unscrewed for installation purposes. (Note: The kafer fitting has been largely replaced by the no-hub pipe and fittings.)
See page 84

55. The U.S. city where plastic pipe was first used:
Key West, Florida, by the Navy, in 1960.
See page 84

56. Two advantages of using DWV plastic systems:
1) it resists outside soil corrosion
2) the smooth inner surface resists deposit formation (stoppages).
See page 84

57. The advantageous heat transfer characteristic of plastic drainage pipe over metal pipe:
plastic piping retains more heat from hot liquid waste, preventing the cool-down of greasy waste that can cause stoppages.
See page 85

58. Rigid vinyl plastic pipe is noted for its resistance to:
chemical and acid waste.
See page 85

59. In the United States, the precaution one must take before buying plastic DWV material:
check with local authority to find out if plastic DWV systems are approved for use in your area.
See page 85

60. The type of joint used in a DWV plastic system:

    solvent welded (cemented) joints.
    See page 85

61. The familiar type of fitting to which plastic fitting patterns are identical:

    cast iron screwed pipe drainage fittings.
    See page 85

62. The reason plastic pipe and fittings can be used advantageously where access areas are limited:

    the solvent-welded or cemented joints can be made with ease. Also, fitting sockets are thinner than cast iron and can be used in thinner partitions. Installation is fast and efficient.
    See page 85

63. It is possible for one person to work alone, when doing plastic installations because:

    the plastic piping is light in weight, which makes it easy to handle.
    See page 85

64. The reason it is impossible to remove a plastic fitting once it is cemented to the pipe:

    the joint created becomes a "welded" joint and therefore is actually a part of the pipe itself.
    See page 85

65. The lengths into which rigid plastic pipe is manufactured:

    20 feet.
    See page 85

66. The precaution that must be taken when using plastic pipe for building sewers:

    the bottom quarter of pipe must be continuously and uniformly supported by the trench bottom. There must be 4 inches of fine materials that can pass through a ¼-inch screen. (Some codes permit using up to ¾-inch material.) Hub and coupling projections must be excavated as shown in Figure 10-7 so that no part of the pipe load is supported by the hub or coupling. Supporting material has to extend 4 inches on each side of pipe.
    See pages 85, 88, Figure 10-14

67. The minimum depth below ground required by code for a plastic building sewer:

    12 inches.
    See page 88, Figure 10-14

68. The minimum size for plastic pipe and fittings in a DWV system:

    1¼ inches.
    See page 88

69. The most common method for cutting cast iron pipe for small jobs (having few cuts):

    the hammer and chisel method.
    See page 88, Figure 10-15

70. To prevent cast iron pipe from breaking unevenly, the following steps must be taken:

    where desired cut is to be made, pipe must be supported firmly by a 2 x 4 board or earth mound.
    See page 89

71. For the inexperienced plumber cutting pipe, a chalk mark around the circumference of pipe being cut is recommended because:

    the cut will be square.
    See page 88

72. A soil pipe cutter must be used to cut cast iron pipe that is rigidly in place because:

    it cannot be turned. The hammer and chisel method will not work. Therefore a cutting tool (soil pipe cutter) that can be rotated around the pipe is one of the best tools still available for this purpose.
    See page 89, Figure 10-16

73. The device that's used so that plastic pipe can be cut square:

    a miter box.
    See page 89, Figure 10-17

74. The most common tools used to cut plastic pipe:

    a fine-tooth hacksaw or a regular hand-saw. (Plastic pipe cutters are also available.)
    See page 89

75. According to code the depth of lead required in a lead and oakum joint:

    1 inch at the outer end of the hub.
    See page 89

76. The approximate amount of lead required to pour a 4-inch joint:

    4 pounds.
    See page 89

77. The following may happen if a cold or damp ladle is plunged into a pot of hot lead:

    it will probably explode.
    See page 90

78. The tool that's used first in making a lead and oakum joint:

    a yarning iron.
    See page 90

79. The caulking iron that should be used first in caulking a vertical lead and oakum joint:

    an outside caulking iron.
    See page 90

80. The tool needed to retain hot lead poured into a horizontal joint:

    a joint runner clamped around the pipe until it cools.
    See page 92, Figure 10-23

81. In a no-hub installation where access areas are limited, it is important to position the heads of the retaining clamps in a certain direction because:

    they must be accessible for tightening.
    See page 92

82. The name of the tool used to assemble no-hub pipe and fittings:

    a torque wrench.
    See page 93

83. It's best to work with plastic joints when the air temperature is above 40 degrees F. because:

    the bonding process is not interfered with. (If the air is too cold, it slows the action of the solvent cement.)
    See page 93

84. That which must be kept visible when installing plastic pipe and fittings:

    the identifying marks so the inspector can see them.
    See page 93

85. The following preparation of plastic pipe and fittings for application of solvent-cement must be made:

    surfaces must be cleaned of any foreign material; the gloss must be removed by emery cloth or special cleaner.
    See page 93

86. When a bead of cement shows around the plastic fitting, the indication is:

    that the proper amount of cement was used in making the joint.
    See page 94

87. When no bead shows around the fitting joint, it means:

    not enough cement was used.
    See page 94

88. The types of joint sealer that should not be used on plastic threaded joints because of the injurious elements they contain:

    conventional pipe compounds, which contain oils.
    See page 94

89. When installing underground or above ground horizontal drainage, waste and vent piping, provision must be made for:

    adequate support to keep the pipe in alignment and prevent sagging.
    See page 95

90. The minimum clearance from the top of a horizontal pipe to the bottom of the footing of a building:

    2 inches.
    See page 95, Figure 10-27

91. When piping must pass through cast-in-place concrete, the amount of annular space around the entire circumference of the pipe must be:

    ½ inch.
    See page 95

92. The procedure that must be followed when excavating a trench parallel to and deeper than the building's foundation:

    the drainage piping cannot be placed within a 45-degree angle of pressure from the foundation of the building, unless a special design is approved by the building official.
    See page 95, Figure 10-28

93. The maximum distance between hangers for horizontal cast iron pipe in 10-foot lengths:
    10 feet.
    See pages 94, 96, Figure 10-26

94. The maximum distance for a hanger from a no-hub cast iron joint:
    must be adjacent to each joint if the developed length exceeds 4 feet.
    See page 96

95. The maximum distance between hangers for 3-inch plastic pipe installed in 20-foot lengths:
    must be supported at intervals not exceeding 4 feet. (The pipe size is irrelevant.)
    See page 96

96. The common support used to keep vertical pipe in alignment:
    a floor clamp.
    See page 96

97. The terminal height of vent pipes above the roof:
    at least 6 inches above building roof.
    See page 96

98. The minimum distance from a window a vent pipe opening may be installed:
    10 feet.
    See page 97

99. To weathertight a vent stack where it penetrates the roof of a building:
    flashing must be used.
    See pages 97-98

100. The result when vent pipes are improperly installed:
    the air circulation within the system is curtailed.
    See page 98

101. The result when moist air is trapped in a horizontal vent pipe:
    air circulation is restricted, pipe corrosion is accelerated and the pipe's lifetime is reduced.
    See page 98

102. The connection of a vent pipe to a horizontal soil or waste pipe should be:
    above the center line of a horizontal soil or waste pipe.
    See page 98, Figure 10-32

Chapter 11

1. The circumstances under which a cesspool may be used:
   when first approved by the local authority (and then, only temporarily).
   See page 103

2. The term used when drinking water becomes contaminated:
   a cross-connection (i.e., when untreated sewage enters the water supply).
   See page 103

3. Two diseases that can be caused by lack of proper sanitary facilities:
   cholera and typhoid.
   See page 103

4. The most acceptable method for sewage disposal when public sewers are not available:
   the septic tank system.
   See page 103

5. The portion of a septic tank to which the installer connects the building sewer or drain:
   to the inlet tee.
   See page 103

6. The two substances from which a septic tank must be kept free if it's to be maintained in good working order:
   sludge (on the bottom) and scum (on the top).
   See page 103

7. To correct a drainfield when it ceases to function:
   it must be dug up and the old drain pipe and gravel removed and replaced with new.
   See pages 103-104

8. That which a septic tank is designed to do:
   separate solids from liquid wastes.
   See page 104

9. Anaerobic bacteria live and feed within a septic tank because of the absence of:
air or free oxygen.
See page 104

10. The form into which bacteria living in a septic tank transform the solid matter:
into gases and harmless liquids.
See page 104

11. That which takes place inside a septic tank when new sewage is discharged into it:
the gases are forced up and through the drainage vent pipes and into the atmosphere above the building roof.
See page 104

12. The approximate number of hours of anticipated sewage flow for which a septic tank must be sized:
24 hours.
See page 104

13. After the liquid has been processed, it is called:
effluent.
See page 104

14. The type of pipe of the subsurface system into which the liquid enters:
open-joint or perforated piping.
See page 104

15. The substance that's permitted to penetrate to the bedrock of a properly-designed drainfield:
air (free oxygen).
See page 104

16. The reason a drainfield should be located in full sunlight:
to allow proper oxidation and evaporation of the waste.
See page 105

17. The result when trees or shrubbery are planted over or near a drainfield:
the roots will penetrate the drainfield and reduce its useful life.
See page 105

18. The size of a septic tank is determined by:
the number of bedrooms.
See page 105

19. The size of a drainfield is determined by:
the number of bedrooms and the required drainfield absorption area.
See page 106, Figure 11-9

20. The closest a septic tank can be located to a building:
no closer than 5 feet.
See page 105

21. The closest a septic tank can be located to a private (potable) well:
no closer than 50 feet.
See page 105

22. From the base of an existing structure, a septic tank excavation cannot be made within:
a 45-degree angle of pressure from the base of the existing structure.
See page 105

23. The way air circulation within a septic tank and drainfield is accomplished.
through the inlet and outlet tees of the septic tank only.
See page 105

24. When a septic tank is abandoned, the code requires that:
1) it be pumped dry (by certified personnel)
2) the bottom of the tank be broken in
3) the tank be filled with clean dirt.
See page 105

25. Codes usually require that a septic tank be cleaned by:
a certified professional.
See pages 103, 105

26. The types of material/construction not usually permitted in the building of septic tanks:
blocks, brick, sectional tanks.
See page 105

27. The material most often used in septic tank construction because it's accepted by most codes:
precast or cast-in-place concrete.
See page 105

28. The minimum distance the outlet tee invert can be below the inlet tee:
not more than 1 inch.
See page 106

29. The place where inlet and outlet tees must be installed in relation to each other:
at opposite ends of the septic tank.
See page 106

30. To protect metal septic tanks from corrosion:
the code requires that the inside and outside of such tank have a bituminous coating applied.
See page 106

31. The two ways the lid of a septic tank can be constructed to permit access for cleaning:
1) two 20-inch manholes, one located over inlet and one over outlet tee
2) tank lid to be sectional type (so it can be removed).
See page 106

32. The essential characteristic of soil suitable for septic tank installation:
must be able to absorb the effluent properly.
See page 106

33. The two factors necessary for sizing drainfields:
1) topography (the lay of the land)
2) the absorption capacities of soils.
See page 106

34. The purpose of underdrains:
to help prevent saturation of the drainage area due to poor percolation.
See page 106

35. Requirement where soil porosity appears to be less than usual:
a percolation test before construction begins.
See page 106

36. When you need more than 100 SF (150 SF in some codes) of absorption area per bedroom:
when the percolation rate is above the three-minute rate to absorb 1 inch of water.
See page 106

37. To meet code requirements, the end of each distribution line in a drainfield must:
be sealed (by capping or by cementing a block to it).
See page 110

38. The minimum inside diameter of drainfield tile:
4 inches.
See page 110

39. The most acceptable downward slope for all types of distribution lines:
no more than ½ inch per 10 feet.
See page 110

40. The reason the entire width of filter material should be covered with untreated paper:
to prevent sand and other small particles from filtering down and through the washed rock when the trench is backfilled.
See page 110

41. The minimum center-to-center spacing for individual trenches:
6 feet.
See page 111

42. The sizing of reservoir-type drainfields is based on the following rule:
the length times width of the reservoir.
See page 111

43. The minimum depth of rock required under drain units of a filter bed:
12 inches.
See page 111

44. To avoid drain spaces (¼ inch) between the ends of each unit, the drain units must be equipped with:
fixed openings (in the bottom of each unit).
See page 111

45. For reservoir-type drain units the maximum distance between centers of distribution lines is:
4 feet (generally acceptable).
See page 111

46. The pipe used to connect the septic tank outlet tee to the distribution box must be:
a tight-jointed pipe (a pipe with no open joints or fixed openings).
See page 111

47. No seepage space is needed when plastic tubing is used because:
the tubing is manufactured with adequate

fixed openings for drainage.
See page 112

48. The maximum distance between the centers of drain lines in a plastic distribution system:

    72 inches (may vary with some codes).
    See page 112

49. The maximum permitted length of a drain line:

    100 feet.
    See page 112

50. The minimum depth of earth cover for a drainfield:

    12 inches.
    See page 112

51. The minimum separation between a drainfield and a basement wall:

    8 feet.
    See page 112

52. The minimum separation between a drainfield and a water supply line:

    5 feet (10 feet in some codes).
    See page 112

53. The solid wastes that should not be disposed of through a drainage system:

    rags, sanitary napkins, heavy paper, cooking fats, coffee grounds, food leftovers.
    See page 112

54. The kinds of chemicals that should not be disposed of through a drainage system served by a septic tank:

    paint thinners, chemical cleaners, photographic chemicals.
    See page 112

55. To help the growth of bacteria in a septic tank:

    flush a yeast cake through the water closet once a week or so.
    See page 113

Chapter 12

1. The sources of water for public water supply systems:

    lakes, rivers, deep wells.
    See page 115

2. Potable water is:

    drinkable water. This means that unpleasant tastes, odors and impurities have been removed.
    See page 115

3. The approximate percentage of households in the United States that depend on a private water supply system:

    approximately 18 percent.
    See page 115

4. Most plumbing fixtures operate satisfactorily when the water main pressure is:

    40 to 55 pounds per square inch.
    See page 115

5. A pressure-reducing valve should be used on a water service line when:

    water main pressure exceeds 80 psi.
    See page 115

6. When public water is available, the law usually requires owners of private water supply systems:

    to connect to the public water supply system.
    See page 115

7. Three factors affecting the sizing of water supply piping:

    1) the type of flush device used
    2) the pressure of the water supply
    3) the length of the pipe in the building
    4) the number and kinds of fixtures installed
    5) the number of fixtures used at a given time.
    See pages 116-117

8. The number of gallons per minute flow which each water supply fixture unit represents:

    7½ gallons per minute.
    See page 117

9. The part of a water system which must be sized first:

    the water service pipe for a building.
    See page 118

10. The minimum practical size for the water service line:

    ¾ inch.
    See page 118

11. The point at which a building main may be reduced in size:

    when the demand placed on the line decreases, as it progresses through the building.
    See page 118

12. The rule of thumb in sizing the water piping in a building when water main pressure is less than 50 psi:

    use the next larger size pipe listed in the minimum water service pipe size table.
    See page 118

13. The number of fixtures that may connect to a ½-inch cold water supply branch:

    two.
    See page 118

14. Hot water piping need not be considered when sizing cold water supply piping because:

    it does not increase the water demand (i.e., does not add to the number of gallons used).
    See page 118

15. The minimum cold water pipe size needed to supply a hot water heater:

    ¾ inch.
    See pages 117, 118, Figure 12-3

16. Acceptable preventive measures to protect metal water pipes where hydrogen sulfide gas or other injurious elements exist:

    sleeve it in polyethylene plastic tubing, or paint it with asphaltum paint. (There may be other approved protective measures in your area.)
    See page 120

17. Corrosion affects the outside of the pipes, yes – but:

    it also affects the inside of the pipes.
    See pages 120-121

18. Galvanic corrosion is caused by:

    weak electric currents generated by using two different types of metal (such as galvanized steel and copper).
    See page 120

19. The type of corrosion that occurs as a result of weak electric currents:

    electrolytic action, or galvanic corrosion.
    See page 120

20. The type of corrosion that will occur inside a water line where oxygen and iron are present:

    (galvanized type) rust formation.
    See page 121

21. Two other causes of corrosion in a water supply system:

    1) *velocity of water moving through the pipes*
    2) *the presence of large amounts of calcium and magnesium compounds that form scales inside pipes (occurs in hard water).*
    See page 122

22. The most common way of reducing corrosion in a plumbing system:

    use dielectric unions to stop the weak flow of electric current from water tank to piping.
    See page 122

23. The reason it's necessary to isolate ferrous water distributing pipes within a building from other metals (ferrous pipes, electrical conduit, building steel and reinforcing steel):

    to prevent electrolytic action at point of contact.
    See page 122

24. Where water velocity is high, the way a water system can be protected from corrosion:

    install a reducing valve in the building's water service pipe.
    See page 122

25. Water containing large amounts of calcium and magnesium affects a water system by:

    forming a scum which is deposited on the inside of pipes. The deposits harden and form scale.
    See page 122

26. Two ways of reducing the formation of scale within pipes due to hard water (any two of the following three are acceptable):
    1) *use copper, CPVC (or polybutylene where permitted) plastic pipe and fittings*
    2) *use a water softener*
    3) *regularly flush the hot water heater of sediments.*
    See page 122

27. In climates where water service supply pipe is subject to freezing, the trench depth should be:
    below the frost line.
    See page 122

28. Maintenance precautions that should be taken where water pipes are not protected from the cold:
    they must be thoroughly insulated to prevent freezing.
    See page 122

29. The cause of a pipe's bursting when water freezes within a pipe:
    ice takes up one-twelfth more space than the same amount of water.
    See page 122

30. When water freezes within a pipe, the place where the pipe will most likely burst:
    the weakest point.
    See page 122

31. Two common insulation materials used to protect pipes against freezing:
    1) *woolfelt pipe covering*
    2) *frostproof insulation (five layers of felt).*
    See page 123

Chapter 13

1. The major function of valves and faucets:
   to regulate the flow of water in a plumbing system.
   See page 125

2. The place where gate valves are usually required:
   where piping lines are to remain com- *pletely open or closed most of the time.*
   See page 125

3. A gate valve should be installed where little use of a valve is expected because:
   it limits the amount of wear on the metal gate and seat.
   See page 125

4. The ways in which the flow controls of a gate valve and a globe valve differ from each other:
   1) *a gate valve controls the flow with a wedge-shaped or tapered metal disc,*
   2) *a globe valve controls the flow with composition or fiber washer compressing against a ground metal seat. Since the seat is at right angles to the flow direction, it changes the direction flow passing through it.*
   See page 125

5. As to whether a gate valve obstructs the flow of water:
   no, an open gate valve obstructs the water flow much less than most other types of valves.
   See page 125

6. The direction the wheel of a valve is turned to stop the flow through it:
   clockwise.
   See page 125

7. The outside of a globe valve looks very similar to:
   a gate valve.
   See page 125

8. The direction of the seat in a gate valve:
   it is not horizontal.
   See page 125, Figure 13-1

9. The direction of the seat in a globe valve:
   it is horizontal.
   See page 125, Figure 13-1

10. The globe valve changes the direction of flow in the following way:
    to a right angle.
    See page 125

11. Globe valves are usually installed:
    1) *where an adjustment of flow rate is desired,*

2) *where a tight shut-down is required at some time.*
See page 125

12. A globe valve must be installed in the line in this manner:

    so that the inlet side will carry the pressure (from the water source) when the valve is closed.
    See page 125

13. The main reason for using a stop and waste valve:

    to protect exposed piping from freezing.
    See page 127

14. The thing that makes a stop and waste valve different from other valves:

    it has a small drain opening on the outlet side of the valve body.
    See page 127

15. The conditions under which stop and waste valves are generally used:

    in cold climates so the building water system can be drained when the main valve is closed.
    See page 127

16. Is it permissible to install a stop and waste valve in the same trench with the water service pipe:

    no – never.
    See page 127

17. The two main purposes an angle valve serves in a water line:
    1) to control the flow
    2) to change the direction of a line.
    See page 127

18. The working principle of a check valve:
    to prevent reversal flow in a line.
    See page 127

19. (a) The place where a check valve is most generally used in residential construction:
    on domestic or irrigation wells.
    (b) It is located:
    on the suction line between the well and the pump.
    See page 127

20. The type of faucet most commonly used for water outlets:

the compression type.
See page 127

21. The advantage of using a mixing faucet instead of a compression faucet:

    water temperature and flow are more easily controlled without having to grind the washer against a metal seat.
    See page 129

22. The working principle of a compression faucet:

    a washer is compressed against a ground seat when faucet is closed.
    See page 127

23. To prevent water from leaking out around the stem of a compression faucet:

    a packing washer or ring fits snugly around the stem.
    See page 128, Figure 13-2

Chapter 14

1. The definition of a hot water system:
   the part of the plumbing installation that heats water and distributes it through pipes to fixtures dispensing hot water.
   See page 130

2. Copper pipe and tubing is the most popular material for hot water systems because:
   it has the ability to resist corrosion (which increases as the temperature of the water increases).
   See page 130

3. The type of plastic pipe that may be used for hot water piping, under limited circumstances:
   CPVC plastic or polybutylene (PB) plastic pipe.
   See page 130

4. Safety equipment is necessary for a hot water system:
   to prevent damage to property and injury to persons using the plumbing fixtures.
   See page 130

5. The sizing of hot water distribution piping is governed by:

good engineering practice.
See page 130

6. The thing that a hot water installation begins with:

   a water heating device.
   See page 130

7. Two principal objectives that must be met in designing a good hot water system:

   1) *the system must satisfy the hot water demand for a particular need,*
   2) *safety features must be built into the system to prevent excessive pressure and temperature.*
   See page 131

8. The recommended temperature setting for the hot water heater thermostat when a dishwasher is used in a residence:

   150 degrees F.
   See page 131

9. The two most common sources of energy for residential heating units:

   1) *electricity*
   2) *fuel-burning substances (gas, etc.).*
   See page 131

10. Another type of energy source for residential heating units:

    solar energy.
    See page 131

11. Electric water heaters are the most popular of all hot water heating units because:

    they are clean, attractive and can be installed nearly anywhere within a building.
    See page 131

12. Two provisions that must be made for gas- or oil-fired water heaters which are not necessary when installing other heating units:

    1) *they must be installed in well-ventilated areas*
    2) *they must have flues to carry away combustion gases.*
    See page 131

13. The possible locations of solar hot water storage tanks are limited by:

    1) *the large size of the tank*
    2) *the type of system planned.*
    See page 131

14. It is not necessary to circulate hot water lines for small jobs because:

    the system has small pipes and short runs. Therefore hot water is available again shortly after the faucet is opened.
    See page 132

15. When designing a hot water piping system, secure pipes with:

    the conventional method of straps. Water expands when heated but in most residential and simple buildings with short piping runs, expansion should not be a problem.
    See page 132

16. Residential hot water heaters may be grouped into three categories:

    1) *electrical*
    2) *fuel-burning*
    3) *solar energy.*
    See page 131

17. The two most common types of water heaters are:

    1) *electric*
    2) *fuel-burning.*
    See page 131

18. The peak draw period for hot water use in most dwelling units is assumed to be:

    one hour.
    See page 132

19. The percentage of a hot water storage tank capacity used in sizing the tank for the peak draw period is:

    75 percent
    See page 132

20. The maximum hourly use of hot water per person is considered to be:

    8 to 10 gallons.
    See page 132

21. The "maximum load" for a hot water heater means:

    the amount of storage tank capacity normally required to provide hot water in any given hour.
    See page 132

22. The "working load" of a hot water heater means:

    the percentage of the maximum load (storage tank capacity) normally needed for any given hour. For residential buildings, that's approximately 35 percent.
    See page 132

23. A 20-gallon water heater for a peak draw period of one hour and a 50 percent recovering rate is expected to supply:

    22.5 gallons per hour.
    See page 132

24. The structural factor that determines the quantity of hot water used in a residential unit:

    the number of bedrooms.
    See page 132

25. When a special demand requirement is not considered, the recommended size of an individual storage tank capacity for a three-bedroom residence is:

    42 gallons.
    See page 133, Figure 14-2

26. Water heaters must be installed where they are:

    accessible for servicing or replacement without removing any permanent part of the building.
    See page 132

27. The rating of a temperature relief valve is determined by:

    its Btu capacity.
    See page 132

28. In addition to the Btu capacity, another special requirement a combination pressure and temperature relief valve must have is:

    must be the reseating type
    See page 132

29. The older fuse-type relief valve is no longer acceptable in new construction because:

    when the water overheats, the fuse melts and the water will continue to run until a new fuse is reinserted.
    See page 133

30. For a relief valve to be most effective it must be installed:

    in the top one-eighth of the tank.
    See page 133

31. The code prohibits the installation of a check valve or shutoff valve between the relief valve and the hot water storage tank for the following reasons:

    if the check valve fails to operate or if someone accidentally closes the shutoff valve, the relief valve would be useless. The tank could then rupture or explode.
    See page 133

32. Relief valve drip pipes should not be connected directly to any plumbing drainage system because:

    1) this could cause contamination (cross-connection)
    2) a continuous discharge of hot water is concealed from view.
    See pages 133-134

33. On a single family residence, a relief valve drip pipe should terminate:

    where it is observable outside the building and extend to within 6 inches of the ground.
    See page 134

34. Relief valve drip pipes should not be threaded in order that:

    other piping cannot be connected to them, accidentally or otherwise.
    See page 134

35. The size of relief lines is determined by:

    the Btu rating of the appliance.
    See page 134

36. The required type of fitting used to secure a relief line to the female thread of a relief valve is:

    a male flare or compression adapter of the same size.
    See page 134

37. The drain cock on a hot water heater must be located:

    in an accessible location.
    See page 135

38. The purposes served by a hot water heater drain cock are:

to permit flushing the tank of sediment and to empty it for repairs or replacement.
See page 135

39. The acceptable minimum size cold water line to any hot water heater is:

    ¾-inch diameter.
    See page 135

40. Using solar energy for heating domestic water is not considered new because:

    it has been used successfully in many areas of the world since before 1900, notably in Florida.
    See page 135

41. The three major components in any solar heat collecting system:

    1) the solar heat collector
    2) the circulation system
    3) the solar storage tank.
    See page 135

42. Some of the obvious features that make the solar hot water system unique are:

    1) it's the most practical and least expensive for residential use
    2) it can produce water temperature up to 200 degrees F.
    3) the collector plate absorbs heat from the sun (which is almost always available) and transfers it to the liquid in the tubing.
    See page 135

43. The type of solar collector that's most practical for residential use is:

    the flat plate solar collector.
    See page 135

44. The three materials most used to build heat decks for solar collectors:

    copper, aluminum, and steel.
    See page 135

45. The thermal difference between the three heat deck materials is:

    none. Thermally there is no difference between copper, aluminum or steel when used as heat deck materials.
    See page 135

46. The only solar installation type in which aluminum tubing may be used is:

    closed-type system with a heat exchanger — and then only with authority approval.
    See page 140

47. The reason the heat collector box should be well insulated:

    to shield the heat deck plate from the weather and to reduce heat loss.
    See page 135

48. The provision needed to prevent heat loss in cold climates:

    the heat collector box should have a double layer of glass installed.
    See page 136

49. The number of gallons of water per day provided by a 4 x 12-foot solar heat collector:

    80 gallons.
    See page 136

50. The minimum size solar storage tank recommended for a family of four is:

    80 gallons.
    See page 136

51. The direction in which a flat collector should be tilted to be the most efficient:

    in the general direction of the sun's path across the sky.
    See page 136

52. It is important to install solar piping to drain dry in climates subject to freezing temperatures because:

    it protects the water from freezing and bursting the pipes.
    See page 137

53. Some of the recommendations made by roofing contractors about mounting collectors:

    the collector should have a structural frame; the frame should be securely bolted to the main roof structure; flashing and rain collar should be put around the pipes and collector; then it can be fitted into the frame and anchored securely to it.
    See page 137

54. The description of how a natural thermosyphon solar water heating system works:

    the hot water circulates through the system on its own, requiring no external energy source, no pumps, no controls or other moving parts.
    See page 138

55. The minimum size piping that must be used in a thermosyphon circulation system:

    ¾ inch.
    See page 139

56. The location of the hot water storage tank in a pumped solar water heating system may be:

    in any convenient place (possibly the garage or utility room).
    See page 139

57. The most common fluid that may be used in a closed solar energy collection system:

    a fluid such as antifreeze.
    See page 139

58. It is not practical to use CPVC plastic pipe in a solar circulation system because:

    plumbing standards forbid the use of plastic pipe where temperatures could exceed 180 degrees F.
    See page 140

Chapter 15

1. The three aspects of potable water piping that are regulated by the plumbing code:

    1) materials
    2) sizing
    3) installation methods.
    See page 144

2. It is important for the code to establish safeguards for public water supply systems in order to:

    avoid contamination of public water supply due to backflow or cross-connection.
    See page 144

3. The quality of a community's water supply affects the plumber's choice of piping materials in the following way:

    the water in some communities corrodes or leaves deposits on the interior walls of pipes. This naturally affects the plumber's choice of piping materials.
    See page 144

4. The way various soils affect the choice of piping materials:

    if a certain type soil corrodes the exterior of the pipe, of course the plumber would choose a piping material that resists corrosion.
    See pages 144, 145

5. It is important not to reuse gas piping and fittings in a potable water supply system because:

    it can leave toxic substances in the water supply.
    See page 144

6. The substance with which galvanized steel pipe has been coated is:

    zinc.
    See page 144

7. Galvanized steel pipe is manufactured in the following standard lengths:

    21-foot lengths.
    See page 144

8. The weight or strength of galvanized steel pipe most often used in a plumbing water piping system:

    standard weight.
    See page 144

9. The type of fitting which galvanized steel pipe usually has attached to one end:

    a pipe coupling.
    See page 145

10. The particular use for which galvanized steel pipe is especially desirable:

    for outside use in water supply piping (because of its resistance to trench loads).
    See page 145

11. Water with a high acid content affects galvanized pipe by:

    rusting the inside of the pipe.
    See page 145

12. The substances which water contains that cause it to be considered "hard" are:

    calcium and magnesium compounds.
    See page 145

13. The four conditions that limit the use of galvanized steel pipe are:

    1) it's more subject to interior and exterior corrosion
    2) it collects deposits when water is hard (high in calcium and magnesium)

3) a high acid water content will rust inside of pipe

4) if water pressure is low, it creates more friction than copper or plastic.
See page 145

14. Galvanized malleable iron threaded fittings are usually used:
in water supply systems.
See page 145

15. In a water piping system a tee provides:
an opening to connect a branch pipe at right angles to the through pipe or main pipe run.
See page 145

16. A tee that has three openings of the same size is called:
a straight tee.
See page 145

17. A tee that has openings of different sizes is known as:
a reducing tee.
See page 145

18. In a rigid pipe line, the purpose served by elbows is:
to change the direction of rigid pipe lines.
See page 145

19. A street elbow is different from a standard elbow in that:
it has both male and female threads and is used in special situations.
See page 145, Figure 15-1

20. A union used in existing lines:
permits removing a section of pipe to install the necessary fitting for the branch opening without disturbing any of the existing pipe or fittings.
See page 145

21. In a piping installation, using a coupling:
makes a straight run when joining two lengths of the same pipe size.
See page 146

22. The name of the coupling that has both male and female threads:
an extension piece.
See page 146

23. Pipe nipples are used for the purpose of:
making an extension from a fitting or joining two fittings.
See page 146

24. A nipple that is threaded almost its entire length is called:
a short or shoulder nipple.
See page 146

25. Two common lengths for nipples used in a plumbing piping system are:
1) a close nipple (the shortest nipple) and
2) a 6-inch nipple (the longest nipple). (There are, of course, other lengths in between.)
See pages 146-147

26. Plugs are generally used for the purpose of:
closing openings in other fittings or sealing the end of a pipe.
See page 147

27. The plug most widely used in plumbing is:
the square head plug.
See page 147

28. The two purposes of caps in a plumbing system are:
1) to plug water outlets for testing purposes
2) to create an air chamber to eliminate water hammer.
See page 147

29. The purpose of a bushing:
to reduce the opening of a fixed fitting to receive a smaller pipe.
See page 147

30. The place where a faced bushing is usually used:
for very close work (though most codes prohibit its use).
See page 147

31. The action which copper is known to resist:
corrosion.
See page 148

32. The lengths in which rigid copper pipe is usually manufactured:
20-foot lengths.
See page 148

33. Flexible copper tubing is known as:
    soft tempered.
    See page 148

34. One main advantage in using flexible copper tubing is:
    it is easily bent (even when cold), thus many fittings can be eliminated.
    See page 148

35. The three weights or types of copper pipe or tubing used in piping systems are:
    1) Type K
    2) Type L
    3) Type M.
    See page 148

36. Two types of copper fittings that may be used in a water piping installation are:
    1) the solder type
    2) the flare type.
    See page 148

37. The portion of the water supply in which PVC plastic pipe may be used (where permitted):
    only for outside installations.
    See page 148

38. The minimum working pressure required for plastic pipe used in a water piping installation:
    100 psi.
    See page 149

39. The additional advantage(s) that PB pipe has over CPVC is/are:
    it's not affected by freezing (the pipe expands to accommodate ice, yet doesn't split) and it's flexible so that temperature changes are largely absorbed. This tends, then, to eliminate water hammer.
    See page 149

Chapter 16

1. When installing a water service pipe in the same trench with a building sewer, the two code-required procedures:
    1) the water service pipe must be located above the sewer pipe on a solid shelf excavated to one side,

    2) the bottom of the water service pipe must be at least 12 inches above the top of the sewer line.
    See page 152

2. The minimum separation between a water service pipe and a sewer line installed in separate trenches:
    5 feet.
    See page 152

3. The minimum separation between a water service pipe and a septic tank or drainfield:
    5 feet.
    See page 153

4. It's important to support water service supply piping for its entire length:
    to prevent sagging, misalignment and breaking.
    See page 153

5. Young trees should not be planted near a water service pipe because:
    the root system could cause buckling, bending or breaking the water line.
    See page 153

6. The characteristics that make galvanized steel pipe superior to use in outside trenches:
    it has strength, durability and is resistant to trench loads.
    See page 153

7. What you should avoid when backfilling a trench containing water service piping:
    large boulders, rocks, cinder fill, or other materials which might physically damage or encourage corrosion of the pipe.
    See page 153

8. The depth of a water service piping trench should be:
    deep enough to protect the pipe against freezing.
    See page 153

9. Why PVC plastic pipe must be treated carefully during installation:
    it is fragile and much more subject to damage than metal pipe.
    See page 153

10. Precautions that should be taken when backfilling over PVC plastic pipe in water piping trenches:

    should be supported continuously and protected with 4 inches of fine uniform fill which should pass through a ¼-inch screen.
    See page 153

11. The clearance required between the top of a water service supply pipe and the bottom of a building footing:

    2 inches.
    See page 153

12. The clearance required around the circumference of a water supply pipe passing through cast-in-place concrete:

    ½ inch.
    See page 153

13. When the source of water for a lawn sprinkler system is taken from a potable water service supply pipe, the device used to prevent a cross-connection:

    a backflow preventer located on the discharge side of the control valve.
    See page 154

14. The location of the backflow preventer above the highest sprinkler head must be:

    at least 6 inches.
    See page 154

15. The code-required device which must be provided on each building water service supply pipe:

    a separate water control valve.
    See page 154

16. A water supply control valve must be located:

    at or near the foundation line outside the building, either above ground or in a separate approved box with a cover.
    See page 154

17. A drip valve is required in cold climates in order to:

    drain off water to prevent freezing and bursting of pipes.
    See page 154

18. The type of pipe straps required to secure copper water lines to wooden partitions.

    copper or copper plated.
    See page 155

19. The problem that might occur if water line pipes are not properly strapped:

    they will move to and fro. This is commonly known as water pipe creep and will make an audible irritating sound.
    See page 155

20. The maximum depth a wooden partition may be notched to conceal water piping is:

    40 percent of its depth.
    See page 155

21. The standard acceptable method to provide openings for the installation of water piping in a load-bearing partition:

    to drill holes through the center of the studs. Holes should not exceed 40 percent of the studs' depth.
    See page 155

22. Soft piping installed in notched wooden partitions should be protected from lath nails:

    by installing metal stud guards.
    See page 155

23. The more common way to protect piping installed in ground where hydrogen sulfide gas is known to be present:

    paint the pipe with two coats of asphaltum paint.
    See page 155

24. The way to protect metal water piping from other metals or ferrous pipes within a building:

    the water piping must be electrically isolated from all ferrous pipes, electrical conduit, building and reinforcing steel. It must not come in physical contact with any good conductor of electricity.
    See page 155

25. When hot and cold metal water piping within a building make contact:

    There is a transfer of heat from the hot water pipe to the cold water pipe.
    See page 155

26. The areas in a building subject to freezing where water piping should not be installed:

    in crawl spaces or other unheated areas.
    See page 157

27. Water pipe installations are protected from water hammer by:

    locating air chambers properly or by installing other approved devices.
    See page 157

28. Water supply systems must be installed to drain dry:

    for the purpose of making repairs, to replace lost air in air chambers, and to prevent freezing.
    See page 157

29. A cross-connection is:

    a direct arrangement of a piping line that allows the potable water supply to be connected to a line which contains a contaminant.
    See page 157

30. Two more common causes of cross-connection:

    1) a garden hose attached to a hose bibb with the end of the hose lying in contaminated water,
    2) failure to keep the correct air gap between fixtures and faucets. (Note: Any 2 of the 5 listed on p. 157 are correct.)
    See page 157

31. Back-siphonage can be triggered when:

    for any reason, water pressure drops in a public main.
    See page 157

32. Except for clothes washing machines, each threaded water outlet (to prevent cross-connection) must have:

    approved backflow preventers.
    See page 158

33. The minimum air gap required from water supply outlets to the overflow rim of each fixture:

    1 inch.
    See page 158

34. Below-rim water supplied fixtures are prohibited because:

    they will permit back-siphonage of polluted water if a negative pressure develops in the supply piping.
    See page 158

35. The two basic types of pipe measurement for plumbing installation:

    1) the straight run,
    2) the offset.
    See page 158

36. End-to-center measurement means:

    the measurement taken from the threaded pipe end to the center of fitting attached to opposite end.
    See page 160, Figure 16-13

37. To find the length of pipe between two 45-degree ells, the mathematical figure one must know is:

    1.4142
    See page 161, Figure 16-14

38. The two common types of pipe vises used to hold pipe:

    1) the hinged vise (yoke vise),
    2) the chain vise.
    See page 162

39. The direction one must turn a vise handle to tighten the jaws on the pipe:

    clockwise.
    See page 162

40. When a vise is used to secure a chrome pipe, the finish should be protected in this manner:

    wrap the pipe with adhesive tape (or some other equally effective tape).
    See page 162

41. When a pipe cutter is used, that which is left inside a pipe:

    a burr (ridge)
    See page 165

42. The way you begin the operation in order to cut steel pipe square with a pipe cutter:

    place cutter wheel exactly on the mark. The rollers of the single-wheel cutter ensure a square cut as cutter body is rotated around pipe.
    See page 163

43. The result when tightening the handle of a pipe cutter too rapidly:

    it can break the cutter wheel or spring the cutter frame.
    See page 163

44. The possible adverse effect in a water pipe if the burr is not removed:

mineral deposits can collect at the burr and cause a clogged line.
See page 165

45. The name of the tool used to remove a burr from inside a water pipe:

a reamer.
See page 165

46. The two most common types of pipe threaders:

a three-way threader, a ratchet threader.
See page 165

47. A three-way threader and a ratchet threader differ in that:

   1) *the three-way threader has three pipe-size dies as part of the stock.*
   2) *the ratchet threader has dies that are interchangeable with a single stock.*
See page 165

48. The type of pipe threader that can be used only where there is enough room to make a complete circumference:

the three-way threader.
See page 165

49. The taper per foot of standard pipe threads is:

¾ inch.
See page 165

50. The first thing you must do before threading the pipe:

fasten the pipe securely in the vise.
See page 165

51. The other steps in threading pipe:

slide stock over end of pipe. Push die against pipe with heel of hand; take three or four short, clockwise strokes. Apply plenty of thread-cutting oil. Continue to turn the stock with downward strokes on handle. Apply oil often. Clear away pipe chips after each full turn forward by backing off about one-quarter turn.
See page 165

52. To reclaim used thread-cutting oil:

place a shallow pan covered with a wire screen under the threader.
See page 166

53. The way joint pipe compound should be used when joining an elbow (or other fittings) to a threaded pipe:

apply pipe joint compound evenly over the male threads of the pipe. (Do not put compound on the female threads of the fittings.)
See page 166

54. It's important to select the right size pipe wrench when joining fittings to threaded pipe for the following reasons:

When wrench is too small you might strain your hands, arm or back. When wrench is too large, you'd force the fitting too far onto the threaded pipe. This overtightening could result in a bad joint or a cracked fitting.
See page 167

55. When a piping system is complete one should:

open a hose bibb and house valve to flush foreign matter from the lines.
See page 168

56. The procedure used in measuring copper pipe is:

the same as for threaded pipe.
See page 168

57. Soft-tempered copper tubing is desirable for some uses because:

it is easily bent, thus saving many fittings.
See page 168

58. The allowances that must be considered when measuring copper pipe for cutting are:

   1) *the fitting dimensions,*
   2) *the distance the pipe or tubing is inserted into the fitting.*
See page 168

59. A fine-toothed saw blade is considered to have:

24 teeth per inch.
See page 168

60. The most common joint used in a copper water system:

a sweat joint.
See page 168

61. Two other types of joints frequently used in a copper system:
    1) *flare joints*
    2) *compression joints.*
    See page 170

62. It's important to solder copper pipe and fittings immediately after they are cleaned:
    to prevent oxidation.
    See page 169

63. It's important to select the proper size tip when soldering copper pipe and fittings because:
    If it's too large it will burn the tubing and fitting. If it's too small it will heat the pipe and fitting unevenly and not draw the solder into the joint. Either way, you'd have a bad joint.
    See page 168

64. The soldering process:
    involves properly cleaning and heating the joints. The surface tension spreads the solder to all parts of the joined surfaces, resulting in a sound joint.
    See page 168

65. The precaution that you should take when soldering needs to be done near combustible material:
    place a metal sheet between the point to be soldered and the combustible material. Have a bucket of water available.
    See page 169

66. It is more difficult to resolder a leaking joint than it was the first time because:
    the heat from the torch turns the moisture in the tubing to steam. The steam makes pinholes in the newly-applied solder before it can harden.
    See pages 169-170

67. If you are unable to keep water away from the joint you're trying to solder, try this "trick of the trade":
    stuff plain white bread (omit the crust) as far into the pipe as possible in the direction from which the water is flowing.
    See page 170

68. The type of seat a galvanized union should have in a water supply system:
    metal-to-metal joints with a brass ground seat.
    See page 170

69. Once a water system is complete you should check for:
    leaks.
    See page 170

70. The type of copper with which flare fittings can be used:
    soft flexible copper tubing.
    See page 170

71. The type of wrench which should be used to tighten flare nuts:
    smooth jaw adjustable wrench.
    See page 170

72. Flare fittings, when used in a copper installation, should never be:
    concealed within walls or other inaccessible places.
    See page 170

73. Compression-type fittings (though seldom used in a water supply system) are usually used:
    to connect fixture supply tubes to an ell or cutoff valve located beneath the fixture.
    See page 170

74. Description of how a compression-type fitting works:
    by tightening a threaded nut which forces a compression ring against a ground joint on the fitting.
    See page 171

75. A tubing cutter should be used to cut supply line tubes to fixtures because:
    it cuts the tubing square and make a smooth cut.
    See page 171

76. A freehand cut should not be used when cutting plastic pipe for the following reason:
    a square end cut is essential for making a good joint.
    See page 171

77. The tool that may be used to ream plastic pipe:
    the reamer attached to the tubing cutter frame.
    See page 171

78. When cutting plastic pipe with a hacksaw, the piece of equipment needed:

a miter box.
See page 171

79. When plastic pipe and fittings are cemented together, the joints may be referred to as:

"welded" joints.
See page 171

80. Liquid cleaner or sandpaper removes from the surface of plastic pipe and fittings:

impurities and gloss.
See page 171

81. After completion of a plastic system, the minimum waiting time recommended before water testing:

at least half an hour.
See page 173

82. To repair a leaking joint in a plastic welded system:

the bad joint must be cut out and replaced with a new fitting. (Welded plastic fittings cannot be reused.)
See page 173

83. The newest plastic pipe now approved for most water supply systems:

polybutylene (PB).
See page 173

84. The type of fitting required to connect plastic pipe to iron pipe valves and fittings:

approved male-threaded plastic adapters.
See page 173

85. The required test pressure for a newly completed water system:

not less than the maximum working pressure under which it is to be used.
See page 174

86. If a water system is to pass final inspection, the thing that must not be present:

water shock or hammer.
See page 174

Chapter 17

1. The approximate number of people in the United States who depend on private water (wells) to supply their needs.

about 47 million.
See page 177

2. The two agencies responsible for approving potable water supply wells:

1) the local health department and

2) Director of Environmental Resource Management (DERM).
See page 177

3. These two agencies are mainly concerned with:

guidelines on well installation that cover depth and separation distance from sources of contamination.
See page 177

4. Well water is generally classified as hard water because:

it has a high mineral content.
See page 177

5. The main reason that well water has a different taste than public water:

it contains no chlorine or other chemicals added as does public water.
See page 177

6. When used for bathing purposes, well water requires more:

soap to make a lather than does public water.
See page 177

7. The thing that causes well water to stain other surfaces a dark reddish brown:

the minerals in well water – especially iron.
See page 177

8. The device that may be used to reduce scale buildup and iron in a plumbing system:

a water softener.
See page 177

9. To provide a continuous supply of water, the depth of a well casing must extend:

into the dry weather water table.
See page 178

10. Most codes require that a potable water supply well and its equipment be installed by:

    certified and licensed well drillers or plumbing contractors.
    See page 178

11. Most authorities require the following separation between a potable water supply well and source of contamination:

    1) a 50-foot separation from a septic tank
    2) a 100-foot separation from a drainfield.
    See page 178

12. Dug or driven wells are classified as:

    shallow wells.
    See page 179

13. Drilled wells are generally classified as:

    deep wells.
    See page 179

14. Water from deep wells is more desirable than water from shallow wells because:

    1) there is less chance of contamination
    2) it is affected very little by dry years.
    See page 179

15. A comparison of the two types of driven wells:

    one has an open end casing and one has a casing equipped with a well point.
    See page 179

16. It is recommended that wells with well points not be installed when:

    the water table is close to the ground surface and a good rock formation is present.
    See page 179

17. Rock and sand may be flushed from a shallow well by the following method:

    Use a pipe smaller than the well casing with a hose attached to one end. Ream the other end and cut it on an angle. Work the smaller pipe up and down inside the larger casing under water pressure. This should flush loose debris out of the well casing.
    See pages 179-180

18. To check a well to determine if water is free of sand:

    catch samples in a glass jar. When no sand or rock fragments show up in the water samples, the well is considered good.
    See page 180

19. The well point recommended when the water reservoir ends in sand:

    one having fine perforations and a fine mesh screen.
    See page 180

20. The well point recommended when the water reservoir ends in gravel:

    one having large perforations and a coarse mesh screen.
    See page 180

21. The material required for a well casing pipe used for residences:

    galvanized steel pipe.
    See page 180

22. Around the top of a well casing you must install:

    a concrete pad at least 4 inches thick and 18 inches from the center of the well.
    See pages 180-181

23. The purposes served by having a tee screwed to the top of a well casing:

    accessibility for inspections, adding disinfecting agents, measuring well depth and testing the static water level.
    See page 181

24. The minimum size suction line permitted for supplying water to plumbing fixtures:

    1 inch.
    See page 182

25. The direction which a suction line should pitch:

    toward the well.
    See page 182

26. For supplying water to a sprinkler system, the suction line should be:

    as large as the pump suction inlet.
    See page 182

27. On all suction lines, as near the well as possible, there must be installed:

    a soft seat check valve.
    See page 181

28. On all screw pipe suction lines as near the pump as possible, there must be installed:

    a union or an approved slip coupling.
    See page 181

29. Some of the advantages of installing a hose bibb on the suction line above grade on an irrigation well pump:

    to prime the pump and to add pesticides or liquid fertilizer.
    See pages 182, 183

30. The minimum size of the discharge pipe from the pump to the pressure tank:

    ¾ inch.
    See page 183

31. The minimum size hydropneumatic tank permitted for a single-family residence:

    42 gallons.
    See page 184, Figure 17-7

32. The size of a hydropneumatic tank is controlled by:

    the fixture units of a building.
    See page 184

33. Two advantages that the newer breather-type pressure tank has over the older hydropneumatic tank:

    1) it gives consistent service and
    2) it requires no air to be injected into the tank.
    See page 184

34. When initially installing the pump and equipment for a single family pressurized system an important consideration that must be made is:

    the equipment must be reasonably accessible for repair or replacement.
    See page 184

35. Possible outcomes when a pump wired for 230 volts is connected to wires supplying 115 volts:

    the pump may not perform properly or if used too long will burn out.
    See page 184

36. Two types of sprinkler heads that may be used for a sprinkler system:

    the flush spray head and the pulsating head.
    See page 186

37. The first factor to be considered in planning and designing a lawn sprinkler system:

    the type of sprinkler heads to be used.
    See page 186

38. The two factors that determine the size pump to be used on an irrigation water well:

    1) the volume of water required and
    2) the depth of the well.
    See page 188

39. Pulsating-type sprinkler heads are usually used in:

    large open areas.
    See page 186

40. The main disadvantage in using pulsating-type sprinkler heads in an average size residential yard:

    they must be in operation for a longer period of time to supply the same amount of water as would be provided by the spray heads.
    See page 186

41. What is accomplished when pressures exceed those listed in Figures 17-12 and 17-13:

    the pressure is greater and the same job is done in less time. (Thus, the pump doesn't have to run as long a time.)
    See page 186

42. The average radius coverage for a ½ flush spray head at 20 psi:

    12 feet.
    See page 187, Figure 17-12

43. The spray pattern flush heads should have when installed along a building wall:

    half heads.
    See page 187

44. It's better not to install flush spray heads in a straight line with other heads because:

    this would prevent the needed overlapping.
    See page 187

45. The depth of the well is an important factor in sizing the well pump because:

    the deeper the well, the larger the pump capacity must be.
    See page 188

46. The type of well pump usually installed for irrigation purposes:

    a centrifugal type.
    See page 188

47. The two types of materials used exclusively for sprinkler systems:

    PVC Schedule 160 rigid pipe and fittings and flexible polyethylene plastic tubing.
    See page 189

48. To protect sprinkler heads:

    place a concrete collar around each head.
    See page 190

49. A requirement for a newly installed sprinkler system before putting it in use:

    do not install head at the end of each in-place run of pipe until the lines have been flushed of foreign matter.
    See page 190

50. When a public water supply is not available, the other source of water is:

    a private well.
    See page 177

Chapter 18

1. The approximate number of feet of concealed piping in a typical single-family residence is:

    300 feet.
    See page 200

2. The difference in meaning of "roughing-in" and "rough plumbing":

    "Roughing-in" means the portion of the rough plumbing system that brings the waste and water pipe through to the wall and floor lines where connections are made to the fixtures. "Rough plumbing" is the term used for all concealed waste and water piping used in a building (i.e., the drainage and waste piping, the vents and the hot/cold water lines).
    See page 200

3. It's crucial that a plumber know installation dimensions when roughing-in the plumbing fixtures because:

    the fixtures must be properly spaced and installed for their intended purpose; also they must be accessible for cleaning and repairs.
    See page 200

4. The major consideration when installing plumbing fixtures:

    to observe the minimum clearances required by code.
    See page 200

5. The center of a water closet bowl to any finished wall must be:

    at least 15 inches.
    See page 200

6. The minimum center-to-center spacing for a bidet installed next to a water closet:

    30 inches.
    See page 200

7. The minimum clearance between the center of a water closet bowl and the edge of a bathtub:

    12 inches.
    See page 200

8. The minimum clearance from the front of most fixtures to any finished wall:

    21 inches.
    See page 200

9. Minimum center-to-center spacing is not applicable to a lavatory because:

    lavatories are manufactured in many various designs and widths.
    See page 200

10. The standard height from finished floor to the overflow rim of a lavatory:

    31 inches.
    See page 206, Figure 18-5

11. The distance from the center of most water closet bowls to the center of the water supply outlet:

    approximately 6 inches.
    See page 207, Figure 18-5

12. The minimum distance from the edge of a lavatory to the nearest obstructions:

    4 inches.
    See page 200

13. The standard height from finished floor to the center of a waste outlet for a wall hung lavatory:

    approximately 18½ inches.
    See page 206, Figure 18-5

14. The standard measurement, center-to-center, of hot and cold water outlets for a lavatory:

8 inches.
See page 206, Figure 18-5

15. The minimum clearance from the nearest obstruction for the entry/exit point of a shower:

24 inches.
See page 201

16. The minimum floor area acceptable for a shower:

1024 square inches.
See page 201

17. The standard roughing-in measurement for a water closet from the finished wall to the center of the waste outlet:

12 inches.
See page 201

18. The standard height from the finished floor to the top of a water closet bowl:

14¼ inches.
See page 207, Figure 18-5

19. The standard height from the finished floor to the overflow rim of a kitchen sink:

36 inches.
See page 205, Figure 18-5

20. When the waste outlet for a water closet is roughed in too close to a finished wall, the error should be corrected in the following manner:

use a special 10-inch rough-in water closet. (An offset closet flange is not permitted by code.)
See pages 201-202

21. The conditions to which bathrooms are more susceptible than probably any other room:

unsanitary conditions.
See page 202

22. To support off-the-floor water closets:

a water closet carrier must be used.
See page 202

23. Two newer piping materials that are compatible with specially-designed residential carriers:

PVC plastic and no-hub cast-iron pipe and fittings.
See page 202

24. The size drain with which a bidet is usually equipped:

1¼-inch O.D. drain.
See page 208, Figure 18-5

Chapter 19

1. Plumbing fixture standards have been developed over the years for this reason:

to control quality and design of plumbing.
See page 210

2. Plumbing fixtures must be free from:

defects and concealed fouling surfaces.
See page 210

3. Plumbing fixtures must be manufactured of:

non-absorbent materials.
See page 210

4. When fixtures are constructed of pervious materials, the specific condition that's not permitted:

waste outlets which retain water.
See page 210

5. The potential danger that exists when bathrooms do not have adequate lighting or ventilation:

likelihood of unsanitary conditions.
See page 210

6. The minimum fixture requirements for a single family residence:

one kitchen sink, one water closet, one lavatory, one bathtub or shower. (Hot water may or may not be required, depending on local code.)
See page 210

7. The minimum number and type of fixtures required by code are determined by:

the type of occupancy and the number of people expected to use the toilet facilities.
See page 210

8. Besides the usual kitchen sink, water closet, lavatory and bathtub (or shower)

required in most living units, the additional fixture required by code:

a clothes washing machine.
See page 210

9. The number of toilet facilities in light commercial buildings is determined by:

the number of employees.
See page 211

10. In addition to the regular toilet facilities, a small office building with 100 employees must also provide:

a service sink on each floor.
See page 211

11. To ensure decency and privacy, a toilet room connected to a public use area (regardless of the number of fixtures) must have:

a vestibule or a privacy screen arrangement.
See page 211

12. The design of toilet bowls for public use must be:

the elongated type.
See page 211

13. The special plumbing fixture required in a place employing more than 5 males:

a urinal.
See page 212

14. The way of determining minimum toilet facilities for a small restaurant:

by the maximum number of persons that can be served at one time.
See page 212

15. The determining factor for minimum toilet facilities in a fast food drive-in restaurant:

the number of parking spaces.
See page 212

16. The required material for floors and walls (up to 5 feet) in public toilet rooms:

tile or other impervious materials.
See page 212

17. The type seat required for toilet bowls serving the public:

elongated, with open front.
See page 213

18. The type of eating establishment in which a dishwashing machine or suitable three-compartment sink is required:

one where dishes, glasses or cutlery are to be reused.
See page 212

19. A requirement for employee use in establishments where food is prepared and served to the public:

a hand sink (in addition to the lavatory in restroom).
See page 212

20. The two fixtures that are generally provided with overflows:

bathtubs and lavatories.
See page 213

21. The two purposes served when plumbing fixtures are designed with overflows:

1) they provide secondary protection against self-siphonage
2) they let excess water escape below the flood-level rim of the fixture.
See page 213

22. The overflow passageway from a fixture must be connected to:

the inlet side of the fixture trap.
See page 213

23. The size of a fixture strainer is determined by:

the fixture waste outlet it serves.
See page 213

24. Two materials of which modern bathtubs are made:

enameled pressed steel, enameled cast-iron or gel-coated fiberglass.
See page 213

25. Bathtubs recessed into the finished walls must have:

waterproof joints.
See page 213

26. The minimum sized for a bathtub waste and overflow:

1½ inches.
See page 214

27. The three common types of bathtub waste and overflow most often used today:

1) the tip-toe

2) *the trip waste*

3) *the pop-up.*

See pages 214-215

28. The type of shower found in most homes today:

    the tiled (or possibly marble) shower designed and sized to the owner's taste.
    See page 218

29. The way in which water is transferred from a tub having a diverter spout to the shower head:

    by pulling up the knob located on the tub spout.
    See page 218

30. The advantages of having a shower stall:

    it takes up little space and uses less water.
    See page 218

31. The two general types of shower stalls used in today's plumbing:

    1) porcelain-coated steel
    2) fiberglass.
    See page 218

32. Accuracy is very important in roughing-in the waste opening for a stall shower because:

    a stall shower is usually installed after the floor is poured and the rough partitions are in place.
    See page 218

33. Two advantages of a tiled shower over a stall shower:

    1) the accuracy of waste opening is not as critical
    2) it can be designed and sized to owner's taste.
    See pages 218-219

34. The most important requirement for a shower:

    complete waterproofing.
    See page 219

35. The minimum size waste outlet for a shower compartment:

    2 inches.
    See page 219

36. The minimum floor space required for any shower compartment:

    1,024 square inches.
    See page 219

37. The minimum code required weight per square foot for lead shower pans:

    4 pounds.
    See page 219

38. Before a shower pan can be installed a carpenter must provide:

    the wall framing and the curb.
    See page 219

39. Design requirements for shower strainers:

    a minimum 2 inch waste outlet; a minimum free area for shower strainer of 3½ square inches; strainer must be removable for easy cleaning of shower trap.
    See page 219

40. To protect lead or copper shower pans installed on concrete floors:

    they must be painted with asphaltum paint inside and outside.
    See page 219

41. When securing the shower pan material to the partition studs, nails or screws should be placed a maximum distance of:

    1 inch from top of pan's turnup.
    See page 219

42. The sides of a shower pan should extend above the finished curb:

    at least 3 inches.
    See page 219

43. The point in a building's construction at which a shower pan should be prepared for inspection:

    at the same time the tub and water pipe are inspected.
    See page 219

44. Shower pans may be omitted when:

    the compartment is built on a concrete slab on the ground floor, and the bottom, sides, and curbs are poured at the same time as the slab.
    See page 219

45. The height of shower rods above the finished floor is usually:

    6½ feet.
    See page 220

46. Shower doors are generally manufactured from:
plastic or safety glass.
See page 220

47. Two common materials used in manufacturing lavatories: (Any two of the following are acceptable.)
enameled cast iron, vitreous china, enameled pressed steel, stainless steel, acrylics.
See page 223

48. Two types of lavatories most often installed:
wall-hung and countertop.
See page 223

49. The standard height from the finished floor to the overflow rim of a lavatory:
31 inches.
See page 223, Figure 18-5

50. On a lavatory, the thing that determines which type of faucet is to be used is:
the design of the faucet openings; i. e., whether a combination faucet or a separate hot and cold water (two faucets) arrangement is to be used.
See page 225

51. The disadvantage of having a single hot and cold water faucet on a lavatory:
it's impossible to mix the water before it enters the lavatory.
See page 225

52. The reason why some commercial lavatories use self-closing faucets:
to save water.
See page 225

53. The two types of drain assemblies used on lavatories:
the chain and rubber stopper, the pop-up stopper.
See page 226

54. Lavatory faucets and waste assemblies should be installed before the lavatory is in place because:
it's much easier to do on the floor.
See page 227

55. Wall-hung lavatories at point of contact with finished wall surfaces should be:
sealed with white cement or other suitable material.
See page 227

56. Three types of flushing action for today's water closets:
1) reverse trap,
2) siphon jet,
3) siphon action.
See page 228

57. The part of a combination water closet that's installed first:
the bowl. The tank is installed second (last).
See pages 230, 231

58. The type closet flange required in a plastic drainage system:
manufactured of plastic so it can be cemented to the plastic stub.
See page 230

59. To make a tight joint between a closet bowl outlet and the building waste pipe opening, use:
a preformed wax setting seal.
See page 230

60. A level should be used when installing a closet bowl and tank because:
it must be level in each direction.
See page 231

61. Tank-type water closets are seldom used in public toilet rooms because:
it takes them a longer time to refill the tank after each flushing.
See page 231

62. The two types of low closet tanks are:
the wall-hung, the close coupled.
See page 231

63. The purpose of a refill tube:
to automatically restore the closet bowl water seal.
See page 231

64. The purpose of an overflow tube in a water closet tank:
to prevent tank overflow by removing excess water at the rate it enters the tank.
See page 232

65. In toilet rooms serving large numbers of people, flushometers are generally required because:
the flushing action is quick and automatic.
See page 232

66. The other commercial fixture that uses the flushometer valve is:
the urinal, either wall-hung or floor mounted.
See page 232-234

67. After a water closet is completely installed it should be:
checked by adjusting the float rod for correct water level and checking all connections for possible leaks.
See page 233

68. It is necessary to adjust the float rod in a water closet tank in order that:
the correct water level may be maintained.
See page 233

69. Toilet seats must be made of:
smooth nonabsorbent materials.
See page 234

70. The two most common urinal designs are:
the wall-hung, the floor mounted stall type.
See page 234

71. The minimum size waste outlets for urinals:
2 inches.
See page 234

72. Most wall-hung urinals do not require a separate trap because:
they are designed with an integral trap.
See page 234

73. Stall urinals must be slightly recessed below the finished floor for the purpose of:
providing drainage.
See page 234

74. The location in the waste pipe for installation of a trap to serve a stall urinal:
8½ inches from the finished wall to center of waste pipe.
See page 234, Figure 18-5

75. A flushometer for a urinal must deliver water at a certain rate in order to:
flush all surfaces of the urinal.
See page 234

76. One outlet and one trap may be used to connect a two compartment laundry tray by:
using a continuous waste.
See page 235

77. The size trap that must be used on a kitchen sink waste pipe:
1½ inches.
See page 235

78. An optional feature of a sink faucet:
a spray for rinsing purposes.
See page 235

79. The minimum size waste opening for a domestic sink is:
1½ inches.
See page 235

80. When a food waste disposer is used according to the manufacturer's instructions:
the drainage system will not clog up.
See page 236

81. When a waste disposer is newly installed in a two-compartment sink, some codes require:
the disposer waste to discharge through a separate trap and waste line.
See page 236

82. The fitting that is required when a waste disposer is installed on an existing two-compartment sink:
a directional tee must be installed to flush garbage away from the other sink compartment.
See page 237, Figure 19-31

83. The trap size that must be used to connect the waste pipe to a food disposer:
1½ inches.
See page 237

84. The side of the sink on which an under-counter dishwasher may be installed:
either side, but never directly under a sink.
See page 237

85. A dishwasher is served by:
hot water only, which is heated to approximately 150 degrees.
See pages 237-238

86. A rubber hose should not be used to connect the water supply to a dishwasher because:

constant pressure and high temperature will in time rupture this material.
See page 237

87. The preferred type water piping materials to supply water to a dishwasher:

hard or soft copper tubing with brass or copper fittings.
See page 237

88. A solder fitting on the inlet valve of a dishwasher should not be used because:

the heat necessary for soldering purposes may damage the inlet valve.
See page 238

89. It's essential to install the high loop fitting on the drain hose in order to:

prevent back-ups into the dishwasher if the sink should become clogged.
See page 238

90. The place where the high loop fitting of a dishwasher drain line should be installed:

clamped securely to the underside of the cabinet countertop.
See page 238

91. An air gap fitting on a dishwasher drain, when code required, must be installed:

on the sink or countertop.
See page 238

92. The maximum distance between a dishwasher and the sink waste connection, according to most codes, should be:

5 feet.
See page 239

93. Where a food disposal unit is installed in a sink, the waste from the dishwasher must connect to:

the opening provided in the body of the food disposer.
See page 239

94. The two most common types of residential water heaters:

electric and gas-fired.
See page 239

95. Three important criteria in selecting a water heater location:

1) accessibility to water and power,
2) must be close to greatest hot water use (to prevent heat loss through pipes),
3) access panels and drain valves must be accessible.
See page 240

Chapter 20

1. The major cause of clogged drains is:

foreign matter which finds its way into the drainage system.
See page 246

2. The portions of a drainage system where stoppages rarely occur:

in straight horizontal runs, vertical drops or fixture traps (except those in water closets).
See page 246

3. The most likely point in a drainage system where a stoppage may occur:

where two pipes are joined together with a fitting for a change of direction.
See page 246

4. If a local blockage on private property is not the cause of a complete stoppage of a drainage system, other possible causes may be:

overloaded public sewers at peak periods, lift station pump breakdown or power failure.
See page 246

5. In order to make private sewer stoppages accessible for unclogging, most codes require:

a two-way cleanout within 5 feet of the building and another at the property line.
See pages 246, 248

6. If a brass cleanout plug is frozen and will not loosen, another procedure one should try is:

use a 14-inch pipe wrench. Place as much pressure as possible on the wrench and rap the handle hard with a hammer. A soldering torch may be used if necessary.
See page 248

7. Some of the factors determining the right kind of sewer cable (electric, flat steel tape or steel spring cable) for a particular type of stoppage:

    whether sewer pipe is fairly straight, whether it's connected to a septic tank, whether it has fittings for a change in direction.
    See page 248

8. The portion of a sanitary system considered adequate for cleaning purposes when an accessible cleanout is not available:

    the vent pipes extending above the roof.
    See page 249

9. The plumbing fixture most subject to clogging:

    the kitchen sink.
    See page 250

10. A "plumber's helper" is:

    a flat rubber force cup with a wooden handle.
    See page 250

11. The procedure needed when using a "plumber's helper" in a two-compartment sink to prevent loss of pressure:

    plug one waste outlet tightly with a rag.
    See page 250

12. The most likely cause of a clogged shower drain:

    accumulated hair in the trap.
    See page 250

13. The tool one should try first in unstopping a shower drain:

    a "plumber's helper."
    See page 250

14. When the drain stopper of a lavatory or bathtub is not causing the stoppage, the tool you should try first to clear the trap or drain line:

    a "plumber's helper."
    See page 252

15. The passageway of a water closet trap is designed to:

    pass "acceptable" materials no larger than 2 inches.
    See page 252

16. It's important to determine the cause of a stoppage in a water closet trap before proceeding to unclog it because:

    if the blockage is something that will result in additional stoppage, it should not be forced into the drainage system.
    See page 253

17. The two tools most often used to unstop a water closet bowl are:

    the closet auger and the force cup.
    See page 253

18. To clear a closet bowl trap when stoppage is caused by cloth, the tool that should be used is:

    a closet auger.
    See page 253

19. The best way to check to be certain a closet bowl blockage has been removed is:

    flush several fairly heavy loads of toilet tissue through the bowl.
    See page 254

20. The substances contained in "hard" water are:

    high amounts of iron and other dissolved minerals.
    See page 254

21. The flushing rim of a closet bowl is located:

    around the top of the water closet bowl against which the seat rests.
    See page 254

22. When a water closet is not flushing properly one of the first checks you should make is:

    check the holes in the rim of the bowl. If these are partly or totally closed, they must be cleaned.
    See page 254

23. The procedure you should use to clear mineral deposits from a shower head's face plate:

    remove the face plate and soak it in vinegar.
    See page 255

24. The first step to take before shutting off the water supply to an improperly-working water heater:

shut off the electric power or the gas supply.
See page 255

25. To find out whether a water heater is working properly, the following yearly check should be made:
    1) flush sediment buildup in the tank bottom through the drain valve until the water runs clear,
    2) raise the test lever at the top of the temperature-pressure relief valve to make certain the waterway is clear.
    See page 255

26. The two things accomplished when an aerator is used on a faucet spout:
    the aerator blends air with the water, which conserves water and prevents splattering.
    See page 255

27. When reassembling a male-type aerator, the pipe compound should be applied:
    only to the outside threads of the aerator.
    See page 255

28. The main cause of reduced water pressure to a clothes washing machine:
    clogged screens.
    See page 256

29. When a house is to be vacant for a period of time, the property can be protected from possible water damage by:
    turning off the faucets to the washing machine.
    See page 256

30. The most likely cause of a complete loss of hot water in a building is:
    someone closed the valves.
    See page 257

31. The most probable cause of a complete loss of water to a residence in Chicago during January:
    frozen water pipes.
    See page 257

32. The most probable cause of a complete loss of water to a residence in Key West, Florida in January:
    the utility company may have shut off the

water to make emergency repairs.
See page 257

33. The most probable cause of a gradual pressure reduction in a compression-type faucet on the hot water side:
    a washer not designed for hot water use may have been installed.
    See page 256

34. Another possible cause for reduced water pressure on a compression faucet:
    a flat washer is replaced with a beveled washer.
    See page 256

35. Usually the cause of a chattering or whistling noise in a compression-type faucet when the water is turned on:
    a loose screw which holds the washer in place.
    See page 256

36. There are two ways to recover a lost bibb screw that has dropped into the supply tube:
    1) remove the faucet stem assembly and flush the screw into the fixture bowl,
    2) disconnect the supply tube, remove it and the screw should fall out.
    See pages 256-257

37. The major cause of water hammer is:
    air loss from the air chambers.
    See page 257

38. The three steps you should try in eliminating water hammer in a water distribution system (in this order):
    1) drain water from system; this permits air to fill pipes,
    2) blow trapped water (that can't be drained) out of the pipes by placing your mouth over faucet spout opening,
    3) install exposed air chambers in both hot and cold water pipes.
    See pages 257-258

39. The old saying, "An ounce of prevention is worth a pound of cure" applies to protecting water piping system from freezing as follows:
    all exposed piping should be insulated well and checked before winter.
    See page 258

40. The thing that is used to thaw water frozen in pipes:

    heat.
    See page 258

41. The procedure that may be used to thaw a frozen section of pipe where fire is not a danger:

    a blowtorch or a burning twist of newspaper.
    See page 258

42. The procedure that may be used to thaw a frozen section of pipe where fire is a danger:

    wrap the affected section of pipe with rags and pour boiling water over the rags.
    See page 258

43. The procedure used to thaw pipes that are inaccessible:

    a low voltage electrical thawing unit.
    See page 258

44. The thing(s) you should not use to rid plumbing fixtures of heavy stains:

    scouring powder or pads.
    See page 259

45. Bathtub or sink stains may safely be removed:

    by filling the fixture with warm water and adding some chlorine bleach.
    See page 259

46. The best way to avoid fixture stains when the source of water is a well:

    install a water softener.
    See page 259

47. The substance that may be used to remove stubborn mineral stains without damage to the fixture finish:

    vinegar.
    See page 259

48. To avoid damaging the finish when cleaning toilet seats, tile and chrome, it is recommended:

    that you use a soft cloth and mild soap.
    See page 259

Chapter 21

1. The valve types normally used to control water to or in a building are:

    gate and globe valves.
    See page 261

2. Valve repairs are generally limited to *leaking washers* or *leaking packing nuts.*
    See pages 262, 263

3. When a globe valve does not control the water, you correct it by:

    replacing the washer.
    See page 262

4. The first step in disassembling a control valve is:

    to shut off the water and drain the line.
    See page 262

5. When a washer requires frequent replacement the problem may be:

    either the washer may be the wrong type for the seat, or the seat may be so pitted and rough that it scores and wears away the washer prematurely.
    See page 263

6. Replaceable valve seats are identified as follows:

    they either have a square or hexagonal water passage, and the seat removal tool must fit the particular passage.
    See page 263

7. The name of the tool used to remove pits or rough surfaces from valve or faucet seats is:

    a seat dressing tool.
    See page 264

8. The first step in repairing a leak around the stem of a control valve:

    tighten the packing nut.
    See page 263

9. The type of faucet now most often used on plumbing fixtures:

    the ordinary compression faucet.
    See page 263

10. The most common repair required on fixture faucets is:

    dripping.
    See page 263

11. The direction in which a valve handle must be turned to shut off the water:
clockwise.
See page 263

12. A house control valve is usually located:
close to the point of entry to the building.
See page 263

13. When water leaks from under the handle, to repair it usually means:
the stem packing (packing washer) needs replacing.
See page 263

14. To prevent leaks from under the handle, newer type compression faucets have:
"O" rings instead of a packing washer.
See page 264

15. When water drips from the spout, the repair needed is:
replacement of the washer.
See page 264

16. When a "handle puller" is not available, the procedure for removing a frozen faucet handle is:
rap the underside of the faucet handle with the plastic end of a screwdriver.
See page 264

17. When water flowing through a faucet is "noisy," the cause (other than a loose washer) may be:
generally worn threads on the stem and worn receiver threads in the faucet body.
See page 264

18. The part that must be replaced when water leaks around the swing spout of a faucet:
replace the "O" ring or rings.
See page 265

19. The precaution that should be taken when pliers are needed to loosen a chrome swing spout lock nut:
wrap lock nuts with adhesive tape to avoid marring the chrome finish.
See page 265

20. Single handle mixing faucets last longer than other types before repairs are needed because:

the moving parts can control the flow of water without grinding the washer against a metal seat.
See page 265

21. The Delta and Moen faucets differ in controlling water temperature and flow as follows:
Delta faucets and valves are designed to control water temperature and flow with a single ball that rests against a seat assembly. Moen faucets and valves are designed to control water temperature and flow with a single cartridge.
See pages 265, 268

22. The name and size of the two washers in a sink trap:
square cut washers, sized 1½ inches in diameter.
See page 270

23. The U-shaped portion of a P-trap is called:
a "J" bend.
See page 270, Figure 21-13

24. The two parts of a sink basket strainer that usually must be replaced because of age are:
1) the basket portion of the strainer,
2) the strainer body rubber washer.
See page 270

25. The thing that distinguishes a washer designed for a sink tailpiece from one designed for a P-trap is:
the tail piece washer is flat.
See page 272

26. The washer that seals the bottom of a lavatory bowl is called:
a mack washer.
See page 274

27. To stop a leak at the water closet tank bolt, the first thing that should be tried is:
to tighten the bolts.
See page 276

28. It's important not to overtighten water closet tank bolts because:
overtightening may crack the tank.
See page 276

29. The device that provides the waterway between a wall-mounted tank and a closet bowl is:

a flush elbow.
See page 276

30. The location of a water closet spud washer:

it fits over the end of the flush valve.
See page 276

31. The probable cause of a water closet leak at the floor connection is:

the setting seal is not holding.
See page 278

32. A water closet tank sometimes "sweats" because:

cold water entering the tank may chill it enough to cause condensation on the tank's outer surface.
See page 278

33. To stop tanks from "sweating," try this:

insulate the tank with tank liners.
See page 278

34. The valve that controls the water to a water closet tank is called:

a ballcock.
See page 278

35. The three checks that should be made when a trickle of water continues to flow from the tank into the bowl:

1) Shut off the water at the valve below the water closet tank and empty tank by flushing. Then trip the flush handle several times to see if the guide is lined up directly over the center of the flush valve seat.

2) Check to see if corrosive water may have deteriorated the copper or brass lift wires and guide assembly.

3) Using fine steel wool, remove any irregularities that might prevent the flush ball from making a tight seal.
See page 283

36. The three checks that should be made when there is the continual sound of running water coming from a water closet but no visible water entering the bowl:

1) Bend the float rod down slightly until the normal water level is approximately

1 inch below the overflow tube.

2) If the float ball rides low in the tank water, it's probably waterlogged and needs to be replaced.

3) If the float ball is not the cause, the ballcock should be repaired or replaced.
See pages 283-284

37. To repair a ballcock you must be familiar with:

the location and how to replace the two washers that control the running water.
See page 284, Figure 21-23

38. The first thing to be done to repair or replace a ballcock:

shut off the water and drain the tank.
See page 284

39. The guide in a water closet tank serves the following purpose:

it guides the flush ball until it seats properly over the flush valve opening.
See page 284

40. When the nuts are frozen to the water closet seat bolts and won't come loose, the way you remove the old seat is:

use penetrating oil (if time is not a factor), or cut through the bolts with a new fine tooth hacksaw blade without the frame.
See pages 284-285

Chapter 22

1. Matter occurs in three physical states:

solid, liquid, gas.
See page 289

2. One of the very valuable characteristics of gas is that it can be *forced* through very small spaces.
See page 289

3. Gas has neither a fixed shape nor a fixed volume but is made of constantly moving *atoms.*
See page 289

4. The point at which gas particles are liquefied:

when cooled below their boiling point.
See page 289

5. The thing that's added to the natural gas system (so that leaks can be detected) before it enters the pipelines:

a chemical scent.
See page 289

6. Some of the other names for natural gas are:

dry or sweet gas.
See page 289

7. The way natural gas can cause death, although it is not poisonous:

suffocation in a closed space.
See page 289

8. Manufactured gas is chiefly produced from:

coal.
See page 289

9. The substance contained in manufactured gas that makes it poisonous is:

carbon monoxide.
See page 289

10. LP gas is convenient to use as fuel in remote areas because:

it liquefies under moderate pressure, and this makes it easy to transport and store in special tanks.
See page 289

11. The gas service pipe to a building is sized by:

the gas supplier.
See page 290

12. The sizing and installation methods of gas supply piping is governed by:

the local gas code.
See page 290

13. The abbreviation "Btu" stands for:

British thermal unit.
See page 290

14. One Btu is:

the quantity of heat required to raise the temperature of 1 pound of water 1 degree Fahrenheit.
See page 290

15. The number of Btu in each cubic foot of natural gas is assumed to be:

1,000 cu. ft. per hour.
See page 290

16. If you know the maximum Btu rating for an appliance, you convert the Btu into cubic feet as follows:

divide the value in Btu by 1,000.
See page 290

17. When connecting a gas supply pipe to an appliance that has lost its Btu rating plate, the minimum pipe size required is:

one that is not smaller than the appliance inlet pipe.
See page 290

18. The minimum size gas pipe outlet that can be used under any circumstances is:

½ inch.
See page 290

19. The two things you must know before sizing any gas main or branch lines:

1) the total developed length of gas piping

2) the Btu input rating of the appropriate appliance outlet.
See page 291

20. The materials and installation methods for a gas system are regulated by:

your local gas code.
See page 292

21. The most common material used in residential gas piping is:

galvanized steel pipe.
See page 292

22. The condition that might prevent the use of copper piping in a gas system is:

the gas to be used is of the corrosive type.
See page 292

23. The two weights of copper pipe and tubing required for interior gas piping are:

type K or L.
See page 292

24. Copper pipe or tubing may be used outside underground with this exception:

under a concrete slab.
See page 292

25. When joints are necessary in a copper piping system, the type of solder that must be used:

    hard solder, usually a silver solder.
    See page 292

26. Flare fittings may be used in a copper gas piping system only if they are *not concealed* and *if the tubing is continuous and of one piece.*
    See page 292

27. Two things that should be done to protect gas piping when it's installed underground in trenches:

    1) *Place it at least 12 inches deep to protect the pipe from damage by sharp tools.*
    2) *In corrosive soils, the pipe should be protected with an approved wrapping or one or two coats of asphaltum paint.*
    See page 292

28. When backfilling a trench containing gas piping, the type of backfill that should be used is:

    fine material.
    See page 294

29. The three steps that must be taken when gas piping must be installed under a slab:

    1) *Encase the pipe completely in conduit.*
    2) *The termination of the conduit above the floor must be sealed.*
    3) *The termination of the conduit outside the building must be tightly sealed with a vent extending above grade.*
    See page 294

30. When a gas appliance is located in the center of a room, the gas piping must be installed:

    in an open channel cut into the concrete floor.
    See page 294

31. To protect gas piping installed in vertical masonry walls:

    provide adequate chases.
    See page 294

32. It might be necessary to drill a hole in the center of the partition for horizontal gas piping when:

    short runs are required.
    See page 294

33. You must not notch a partition deeper than 40 percent the width of the stud in order to:

    avoid weakening the stud.
    See page 294

34. To secure gas piping installed in metal stud partitions:

    tie wire is generally used.
    See page 295

35. A gas main must be *installed* to drain dry.
    See page 295

36. The thing that must be installed for certain gases which contain moisture:

    a drip pipe.
    See page 295

37. Gas branch pipes should be taken only from the top or side of a gas feeder pipe in order to:

    avoid condensate from filling and obstructing the flow of gas.
    See page 295

38. The purpose of installing a shutoff valve near the gas meter is:

    to shut off gas to the building.
    See page 296

39. The type of shutoff valve each gas appliance in a building should have is:

    a manually-operated one.
    See page 296

40. The two types of shutoff valves manufactured for appliances are:

    the straight and angle patterns.
    See page 296

41. The maximum distance allowed from a shutoff valve to the appliance it serves is:

    6 feet.
    See page 296

42. The time when left and right threaded couplings in a concealed gas piping system are permitted:

    never.
    See page 296

43. The thing that must be done to prevent a union in an existing concealed gas line from working loose:

punch the center nut on the joint.
See page 296

44. The time when a new connection in an existing concealed copper tubing is allowed:
never.
See page 296

45. The standard to which the threads for gas piping must conform:
the standards adopted by the American Standards Association.
See page 296

46. The procedure for preparing threads for gas piping is the same as for *water piping.*
See page 296

47. The last major step that must be taken before gas piping can be concealed is:
it must be pressure tested.
See page 296

48. The best and safest way to check gas piping for leaks is:
brush a liquid soap around each joint to see if bubbles appear.
See page 296

49. The minimum height above the garage floor that the combustion chamber for a gas water heater may be set is:
18 inches.
See page 297

50. When a gas water heater is installed in a separate room, the type of ventilation that must be provided:
permanent openings.
See page 297

51. The place where gas water heaters must never be installed:
in living areas which may be closed.
See page 297

52. The thing that may be omitted when a gas appliance input rating does not exceed 30 Btu per hour per cubic foot of room space:
the vent.
See page 297

53. The minimum separation between any combustible material and a gas water heater with an insulated jacket is:
2 inches.
See page 297

54. The size vent pipe required for a 30-gallon gas water heater with a 4-inch draft hood:
a vent no smaller than the opening of the draft hood (4 inches in this case).
See page 297

55. The two types of vent piping materials acceptable for installations:
the double wall metal pipe and single wall metal piping.
See page 297

56. Required provision for vent pipes installed in partitions constructed of combustible material:
an approved metal spacing device.
See page 297

57. The required gauge for metal straps or hangers used to support horizontal gas vent piping:
at least 20 gauge sheet metal.
See page 297

58. The thing required for all gas vent pipes terminating above a roof:
UL-approved cap.
See page 297

Chapter 23

1. A pool that is considered permanent cannot be *disassembled.*
See page 301

2. The most common residential pool is:
the recirculating type.
See page 301

3. A good filtration system protects pool users by:
providing water that is clear of organic matter and safe from harmful bacteria.
See page 301

4. A hose bibb when used for adding water to a swimming pool must be provided with:
a vacuum breaker.
See page 301

5. When water is added to a swimming pool by means of a direct water supply connected to public water, a cross connection can be prevented by:
installing the fill spout with an air gap above the overflow rim of the pool.
See page 301

6. When well water is used to supply a swimming pool, special problems that may be encountered are:
if well water doesn't meet the requirements for domestic water supply, a filtration system is needed.
See page 301

7. Two acceptable ways of disposing of swimming pool waste water:
(Any two of the following are acceptable.)
expel it into an adequately sized drainfield, puddle it onto private property (with certain limitations), direct it into a sprinkler system to be used for irrigation purposes.
See pages 301-302

8. The part of a pool in which the main drain must be located:
at the pool's lowest point.
See page 302

9. Other than recirculating pool water, the main drain serves this purpose:
it drains the pool dry.
See page 302

10. The grate over a main pool drain must be securely fastened so that tools are required for its removal because:
the suction from the drain pipe could hold a small person under water and cause drowning.
See page 302

11. The location of an interceptor in a pool piping system:
in the suction line ahead of the pump.
See page 302

12. It's important to size and space pool inlets properly in order to:
produce uniform circulation of incoming water throughout the pool.
See page 302

13. The square footage of water surface which one pool inlet may adequately serve:
350 square feet.
See page 302

14. The minimum pipe size connected to a main drain is:
2 inches.
See page 303

15. The minimum diameter size of a pool vacuum fitting is:
2 inches.
See page 303

16. The maximum depth below the water line to which a vacuum fitting may be installed is:
18 inches.
See page 303

17. The following provision must be made for the continued operation and maintenance of valves, pumps, filters and other pool equipment:
they must be readily accessible.
See page 303

18. The types of filtration equipment accepted by most codes are:
sand filters, diatomaceous earth filters, cartridge filters.
See pages 304-307

19. The two methods commonly used for surface skimming a pool are:
1) skimmers built into the pool wall
2) an overflow gutter at the end of the pool.
See pages 307-308

20. The required height above surface level for bottom of motor for swimming pool pump:
4 inches.
See page 308

21. The most commonly-used materials for swimming pool piping are:
thermoplastic, PVC, polyethylene and polybutylene pipe and fittings.
See page 308

22. The way pool piping should be supported if it's not supported directly on the ground:
supported directly to the pool structure by means of pipe hangers or heavy duty strap iron.
See page 308

23. The type of ell that should be used underground on any swimming pool suction line is:

    the long radius type.
    See page 308

24. The installation of pool water heating equipment is the same as for:

    domestic water heaters.
    See page 308

25. Pool piping must be water-tested to prove the system is tight by a pressure of:

    40 psi.
    See page 308

26. Before installing and repairing pool piping, this is required:

    a plumbing permit.
    See page 308

27. Most spas and hot tubs differ in the method of construction as follows:

    spas are of permanent construction; hot tubs may be disassembled and reassembled.
    See page 308

28. The general use of a spa is:

    recreational or therapeutic.
    See page 308

29. Gas-fired swimming pools must meet whose standards?

    American Gas Association.
    See page 308

30. What is the most popular energy source for a swimming pool?

    Natural or LP gas.
    See page 309

Definitions of Terms

The terms included here are found in most plumbing codes. Some words in the code have become so descriptive and specialized that their meaning is different from what you'll find in a standard dictionary.

Two words that appear repeatedly in every code are *shall* and *may*. Every plumber should understand the specialized meanings of these words.

Shall means that compliance is mandatory and that the procedure or condition specified must be performed without deviation. For example, part of the requirements for waste disposal is that "sewage and liquid waste *shall* be treated and disposed of as hereinafter provided."

May is a term of permission. When used in the code, it means "allowable" or "optional," but not required. For example: "Drinking fountains *may* be installed with indirect waste only for the purpose of resealing required traps of floor drains."

Building drain and building sewer are two terms often used improperly. Many professionals assume that both terms apply to the same part of the drainage system. However, note the code definition for each term:

Building drain is the main horizontal collection system within the walls of a building which extends to 3 feet beyond the building line. (This distance may vary in some codes.)

Building sewer is defined as that part of the horizontal drainage system outside the building line which connects to the building drain and conveys the liquid waste to a legal point of disposal.

Effective and constructive code interpretation is possible only when the words and terms used in the code are understood. It is possible to have some variation and still remain within the intent of the code definitions. Isometric illustrations help to provide an understanding of code definitions.

Absorption Drainfield absorption area.

Air gap (in a water supply system) The unobstructed vertical distance through the free atmosphere between the lowest opening from any pipe or faucet supplying water to a tank, plumbing fixture, or other device, and the flood level rim of the receptacle.

Anaerobic Living without free oxygen. Anaerobic bacteria found in septic tanks are beneficial in digesting organic matter.

Anchors See *Supports*.

Approved Approved by the plumbing official or other authority given jurisdiction by the code.

Area drain A receptacle designed to collect surface or rain water from an open area.

Backfill That portion of the trench excavation up to the original earth line which is replaced after the sewer or other piping has been laid.

Backflow The flow of water or other liquids, mixtures, or substances into the distributing pipes of a potable supply of water and any other fixture or appliance, from any source, which is opposite to the intended direction of flow.

Backflow connection Any arrangement that can allow backflow to occur.

Backflow preventer A device or means to prevent backflow into the potable water system.

Back siphonage The flow of water or other liquids, mixtures or substances into the distributing pipes of a potable supply of water or any other fixture, device, or appliance, from any source, which is opposite to the intended direction of flow due to negative pressure in the pipe.

Base The lowest point of any vertical pipe.

Battery of fixtures Any group of two or more similar adjacent plumbing fixtures which discharge into a common horizontal waste or soil branch.

Boiler blow-off An outlet on a boiler to permit emptying or discharging of water or sediment in the boiler.

Branch Any part of the piping system other than a main, riser or stack.

Branch interval A length of soil or waste stack (vertical pipe) generally one story in height (approximately 9 feet, but not less than 8 feet) into which the horizontal branches from one floor or story of a building are connected to the stack.

Building drain The main horizontal sanitary collection system, inside the wall line of the building, which conveys sewage to the building sewer beginning 3 feet (more or less in some codes) outside the building wall. The building drain excludes the waste and vent stacks which receive the discharge from soil, waste and other drainage pipes, including storm water.

Building sewer That part of the horizontal piping of a drainage system which connects to the end of the building drain and conveys the contents to a public sewer, private sewer, or individual sewage disposal system.

Building storm drain A drain used to receive and convey rain water, surface water, ground water, subsurface water and other clear water waste, and discharge these waste products into a building storm sewer or a combined building sewer beginning 3 feet outside the building wall.

Building storm sewer Connects to the end of the building storm drain to receive and convey the contents to a public storm sewer, combined sewer, or other approved point of disposal.

Building subdrain Any portion of a drainage system which cannot drain by gravity into the building sewer.

Caulking Any approved method of rendering a joint watertight and gastight. For cast iron pipe and fittings with hub joints, the term refers to caulking the joint with lead and oakum.

Code Regulations and their subsequent amendments or any emergency rule or regulation lawfully adopted to control the plumbing work by the administrative authority having jurisdiction.

Combined building sewer A building sewer which receives storm water, sewage and other liquid waste.

Common vent   The vertical vent portion serving to vent two fixture drains which are installed at the same level in a vertical stack.

Conductor   See *Leader*.

Continuous waste   A drain connecting a single fixture with more than one compartment or other permitted fixtures to a common trap.

Cross connection   Any physical connection or arrangement between two separate piping systems, one containing potable water and the other water of unknown or questionable safety.

Dead end   A branch leading from a soil, waste or vent pipe, building drain or building sewer which is terminated by a plug or other close fitting at a developed distance of 2 feet or more. A dead end is also classified as an extension for future connection, or as an extension of a cleanout for accessibility.

Developed length   The length as measured along the center line of the pipe and fittings.

Diameter   The nominal diameter of a pipe or fitting as designed commercially, unless specifically stated otherwise.

Downspout   See *Leader*.

Drain   Any pipe which carries liquid, waste water or other water-borne wastes in a building drainage system to an approved point of disposal.

Drainage system   All the piping within a public or private premises that conveys sewage, rain water, or other types of liquid wastes to a legal point of disposal.

Drainage well   Any drilled, driven or natural cavity which taps the underground water and into which surface waters, waste waters, industrial waste or sewage is placed.

Durham system   An all-threaded pipe system of rigid construction, using recessed drainage fittings to correspond to the types of piping being used.

Effective opening   The minimum cross-sectional area of the diameter of a circle at the point of water supply discharge.

Effluent   The liquid waste as it flows from the septic tank and into the drainfield.

Fall   See *Grade*.

Fire lines   The complete wet standpipe system of the building, including the water service, standpipe, roof manifold, Siamese connections and pumps.

Fixture branch   The drain from the trap of a fixture to the junction of that drain with a vent. Some codes refer to a fixture branch as a "fixture drain."

Fixture drain   The drain from the fixture branch to the junction of any other drain pipe, referred to in some codes as a "fixture branch."

Fixture unit   A design factor to determine the load-producing value of the different plumbing fixtures. For instance, the unit flow rate from fixtures may be determined to be 1 cubic foot, or 7.5 gallons of water per minute.

Flood level rim   The top edge of a plumbing fixture or other receptacle from which water or other liquids will overflow.

Floor drain   An opening or receptacle located at approximately floor level connected to a trap to receive the discharge from indirect waste and floor drainage.

Floor sink   An opening or receptacle usually made of enameled cast iron located at approximately floor level which is connected to a trap, to receive the discharge from indirect waste and floor drainage. A floor sink is more sanitary and easier to clean than a regular floor drain, and is usually used for restaurant and hospital installations.

Flushometer valve   A device actuated by direct water pressure which discharges a predetermined quantity of water to fixtures for flushing purposes.

Grade   The slope or pitch, known as "the fall," usually expressed in drainage piping as a fraction of an inch per foot.

Hangers   See *Supports*.

Horizontal branch   A drain pipe extending laterally from a soil or waste stack or building drain. May or may not have vertical sections or branches.

Horizontal pipe   Any pipe or fitting which makes an angle of more than 45 degrees with the vertical.

Indirect wastes   A waste pipe charged to convey liquid wastes (other than body wastes) by discharging them into an open plumbing fixture or receptacle such as floor drain or floor sink. The overflow point of such fixture or receptacle is at a lower elevation than the item drained.

Industrial waste   Liquid waste, free of body waste, resulting from the processes used in industrial establishments.

Insanitary   Contrary to sanitary principles; injurious to health.

Interceptor   A device designed and installed to separate and retain deleterious, hazardous, or undesirable matter from normal wastes and permit normal sewage or liquid wastes to discharge by gravity into the disposal terminal or sewer.

Leader   The vertical water conductor or downspout from the roof to the building storm drain, combined building sewer, or other approved means of disposal.

Liquid waste   The discharge from any fixture, appliance or appurtenance that connects to a plumbing system which does not receive body waste.

Load factor   The percentage of the total connected fixture unit flow rate which is likely to occur at any point with the probability factor of simultaneous use. It varies with the type of occupancy and the total flow unit above this point being considered.

Loop or *circuit waste and vent*   A combination of plumbing fixtures on the same floor level in the same or adjacent rooms connected to a common horizontal branch soil or waste pipe.

Main   The principal artery of any system of continuous piping, to which branches may be connected.

Main vent   The principal artery of the venting system, to which vent branches may be connected.

May   The word "may" as used in the code book is a term of permission.

Mezzanine   An intermediate floor placed in any story or room. When the total area of any such mezzanine floor exceeds 33⅓ percent of the total floor area in that room or story, it is considered an additional story rather than a mezzanine.

Pitch   See *Grade*; also referred to as *slope* or *fall*.

Plumbing   Includes any or all of the following: (1) the materials including pipe, fittings, valves, fixtures and appliances attached to and a part of a system for the purpose of creating and maintaining sanitary conditions in buildings, camps and swimming pools on private property where people live, work, play, assemble or travel; (2) that part of a water supply and sewage and drainage system extending from either the public water supply main or private water supply to the public sanitary, storm or combined sanitary and storm sewers, or to a private sewage disposal plant, septic tank, disposal field, pit, box filter bed or any other receptacle or into any natural or artificial body of water or watercourse on public or private property; (3) the design, installation or contracting for installation, removal and replacement, repair or remodeling of all or any part of the materials, appurtenances or devices attached to and forming a part of a plumbing system, including the installation of any fixture, appurtenance or devices used for cooking, washing, drinking, cleaning, fire fighting, mechanical or manufacturing purposes.

Plumbing fixtures   Receptacles, devices, or appliances which are supplied with water or which receive or discharge liquids or liquid-borne waste, with or without discharge, into the drainage system with which they may be directly or indirectly connected.

Plumbing official inspector   The chief administrative officer charged with the administration, enforcement and application of the plumbing code and all its amendments.

Plumbing system   The drainage system, water supply, water supply distribution pipes, plumbing fixtures, traps, soil pipes, waste pipes, vent pipes, building drains, building sewers, building storm drain, building storm sewer, liquid waste piping, water-treating and water-using equipment, sewage treatment, sewage treatment equipment, fire standpipes, fire sprinklers and related appliances and appurtenances, including their respective connections and devices, within the private property lines of a premises.

Potable water Water that is satisfactory for drinking, culinary and domestic purposes and meets the requirements of the health authority having jurisdiction.

Private or *private use* In relation to plumbing fixtures: in residences and apartments, and in private bathrooms of hotels and similar installations where the fixtures are intended for the use of a family or an individual.

Private property For the purposes of the code, all property except streets or roads dedicated to the public, and easements (excluding easements between private parties).

Private sewer A sewer privately owned and not directly controlled by public authority.

Public or *public use* In relation to plumbing fixtures: in commercial and industrial establishments, in restaurants, bars, public buildings, comfort stations, schools, gymnasiums, railroad stations or places to which the public is invited or which are frequented by the public without special permission or special invitation, and other installations (whether paid or free) where a number of fixtures are installed so that their use is similarly unrestricted.

Public sewer A common sewer directly controlled by public authority.

Public swimming pool A pool together with its buildings and appurtenances where the public is allowed to bathe or which is open to the public for bathing purposes by consent of the owner.

Relief vent A vent, the primary function of which is to provide circulation of air between drainage and vent systems.

Rim In code usage, an unobstructed open edge at the overflow point of a fixture.

Rock drainfield Three-quarter-inch drainfield rock of which 100 percent passes a 1-inch screen and a maximum of 10 percent passing a ½-inch screen.

Roof drain An outlet installed to receive water collecting on the surface of a roof which discharges into the leader or downspout.

Roughing-in The installation of all parts of the plumbing system that can be completed prior to the installation of plumbing fixtures; includes drainage, water supply, vent piping, and the necessary fixture supports.

Sanitary sewer A pipe which carries sewage and excludes storm, surface and ground water.

Second hand A term applied to material or plumbing equipment which has been installed and used, or removed.

Septic tank A watertight receptacle which receives the discharge of a drainage system or part thereof, so designed and constructed as to separate solids from liquid, digest organic matter through a period of detention, and allow the liquids to discharge into the soil outside the tank through a subsurface system of open-joint or perforated piping, or other approved methods.

Sewage Any liquid waste containing animal, mineral or vegetable matter in suspension or solution. May include liquids containing chemicals in solution.

Shall As used in the code, a term meaning that compliance is mandatory and that the procedure or condition specified must be performed without deviation.

Slope See *Grade*.

Soil pipe Any pipe which conveys the discharge of water closets or fixtures having similar functions, with or without the discharge from other fixtures, to the building drain or building sewer.

Soil vent See *Stack vent*.

Stack The vertical pipe of a system of soil, waste or vent piping.

Stack vent The extension of a soil or waste stack above the highest horizontal drain connected to the stack. (Sometimes called a *waste vent* or *soil vent.*)

Standpipe system   A system of piping installed for fire protection purposes having a primary water supply constantly or automatically available at each hose outlet.

Storm sewer   A sewer used for conveying rain water and/or surface water.

Subsurface drain   A drain which receives only subsurface or seepage water and conveys it to a place of disposal.

Sump   A tank or pit which receives sewage or liquid waste, located below the normal grade of the gravity system and which must be emptied by mechanical means.

Supply well   Any artificial opening in the ground designed to conduct water from a source bed through the surface when water from such well is used for public, semi-public or private use.

Supports   Devices for supporting and securing pipe and fixtures to walls, ceilings, floors or structural members. (Also known as *hangers* or *anchors.*)

Trap   A fitting or device so designed and constructed as to provide a liquid seal which will prevent the back passage of air without materially affecting the flow of sewage or waste water through it.

Trap seal   The maximum vertical depth of liquid that a trap will retain, measured between the crown weir and the top of the dip of the trap.

Vent stack   A vertical vent pipe installed primarily for the purpose of providing a circulation of air to and from any part of the drainage system.

Vent system   A pipe or pipes installed to provide a flow of air to or from a drainage system or to provide a circulation of air within such system.

Vertical pipe   Any pipe or fitting which is installed in a vertical position or which makes an angle of not more than 45 degrees with the vertical.

Waste pipe   Any pipe which receives the discharge of any fixture, except water closets or fixtures having similar functions, and conveys it to the building drain or to the soil or waste stack.

Waste vent   See *Stack vent.*

Water-distributing pipe   A pipe which conveys water from the water service pipe to the plumbing fixtures, appliances and other water outlets.

Water main   A water supply pipe for public or community use.

Water outlet   As used in connection with the water-distributing system, the discharge opening for the water (1) to a fixture, (2) to atmospheric pressure (except into an open tank which is part of the water supply system), (3) to a boiler or heating system, or (4) to any water-operated device or equipment requiring water to operate, but not a part of the plumbing system.

Water service pipe   The pipe from the water main or other source of water supply to the building served.

Water supply system   Consists of the water service pipe, the water-distributing pipes, standpipe system and the necessary connecting pipes, fittings, control valves and all appurtenances in or on private property.

Wet vent   A waste pipe which serves to vent and convey waste from fixtures other than water closets.

Yoke vent   A pipe connecting upward from a soil or waste stack for the purpose of preventing pressure changes in the stacks.

Abbreviations and Symbols

The abbreviations here are often found on blueprints (building plans) and in plumbing reference books (including the code) to identify plumbing fixtures, pipes, valves and nationally-recognized associations.

A	area
AD	area drain
AGA	American Gas Association
AISI	American Iron and Steel Institute
ASA	American Standard Association
ASCE	American Society of Civil Engineering
ASHRAE	American Society of Heating, Refrigeration and Air Conditioning Engineers
ASME	American Society of Mechanical Engineers
ASSE	American Society of Sanitary Engineering
ASTM	American Society for Testing Materials
AWWA	American Water Works Association
B.S.	bar sink
B	bidet
B.T.	bathtub
Btu	British thermal unit
C to C	center to center
CI	cast iron
CISPI	Cast Iron Soil Pipe Institute
C	condensate line
C.O.	cleanout
C.W.	cold water
cu. ft.	cubic feet
cu. in.	cubic inches
C.W.M.	clothes washing machine
C.V.	check valve
D.F.	drinking fountain
D.W.	dishwasher
E to C	end to center
E.W.C.	electric water cooler
°F	degrees Fahrenheit
F	Fahrenheit
F.B.	foot bath
F.F.	finish floor
F.C.O.	floor cleanout

F.D.	floor drain	NFPA	National Fire Protection Association
F.D.C.	fire department connection	NPS	nominal pipe size
F.E.C.	fire extinguisher cabinet	O	oxygen
F.G.	finish grade	O.D.	outside diameter
F.H.C.	fire hose cabinet	Oz.	ounce
F.L.	fire line	P.D.	planter drain
F.P.	fire plug	P.P.	pool piping
F.S.P.	fire standpipe	psi	pounds per square inch
F.U.	fixture unit	Rad.	radius
GAL.	gallons	R.D.	roof drain
gpm	gallons per minute	Red.	reducer
Galv.	galvanized	R.L.	roof leader
G.S.	glass sink	San.	sanitary
G.V.	gate valve	SH.	shower
G.P.D.	gallons per day	Spec.	specification
H.B.	hose bibb	Sq.	square
Hd or H.D.	head	S.B.	sitz bath
H.W.	hot water	Sq. Ft.	square feet
H.W.R.	hot water return	S.P.	swimming pool
HWT	hot water tank	SS	service sink
in.	inch	Std.	standard
I.D.	inside diameter	SV	service
I.W.	indirect waste	SW	service weight
IPS	iron pipe size	S & W	soil and waste
K.S.	kitchen sink	T	temperature
L. or LAV.	lavatory	U or Urn	urinal
L.T.	laundry tray	V	volume
L	length	Vtr	vent through roof
lb.	pound	W	waste
Max.	maximum	W.C.	water closet
Mfr.	manufacturer	W.H.	water heater
Min.	minimum	XH	extra heavy
M.H.	manhole		
NAPHCC	National Association of Plumbing Heating and Cooling Contractors		
NBFU	National Board of Fire Underwriters		
NBS	National Bureau of Standards		

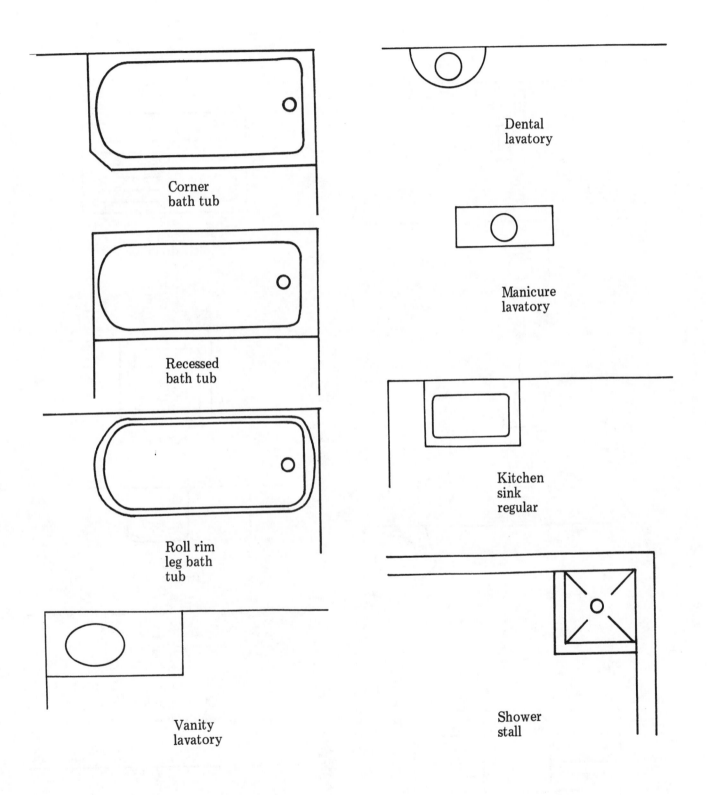

Corner
bath tub

Recessed
bath tub

Roll rim
leg bath
tub

Vanity
lavatory

Dental
lavatory

Manicure
lavatory

Kitchen
sink
regular

Shower
stall

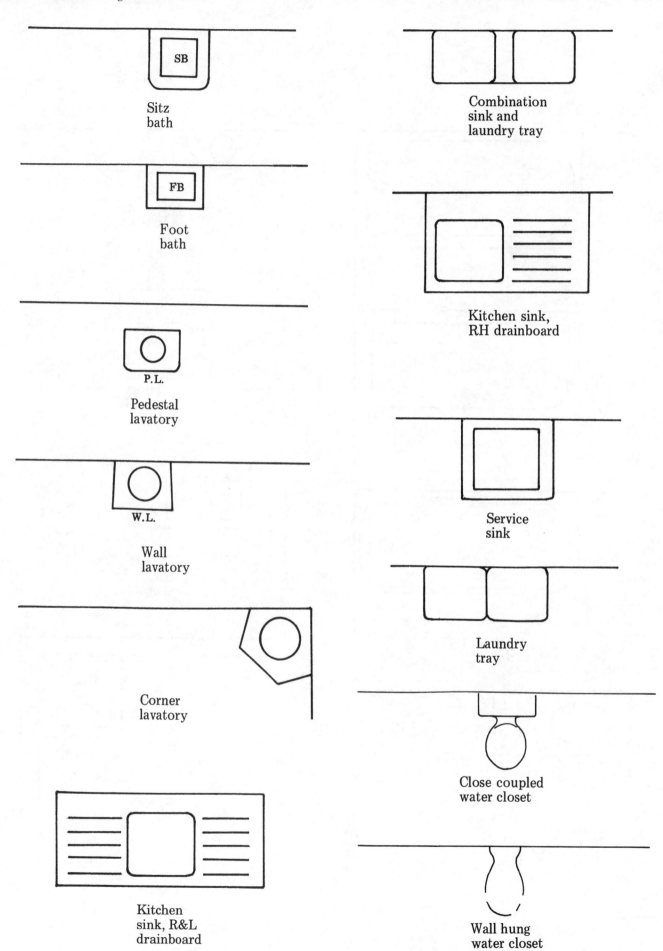

Sitz
bath

Foot
bath

Pedestal
lavatory

Wall
lavatory

Corner
lavatory

Kitchen
sink, R&L
drainboard

Combination
sink and
laundry tray

Kitchen sink,
RH drainboard

Service
sink

Laundry
tray

Close coupled
water closet

Wall hung
water closet

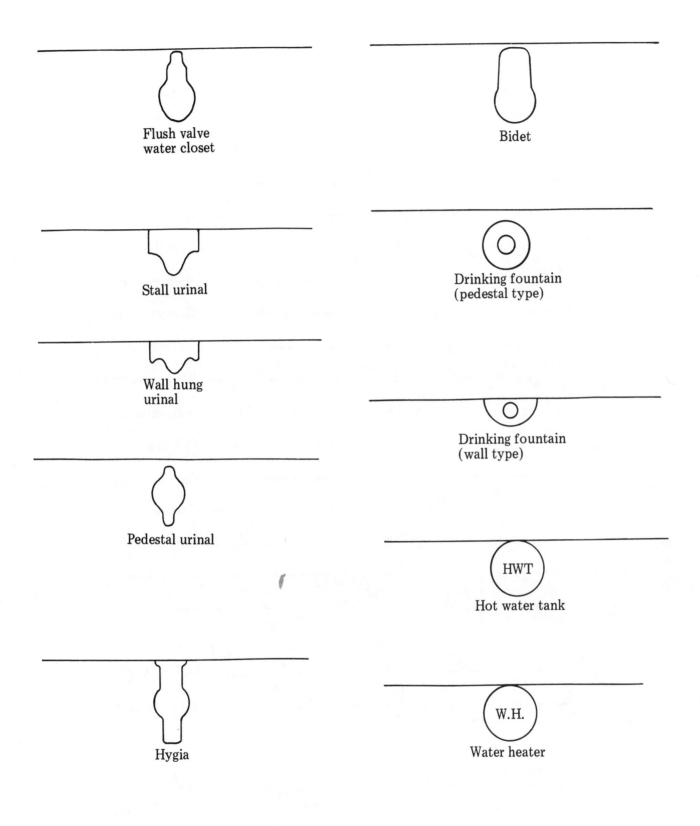

Flush valve
water closet

Bidet

Stall urinal

Drinking fountain
(pedestal type)

Wall hung
urinal

Drinking fountain
(wall type)

Pedestal urinal

HWT
Hot water tank

Hygia

W.H.
Water heater

PLUMBING SYMBOL LEGEND

Symbol		Description
——— — ——— – —	C.W.	——— – — – ——— Cold water
——— —— — —— —	H.W.	——— — — —— — Hot water
——— – — — — —	H.W.R.	——— — — — — ——— Hot water return
———————————	W.L.	——————————— Waste line
– — — — — — —	V.L.	– — – — — – — Vent line
———————————	S S	——————————— Sanitary sewer
———————————	C	——————————— Condensate line
———————————	S D	——————————— Storm drain
———————————	R.W.L.	——————————— Rain water leader
→——→——→	I.W.	→——→ Indirect waste
———————————	F	——————————— Fire line
———————————	G	——————————— Gas line
———⋈———		——————————— Gate valve
———⋈•———		——————————— Globe valve
———————————		——————————— Check valve
———————————	R	——————————— Relief line
———————————		——————————— P&T relief valve
———⊗———	F.C.O.	——————————— Floor cleanout

F.D.　　Floor drain

P.D.　　Planter drain

R.D.　　Roof drain

H.B.　　Hose bibb

A.D.　　Area drain

FITTING SYMBOLS

Bell-and-spigot ends	Return bend
Cap	Tee
Plug	Cross
Union, screwed	True Y- wye
Coupling	Y- Wye single
Expansion joint, sliding	Combination wye and 1/8 bend
Sleeve	Tee union
Expansion joint, bellows	Y- Wye double
Bushing	Double combination wye and 1/8 bend
Reducer	Elbow, 90 degrees
Eccentric reducer	
Reducing flange	
Union, flanged	

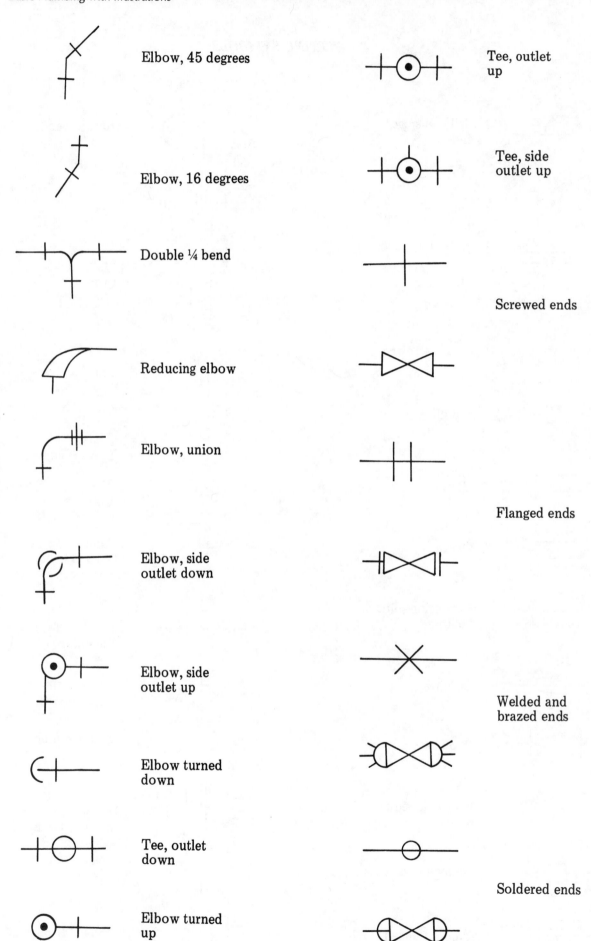

Elbow, 45 degrees

Elbow, 16 degrees

Double ¼ bend

Reducing elbow

Elbow, union

Elbow, side
outlet down

Elbow, side
outlet up

Elbow turned
down

Tee, outlet
down

Elbow turned
up

Tee, outlet
up

Tee, side
outlet up

Screwed ends

Flanged ends

Welded and
brazed ends

Soldered ends

Index

Other Practical References

Basic Engineering for Builders

If you've ever been stumped by an engineering problem on the job, yet wanted to avoid the expense of hiring a qualified engineer, you should have this book. Here you'll find engineering principles explained in non-technical language and practical methods for applying them on the job. With the help of this book you'll be able to understand engineering functions in the plans and how to meet the requirements, how to get permits issued without the help of an engineer, and anticipate requirements for concrete, steel, wood and masonry. See why you sometimes have to hire an engineer and what you can undertake yourself: surveying, concrete, lumber loads and stresses, steel, masonry, plumbing, and HVAC systems. This book is designed to help the builder save money by understanding engineering principles that you can incorporate into the jobs you bid. **400 pages, 8½ x 11, $34.00**

Roofing Construction & Estimating

Installation, repair and estimating for nearly every type of roof covering available today in residential and commercial structures: asphalt shingles, roll roofing, wood shingles and shakes, clay tile, slate, metal, built-up, and elastomeric. Covers sheathing and underlayment techniques, as well as secrets for installing leakproof valleys. Many estimating tips help you minimize waste, as well as insure a profit on every job. Troubleshooting techniques help you identify the true source of most leaks. Over 300 large, clear illustrations  help you find the answer to just about all your roofing questions. **432 pages, 8½ x 11, $35.00**

Contractor's Survival Manual

How to survive hard times and succeed during the up cycles. Shows what to do when the bills can't be paid, finding money and buying time, transferring debt, and all the alternatives to bankruptcy. Explains how to build profits, avoid problems in zoning and permits, taxes, time-keeping, and payroll. Unconventional advice on how to invest in inflation, get high appraisals, trade and postpone income, and stay hip-deep in profitable work. **160 pages, 8½ x 11, $16.75**

National Plumbing & HVAC Estimator

Manhours, labor and material costs for all common plumbing and HVAC work in residential, commercial, and industrial buildings. You can quickly work up a reliable estimate based on the pipe, fittings and equipment required. Every plumbing and HVAC estimator can use the cost estimates in this practical manual. Sample estimating and bidding forms and contracts also included. Explains how to handle change orders, letters of intent, and warranties. Describes the right way to process submittals, deal with suppliers and subcontract specialty work. Includes an electronic version of the book on computer disk with a stand-alone *Windows* estimating program FREE on a 3½" high-density (1.44 Mb) disk. **352 pages, 8½ x 11, $38.25. Revised annually**

Rough Framing Carpentry

If you'd like to make good money working outdoors as a framer, this is the book for you. Here you'll find shortcuts to laying out studs; speed cutting blocks, trimmers and plates by eye; quickly building and blocking rake walls; installing ceiling backing, ceiling joists, and truss joists; cutting and assembling hip trusses and California fills; arches and drop ceilings — all with production line procedures that save you time and help you make more money. Over 100 on-the-job photos of how to do it right and what can go wrong. **304 pages, 8½ x 11, $26.50**

CD Estimator

If your computer has *WindowsTM* and a CD-ROM drive, CD Estimator puts at your fingertips 85,000 construction costs for new construction, remodeling, renovation & insurance repair, electrical, plumbing, HVAC and painting. You'll also have the National Estimator program — a stand-alone estimating program for *Windows* that *Remodeling* magazine called a "computer wiz." Quarterly cost updates are available at no charge on the Internet. To help you create professional-looking estimates, the disk includes over 40 construction estimating and bidding forms in a format that's perfect for nearly any word processing or spreadsheet program for *Windows*. And to top it off, a 70-minute interactive video teaches you how to use this CD-ROM to estimate construction costs. **CD Estimator is $59.00**

National Repair & Remodeling Estimator

The complete pricing guide for dwelling reconstruction costs. Reliable, specific data you can apply on every repair and remodeling job. Up-to-date material costs and labor figures based on thousands of jobs across the country. Provides recommended crew sizes; average production rates; exact material, equipment, and labor costs; a total unit cost and a total price including overhead and profit. Separate listings for high- and low-volume builders, so prices shown are specific for any size business. Estimating tips specific to repair and remodeling work to make your bids complete, realistic, and profitable. Includes an electronic version of the book on computer disk with a stand-alone *Windows* estimating program FREE on a 3½" high-density (1.44 Mb) disk. **416 pages, 8½ x 11, $38.50. Revised annually**

Builder's Guide to Accounting Revised

Step-by-step, easy-to-follow guidelines for setting up and maintaining records for your building business. This practical, newly-revised guide to all accounting methods shows how to meet state and federal accounting requirements, explains the new depreciation rules, and describes how the Tax Reform Act can affect the way you keep records. Full of charts, diagrams, simple directions and examples, to help you keep track of where your money is going. Recommended reading for many state contractor's exams. **320 pages, 8½ x 11, $26.50**

Wood-Frame House Construction

 Step-by-step construction details, from the layout of the outer walls, excavation and formwork, to finish carpentry and painting. Contains all new, clear illustrations and explanations updated for construction in the '90s. Everything you need to know about framing, roofing, siding, interior finishings, floor covering and stairs — your complete book of wood-frame homebuilding. **320 pages, 8½ x 11, $25.50. Revised edition**

The Contractor's Legal Kit

Stop "eating" the costs of bad designs, hidden conditions, and job surprises. Set ground rules that assign those costs to the rightful party ahead of time. And it's all in plain English, not "legalese."
For less than the cost of an hour with a lawyer you'll learn the exclusions to put in your agreements, why your insurance company may pay for your legal defense, how to avoid liability for injuries to your sub and his employees or damages they cause, how to collect on lawsuits you win, and much more. It also includes a FREE computer disk with contracts and forms you can customize for your own use. **352 pages, 8½ x 11, $59.95**

Profits in Building Spec Homes

If you've ever wanted to make big profits in building spec homes yet were held back by the risks involved, you should have this book. Here you'll learn how to do a market study and feasibility analysis to make sure your finished home will sell quickly, and for a good profit. You'll find tips that can save you thousands in negotiating for land, learn how to impress bankers and get the financing package you want, how to nail down cost estimating, schedule realistically, work effectively yet harmoniously with subcontractors so they'll come back for your next home, and finally, what to look for in the agent you choose to sell your finished home. Includes forms, checklists, worksheets, and step-by-step instructions. **208 pages, 8½ x 11, $27.25**

Carpentry Estimating

Simple, clear instructions on how to take off quantities and figure costs for all rough and finish carpentry. Shows how to convert piece prices to MBF prices or linear foot prices, use the extensive manhour tables included to quickly estimate labor costs, and how much overhead and profit to add. All carpentry is covered; floor joists, exterior and interior walls and finishes, ceiling joists and rafters, stairs, trim, windows, doors, and much more. Includes Carpenter's Dream a material-estimating program, at no extra cost on a 5¼" high-density disk. **336 pages, 8½ x 11, $35.50**

National Renovation & Insurance Repair Estimator

Current prices in dollars and cents for hard-to-find items needed on most insurance, repair, remodeling, and renovation jobs. All price items include labor, material, and equipment breakouts, plus special charts that tell you exactly how these costs are calculated. Includes an electronic version of the book on computer disk with a stand-alone *Windows* estimating program FREE on a 3½" high density (1.44 Mb) disk. **560 pages, 8½ x 11, $39.50. Revised annually**

Profits in Buying & Renovating Homes

Step-by-step instructions for selecting, repairing, improving, and selling highly profitable "fixer-uppers." Shows which price ranges offer the highest profit-to-investment ratios, which neighborhoods offer the best return, practical directions for repairs, and tips on dealing with buyers, sellers, and real estate agents. Shows you how to determine your profit before you buy, what "bargains" to avoid, and how to make simple, profitable, inexpensive upgrades. **304 pages, 8½ x 11, $19.75**

Contractor's Guide to the Building Code Revised

This completely revised edition explains in plain English exactly what the Uniform Building Code requires. Based on the newly-expanded 1994 code, it explains many of the changes made. Also covers the Uniform Mechanical Code and the Uniform Plumbing Code. Shows how to design and construct residential and light commercial buildings that'll pass inspection the first time. Suggests how to work with an inspector to minimize construction costs, what common building shortcuts are likely to be cited, and where exceptions are granted. **384 pages, 8½ x 11, $39.00**

Builder's Guide to Room Additions

How to tackle problems that are unique to additions, such as requirements for basement conversions, reinforcing ceiling joists for second-story conversions, handling problems in attic conversions, what's required for footings, foundations, and slabs, how to design the best bathroom for the space, and much more. Besides actual construction methods, you'll also find help in designing, planning and estimating your room-addition jobs. **352 pages, 8½ x 11, $27.25**

Roof Framing

Shows how to frame any type of roof in common use today, even if you've never framed a roof before. Includes using a pocket calculator to figure any common, hip, valley, or jack rafter length in seconds. Over 400 illustrations cover every measurement and every cut on each type of roof: gable, hip, Dutch, Tudor, gambrel, shed, gazebo, and more. **480 pages, 5½ x 8½, $22.00**

Plumber's Handbook Revised

This new edition shows what will and won't pass inspection in drainage, vent, and waste piping, septic tanks, water supply, fire protection, and gas piping systems. All tables, standards, and specifications completely up-to-date with recent plumbing code changes. Covers common layouts for residential work, how to size piping, selecting and hanging fixtures, practical recommendations, and trade tips. The approved reference for the plumbing contractor's exam in many states. **240 pages, 8½ x 11, $18.00**

Construction Surveying & Layout

A practical guide to simplified construction surveying. How to divide land, use a transit and tape to find a known point, draw an accurate survey map from your field notes, use topographic surveys, and the right way to level and set grade. You'll learn how to make a survey for any residential or commercial lot, driveway, road, or bridge — including how to figure cuts and fills and calculate excavation quantities. Use this guide to make your own surveys, or just read and verify the accuracy of surveys made by others. **256 pages, 5½ x 8½, $19.25**

Construction Estimating Reference Data

Provides the 300 most useful manhour tables for practically every item of construction. Labor requirements are listed for sitework, concrete work, masonry, steel, carpentry, thermal and moisture protection, door and windows, finishes, mechanical and electrical. Each section details the work being estimated and gives appropriate crew size and equipment needed. Includes an electronic version of the book on computer disk with a stand-alone *Windows* estimating program FREE on a 3½" high-density (1.44 Mb) disk. **432 pages, 8½ x 11, $39.50**

Drafting House Plans

Here you'll find step-by-step instructions for drawing a complete set of home plans for a one-story house, an addition to an existing house, or a remodeling project. This book shows how to visualize spatial relationships, use architectural scales and symbols, sketch preliminary drawings, develop detailed floor plans and exterior elevations, and prepare a final plot plan. It even includes code-approved joist and rafter spans and how to make sure that drawings meet code requirements. **192 pages, 8½ x 11, $27.50**

Renovating & Restyling Vintage Homes

Any builder can turn a run-down old house into a showcase of perfection — if the customer has unlimited funds to spend. Unfortunately, most customers are on a tight budget. They usually want more improvements than they can afford — and they expect you to deliver. This book shows how to add economical improvements that can increase the property value by two, five or even ten times the cost of the remodel. Sound impossible? Here you'll find the secrets of a builder who has been putting these techniques to work on Victorian and Craftsman-style houses for twenty years. You'll see what to repair, what to replace and what to leave, so you can remodel or restyle older homes for the least amount of money and the greatest increase in value. **416 pages, 8½ x 11, $33.50**

National Construction Estimator

Current building costs for residential, commercial, and industrial construction. Estimated prices for every common building material. Manhours, recommended crew, and labor cost for installation. Includes an electronic version of the book on computer disk with a stand-alone *Windows* estimating program FREE on a 3½" high-density (1.44 Mb) disk.
528 pages, 8½ x 11, $37.50. Revised annually

Estimating Plumbing Costs

Offers a basic procedure for estimating materials, labor, and direct and indirect costs for residential and commercial plumbing jobs. Explains how to read and understand plot plans, design drainage, waste, and vent systems, meet code requirements, and make an accurate take-off for materials and labor. Includes sample cost sheets, manhour production tables, complete illustrations, and all the practical information you need. **224 pages, 8½ x 11, $22.50**

Planning Drain, Waste & Vent Systems

How to design plumbing systems in residential, commercial, and industrial buildings. Covers designing systems that meet code requirements for homes, commercial buildings, private sewage disposal systems, and even mobile home parks. Includes relevant code sections and many illustrations to guide you though what the code requires in designing drainage, waste, and vent systems.
192 pages, 8½ x 11, $19.25

Stair Builders Handbook

If you know the floor-to-floor rise, this handbook gives you everything else: number and dimension of treads and risers, total run, correct well hole opening, angle of incline, and quantity of materials and settings for your framing square for over 3,500 code-approved rise and run combinations — several for every 1/8-inch interval from a 3 foot to a 12 foot floor-to-floor rise.
416 pages, 5½ x 8½, $15.50

Estimating Tables for Home Building

Produce accurate estimates for nearly any residence in just minutes. This handy manual has tables you need to find the quantity of materials and labor for most residential construction. Includes overhead and profit, how to develop unit costs for labor and materials, and how to be sure you've considered every cost in the job.
336 pages, 8½ x 11, $21.50

Plumber's Exam Preparation Guide

Hundreds of questions and answers to help you pass the apprentice, journeyman, or master plumber's exam. Questions are in the style of the actual exam. Gives answers for both the Standard and Uniform plumbing codes. Includes tips on studying for the exam and the best way to prepare yourself for examination day.
320 pages, 8½ x 11, $29.00

Estimating Home Building Costs

Estimate every phase of residential construction from site costs to the profit margin you include in your bid. Shows how to keep track of manhours and make accurate labor cost estimates for footings, foundations, framing and sheathing finishes, electrical, plumbing, and more. Provides and explains sample cost estimate worksheets with complete instructions for each job phase.
320 pages, 5½ x 8½, $17.00